Lecture Notes in Artificial Intelligence 9356

Subseries of Lecture Notes in Computer Science

More information about this series at http://www.springer.com/series/1244

Nathalie Japkowicz · Stan Matwin (Eds.)

Discovery Science

18th International Conference, DS 2015
Banff, AB, Canada, October 4–6, 2015
Proceedings

 Springer

Editors
Nathalie Japkowicz
University of Ottawa
Ottawa, ON
Canada

Stan Matwin
Faculty of Computer Science
Dalhousie University
Halifax, NS
Canada

ISSN 0302-9743   ISSN 1611-3349 (electronic)
Lecture Notes in Artificial Intelligence
ISBN 978-3-319-24281-1   ISBN 978-3-319-24282-8 (eBook)
DOI 10.1007/978-3-319-24282-8

Library of Congress Control Number: 2015948779

LNCS Sublibrary: SL7 – Artificial Intelligence

Springer Cham Heidelberg New York Dordrecht London

Printed on acid-free paper

Springer International Publishing AG Switzerland is part of Springer Science+Business Media
(www.springer.com)

Preface

This year's International Conference on Discovery Science, DS 2015, was the 18th event in this series. Like in previous years, the conference was co-located with the International Conference on Algorithmic Learning Theory, ALT 2015, which is already in its 26th year. Started in 2001, ALT/DS is one of the longest-running series of co-located events in computer science. The unique combination of recent advances in the development and analysis of methods for discovering scientific knowledge, coming from machine learning, data mining, and intelligent data analysis, as well as their application in various scientific domains, on the one hand, with the algorithmic advances in machine learning theory, on the other hand, makes every instance of this joint event unique and attractive. This volume contains the papers presented at the 18th International Conference on Discovery Science, while the papers of the 26th International Conference on Algorithmic Learning Theory are published by Springer in a companion volume (LNCS Vol. 9355).

The 18th Discovery Science conference received 44 international submissions. Each submission was reviewed by at least two committee members. The committee decided to accept 28 papers, of which 16 are long and 12 are short papers. This results in the 36% acceptance rate for long papers. As is the tradition of the Discovery Science and the Algorithmic Learning Theory conferences, invited talks were shared between the two meetings. This year's DS invited talks were "Turning Prediction Tools into Decision Tools" by Cynthia Rudin from MIT, and "Overcoming Obstacles to the Adoption of Machine Learning by Domain Experts" by Kiri Wagstaff from Jet Propulsion Laboratorics, while the ALT invited talks were "Finding Hidden Structure in Data with Tensor Decompositions" by Sham Kakade from Microsoft and the University of Washington, and "Bilinear Prediction Using Low Rank Models" by Inderjit Dhillon form the University of Texas at Austin. Abstracts of all four invited talks are included in these proceedings.

We would like to thank all authors of submitted papers, the Program Committee members, and the additional reviewers for their efforts in evaluating the submitted papers, as well as the invited speakers and tutorial presenters. Support and advice from Randy Goebel, the General Chair of both conferences, were essential every step of the way. We are grateful to Kamalika Chaudhuri, Claudio Gentile, Sandra Zilles, and Csaba Szepesvari for ensuring a smooth coordination with ALT. We are indebted to Jonathan Amyot from the Faculty of Computer Science, Dalhousie University, for putting up and maintaining our website with great competence and efficiency.

We are grateful to the people behind Easychair for making the system available free of charge. It was an essential tool in the paper submission and evaluation process, as well as in the preparation of the Springer proceedings. We are also grateful to Springer for their continuing support of Discovery Science and for publishing the conference proceedings since its inception.

This year, both conferences were held on October 4-6 in the picturesque setting of Banff, Alberta, and were organized by Sandra Zilles and Csaba Szepesvari. We are very grateful to ISM Canada, an IBM company, to the Alberta Innovates - Technology Futures (AITF), to the Canadian Artificial Intelligence Association (CAIAC), and to the Faculty of Computer Science at Dalhousie University for their sponsorship of both the conferences.

October 2015   Nathalie Japkowicz
                                                                              Stan Matwin

Organization

Program Committee

Aijun An	York University, Canada
Vincent Barnabe-Lortie	University of Ottawa, Canada
Colin Bellinger	University of Ottawa, Canada
Sabine Bergler	Concordia University, Canada
Albert Bifet	University of Waikato, New Zealand
Hendrik Blockeel	K.U. Leuven, Belgium
Ivan Bratko	University of Ljubljana, Slovenia
Michelangelo Ceci	Università degli Studi di Bari, Italy
Tapio Elomaa	Tampere University of Technology, Finland
Johannes Fürnkranz	TU Darmstadt, Germany
Dragan Gamberger	Rudjer Boskovic Institute, Croatia
Howard Hamilton	University of Regina, Canada
Geoffrey Holmes	University of Waikato, New Zealand
Diana Inkpen	University of Ottawa, Canada
Aminul Islam	Dalhousie University, Canada
Nathalie Japkowicz	SITE, University of Ottawa, Canada
Ross King	University of Manchester, UK
Svetlana Kiritchenko	NRC, Canada
William Klement	UHN, University of Toronto, Canada
Philippe Langlais	Université de Montréal, Canada
Guy Lapalme	RALI-DIRO, Université de Montréal, Canada
Donato Malerba	Università degli Studi di Bari "Aldo Moro", Italy
Stan Matwin	University of Ottawa, Canada
Robert Mercer	The University of Western Ontario, Canada
Evangelos Milios	Dalhousie University, Canada
Zoran Obradovic	Temple University, USA
Bernhard Pfahringer	University of Waikato, New Zealand
Fred Popowich	Simon Fraser University, Canada
Doina Precup	McGill University, Canada
Marko Robnik-Sikonja	University of Ljubljana, FRI, Slovenia
Mohak Shah	GE Global Research, USA
Marina Sokolova	University of Ottawa and Institute for Big Data Analytics, Canada
Jerzy Stefanowski	Poznań Univeristy of Technology, Poland
Einoshin Suzuki	Kyushu University, Japan
Maguelonne Teisseire	Cemagref - UMR Tetis, France
Herna Viktor	University of Ottawa, Canada

Harry Zhang   University of New Brunswick, Canada
Min-Ling Zhang   Southeast University, China
Nur Zincir-Heywood   Dalhousie University, Canada
Blaz Zupan   University of Ljubljana, Slovenia

Additional Reviewers

Alharthi, Haifa   Fass, Dan   Kurup, Unmesh
Cao, Xi Hang   Fok, Ricky   Lanotte, Pasqua Fabiana
Christoff, Zoe   Grinberg, Nastasiya   Odilinye, Lydia
Corizzo, Roberto   Han, Chao   Olier, Ivan
Davoudi, Heidar   Kirinde Gamaarachchige,   Tofiloski, Milan
Ellert, Bradley   Prasadith

Bilinear Prediction Using Low Rank Models

Inderjit S. Dhillon

Department of Computer Science
University of Texas at Austin, Austin, USA
inderjit@cs.utexas.edu

Linear prediction methods, such as linear regression and classification, form the bread-and-butter of modern machine learning. The classical scenario is the presence of data with multiple features and a single target variable. However, there are many recent scenarios, where there are multiple target variables. For example, predicting bid words for a web page (where each bid word acts as a target variable), or predicting diseases linked to a gene. In many of these scenarios, the target variables might themselves be associated with features. In these scenarios, we propose the use of bilinear prediction with low-rank models. The low-rank models serve a dual purpose: (i) they enable tractable computation even in the face of millions of data points as well as target variables, and (ii) they exploit correlations among the target variables, even when there are many missing observations. We illustrate our methodology on two modern machine learning problems: multi-label learning and inductive matrix completion, and show results on two applications: predicting Wikipedia labels, and predicting gene-disease relationships.

This is joint work with Prateek Jain, Nagarajan Natarajan, Hsiang-Fu Yu and Kai Zhong.

Finding Hidden Structure in Data with Tensor Decompositions

Sham M. Kakade

Microsoft Research, New England, USA
and University of Washington, Seattle, USA

In many applications, we face the challenge of modeling the interactions between multiple observations. A popular and successful approach in machine learning and AI is to hypothesize the existence of certain latent (or hidden) causes which help to explain the correlations in the observed data. The (unsupervised) learning problem is to accurately estimate a model with only samples of the observed variables. For example, in document modeling, we may wish to characterize the correlational structure of the "bag of words" in documents, or in community detection, we wish to discover the communities of individuals in social networks. Here, a standard model is to posit that documents are about a few topics (the hidden variables) and that each active topic determines the occurrence of words in the document. The learning problem is, using only the observed words in the documents (and not the hidden topics), to estimate the topic probability vectors (i.e. discover the strength by which words tend to appear under different topcis). In practice, a broad class of latent variable models is most often fit with either local search heuristics (such as the EM algorithm) or sampling based approaches.

This talk will discuss a general and (computationally and statistically) efficient parameter estimation method for a wide class of latent variable models—including Gaussian mixture models (for clustering), hidden Markov models (for time series), and latent Dirichlet allocation (for topic modeling and community detection)—by exploiting a certain tensor structure in their low-order observable moments. Specifically, parameter estimation is reduced to the problem of extracting a certain decomposition of a tensor derived from the (typically second- and third-order) moments; this particular decomposition can be viewed as a natural generalization of the (widely used) principal component analysis method.

Turning Prediction Tools into Decision Tools

Cynthia Rudin

MIT CSAIL and Sloan School of Management
Building E62-576, 100 Main Street, Cambridge, MA 02142, USA
rudin@mit.edu

Arguably, the main stumbling block in getting machine learning algorithms used in practice is the fact that people do not trust them. There could be many reasons for this, for instance, perhaps the models are not sparse or transparent, or perhaps the models are not able to be customized to the user's specifications as to what a decision tool should look like. I will discuss some recent work from the Prediction Analysis Lab on how to build machine learning models that have helpful decision-making properties. I will show how these models are applied to problems in healthcare and criminology.

Overcoming Obstacles to the Adoption
of Machine Learning by Domain Experts

Kiri L. Wagstaff

Jet Propulsion Laboratory, California Institute of Technology,
4800 Oak Grove Drive, Pasadena, CA 91109, USA
kiri.l.wagstaff@jpl.nasa.gov

The ever-increasing volumes of scientific data being collected by fields such as astronomy, biology, planetary science, medicine, etc., present a need for automated data analysis methods to assist investigators in understanding and deriving new knowledge from their data. Partnerships between domain experts and computer scientists can open that door. However, there are obstacles that sometimes prevent the successful adoption of machine learning by those who stand to benefit most.

We devote a lot of effort to solving technological challenges (e.g., scalability, performance), but less effort to overcoming psychological and logistical barriers. Domain experts may fail to be persuaded to adopt a tool based on performance results that are otherwise compelling to those in machine learning, which can be frustrating and perplexing. Algorithm aversion is the phenomenon in which people place more trust in human predictions than those generated by an algorithm, even when the algorithm demonstrably performs better. Media hype about the dangers of artificial intelligence and fears about the loss of jobs or the loss of control create additional obstacles.

I will describe two case studies in which we have developed and delivered machine learning systems to solve problems from radio astronomy and planetary science domains. While we cannot claim to have a magic wand that ensures the adoption of machine learning systems, we can share lessons learned from our experience. Key elements include progressive integration, enthusiasm on the part of the domain experts, and a system that visibly learns or adapts to user feedback to correct any mistakes.

This work was performed at the Jet Propulsion Laboratory, California Institute of Technology, under a contract with NASA. Government sponsorship acknowledged.

Contents

Resolution Transfer in Cancer Classification Based on Amplification Patterns

Prem Raj Adhikari[1,2]([⊠]) and Jaakko Hollmén[1]

[1] Helsinki Institute for Information Technology HIIT and Department of Information
and Computer Science, Aalto University School of Science,
PO Box 15400, 00076 Aalto, Espoo, Finland
`prem.adhikari@utu.fi`

[2] Department of Physiology and Turku Center for Disease Modeling Institute
of Biomedicine, University of Turku, Kiinamyllynkatu 10, 20520 Turku, Finland
`jaakko.hollmen@aalto.fi`

Abstract. In the current scientific age, the measurement technology has
considerably improved and diversified producing data in different representations. Traditional machine learning and data mining algorithms
can handle data only in a single representation in their standard form.
In this contribution, we address an important challenge encountered in
data analysis: what to do when the data to be analyzed are represented
differently with regards to the resolution? Specifically, in classification,
how to train a classifier when class labels are available only in one resolution and missing in the other resolutions? The proposed methodology
learns a classifier in one data resolution and transfers it to learn the
class labels in a different resolution. Furthermore, the methodology intuitively works as a dimensionality reduction method. The methodology is
evaluated on a simulated dataset and finally used to classify cancers in
a real–world multiresolution chromosomal aberration dataset producing
plausible results.

1 Introduction

Over the years, the measurement technologies have improved considerably
providing an opportunity to measure the finer details of the phenomenon [8].
Multiresolution data is generated when the same phenomenon is measured in
different levels of detail [13]. The older generation technologies measure only
the coarser units of the phenomenon resulting in the data in the coarse resolution while the newer generation technology can measure the finer units of the
phenomenon generating the data in the fine resolution. The fine resolution data
carries more information in the data sample compared to the coarse resolution
data but also has the larger data dimensionality than the coarse resolution data.
The importance of combining multiple data sources, and information within a
single analysis and the availability of multiresolution data in different application areas, such as, image processing, and time–series analysis have given major
impetus to the research in multiresolution data analysis [13].

© Springer International Publishing Switzerland 2015
N. Japkowicz and S. Matwin (Eds.): DS 2015, LNAI 9356, pp. 1–8, 2015.
DOI: 10.1007/978-3-319-24282-8_1

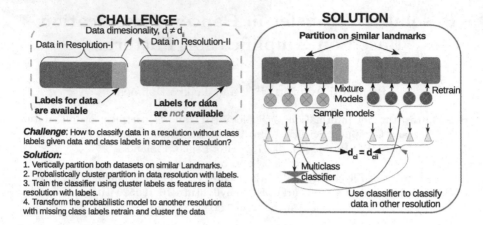

Fig. 1. The research challenge considered in this contribution and the proposed solution.

In this contribution, we address an important challenge that lies in between supervised learning and learning from unlabeled data; which resembles the semi–supervised learning [5] and the transfer learning [10]. The proposed methodology learns the classifier in one resolution where the class labels are available and transforms the classifier to other resolutions where the class labels are not available. The methodology uses a combination of an unsupervised probabilistic clustering and a supervised multiclass classification in a pipeline to address the challenges in classifying the data in different resolutions when class labels are not available in all the resolutions of the data. In this contribution, we do not propose a new probabilistic modeling algorithm or a multiclass classification algorithm. The novelty in the contribution comes from the design of the pipeline for classifying the datasets in different resolutions by resolution transfer of the classifier and intuitive dimensionality reduction achieved through the unsupervised probabilistic clustering. While the clustering results have been used to improve the classification results [7], such a methodological pipeline of resolution transfer has not been proposed in the literature.

The Fig. 1 shows that between the two high dimensional datasets in two different resolutions, only the data in one resolution has the associated class labels; while the class labels are missing in the data in other resolutions. Therefore, the challenge to learn a classifier on the data in the resolution having the class labels and use the same classifier to classify the samples in the data resolution without the associated labels. A simplified approach of using projection methods such as principal component analysis (PCA) would not produce expected results because the representation of data would be lost.

As shown in the bottom panel of Fig. 1, first, we vertically partition both the datasets based on some predefined landmarks in such a way that the number of vertical partitions in each dataset in different resolutions are the same. Second, we train the mixture models in each of the vertical partitions of the dataset with the associated labels. Third, we transform the trained mixture models to another resolution with missing class labels. The transformations are performed using the apriori knowledge of the relationships among different resolutions of the data from the domain ontology. Fourth, we retrain the mixture models with the data partitions in data resolution not consisting of associated labels. The retrained mixture model generates the cluster labels for the data partitions in the data resolution with missing class labels. Fifth, we concatenate the obtained cluster labels in both the data resolutions separately to regenerate the whole genome but with a reduced dimensionality. Sixth, we learn a multiclass classifier for the whole genome in the data resolution having the associated class labels. Finally, we can use the same classifier to classify the data resolution not consisting of associated labels. Since the clusters labels which are used as features are equivalent in both the resolution, the classifier can be used for data in both resolution producing comparable results in both resolutions.

2   Methodology of Multiresolution Multiclass Classification

In our proposed methodology, we first set aside the class labels and vertically partition the feature space in both the given high dimensional datasets on specific landmarks in such a way that specific relationship between different data resolutions can be easily established. Furthermore, vertical partition should be such that data in different resolutions will have equal number of partitions while data dimensionality in each partition can be different. We then use unsupervised probabilistic algorithm, i.e., mixture models, to cluster each partition of the data separately. The number of clustering experiments is equal to the number of vertical partitions in the data. We use model selection to determine the number of clusters in each of the partition of the data separately using ten–fold cross–validation as in [12].

The cluster labels generated by the clustering algorithms are then vertically concatenated forming a new dataset with reduced dimensionality for classification. The newly formed dataset obtained by concatenating the cluster labels emulates the original data but results in the reduced data dimensionality. This is because a cluster label comprise multiple data dimensionality in the vertical partitions of the data thus ameliorating the problem of curse of dimensionality [4].

In our previous research, we have shown that the mixture models learned in a resolution can be transformed to a different resolution provided there exists a well–defined relationship among the different data resolutions [2]. We can use the domain ontology to determine the relationship between the model parameters in different resolutions of data and exploit that to transform the mixture models across different resolutions.

$$\theta_f \sim N(\mu = \theta_c, \sigma = 0.01) = \begin{cases} \theta_f & \text{if } 0 \le \theta_f \le 1 \\ \theta_c & \text{if } \theta_f < 0 \text{ or } \theta_f > 1 \end{cases} \qquad (1)$$

We can re–sample the number of parameters required in the fine resolution from a normal distribution with the mean (μ) equal to the parameter value in the coarse resolution and a small standard deviation (σ); 0.01 in our experiments. Mathematically, we can represent the transformations as in Eq. 1, where θ_c, and θ_f denote the parameters of the mixture components θ in the coarse and the fine resolution. We can also further ensure that the re–sampled parameters obey the laws of probability in such a way that the value of the parameter in the fine resolution is between 0 and 1, i.e., $0 \leq \theta \leq 1$. If the re–sampled value of θ is outside the given range, we replace it with the value of the parameter, θ, in the coarse resolution such that $\theta_f = \theta_c$. Finally, the transformed mixture model is then retrained on the fine resolution data.

We represent the categorical cluster labels as binary features as discussed in [6]. The number of bits in binary features is equal to the number of components in the mixture model for that data partition, i.e., the clusters in the data. For example, if the number of components is four then the clusters one, two, three, and four are represented as: 1000, 0100, 0010, and 0001. We then vertically concatenate the cluster labels in binary representation to represent an entire dataset. This clustering labels can be assumed to be the summary of the patterns present in the data. Finally, a multiclass classifier, e.g., support vector machines, can be trained using on the dataset generated by concatenating the clustering labels.

3   Experiments on Multiresolution Chromosomal Aberrations Dataset

Two chromosomal amplification datasets were available in two different resolutions for our experiments. The data in coarse resolution describing the DNA amplification patterns of 4590 cancer patients were available from [9]. Similarly, data describing the DNA amplification patterns in fine resolution were available from [3]. The coarse resolution data describes the chromosomal amplifications dividing genome in 393 different parts as described in [11]. In contrast, the fine resolution data describes the chromosomal amplifications dividing the genome in 862 different parts. In addition to resolution, another important difference between the datasets is that the coarse resolution data have associated class labels, i.e., the 4590 patients were associated with 73 different cancer types whereas the data in the fine resolution do not have the associated cancer types (class labels).

Data Preprocessing. The number of cancer types in the coarse resolution dataset (73) were too high to learn any credible cancer classifier. Some of the cancers had less than 10 samples making it difficult to learn a classifier that generalizes better on the unseen data [5]. Therefore, we only experimented with top 34 cancer types. The top 34 cancer types were chosen because they covers 90 % of the data. This simplification reduces the number of samples in the data to 4104. The cancer with the highest number of samples is Neuroepithelial tumors with 544 samples. In contrast, the cancer with the minimum number of samples is Pulmonary sarcoma with only 30 samples. The simplified data is then processed chromosome–wise, i.e., the data describing the genome is vertically partitioned into 24 different chromosomes. When the data is divided into chromosomes, some samples in some chromosome do not show any amplification. We remove those samples without amplifications (vectors with all zeros) because they carry no information about the cancer and also further simplify the experimental procedure.

Chromosome–wise Mixture Modeling. After the data have been vertically partitioned into the different chromosomes, we learn the mixture models based on a model selection procedure in a ten–fold cross–validation setting as discussed in [12]. The model selection procedure selects different number of components in the mixture model to fit the data in different chromosomes. Mixture models are generally used to represent the probability distribution of the data. Nevertheless, it can also be used to cluster the data into hard partitions. The number of partitions is equal to the number of components in the mixture model. The cluster labels are then transformed to binary format and chromosomes are concatenated to form the whole genome. We do not use model selection algorithm on Chromosome Y because of lack of data samples. In chromosome Y, the cluster label is 1 if any of the chromosomal regions is amplified; otherwise 0.

Cancer Classification Using SVMs. The cluster labels are transformed to binary format and the chromosomes are concatenated to regenerate the whole genome. In the cancer samples showing no amplifications in specific chromosomes that were ignored during mixture modeling are replaced by all zeros in the binary features. The example, of four clusters discussed in Sect. 2, it would be represented as 0000. We then learn multiclass support vector machines using open–source libsvm software package [6]. The kernel type selected is radial basis function. The parameters of the support vector machines are γ and C which also learned in a ten–fold cross–validation setting by a grid search. The support vector machines are initially learned on the original data with full set of features and also on the vertical concatenation of cluster labels. The original data dimensionality is 393 whereas vertical concatenation of cluster labels results in data dimensionality of 112. The concatenation of features result in reduction of data dimensionality that is less than one third of the features in original data dimension. Naturally, when the data dimension is reduced, the accuracy of classification decreases. Figure 2 shows that decrease in accuracy is negligible when the partitions of data is represented by the cluster labels. The computational and memory efficiency of reduced data dimensionality surpasses the decrease in the classification accuracy.

Resolution Transfer of the Classifier. The crux of this contribution is the resolution transfer of the classifier. We use the knowledge of domain ontology

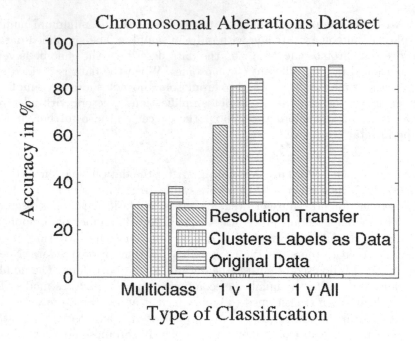

Fig. 2. Comparative study of the accuracy of SVM in different modes of the classification settings.

relations to transform the parameters of the mixture model learned in the resolution having the associated class labels to the data resolution not having the associated class labels as in [2]. The transformed model is retrained on the data in other resolution in such a way that the components are not much different than the model in the original resolution. This requirement is enforced because the clustering algorithm should produce same labels for similar data vectors as we are using the cluster labels as data features for classification. Since the algorithm is trained on the data in other resolution, the features in the concerned resolution must be the same as the features in original resolution. Finally, the data in concerned resolution is classified using the classifier trained in original data resolution.

The data in fine resolution obtained from [3] does not have associated cancer types, i.e., class labels. Therefore, we can use the same classifier to classify cancers but we cannot access the performance of the classification algorithm. Therefore, we transform the data to another resolution using deterministic methods similar to the one suggested in [1]. The data in fine resolution can then be classified using the classifier learned in the resolution having associated cancer labels. Furthermore, the performance can be accessed because the transformed data have labels from the coarse resolution.

Figure 2 depicts the classification accuracy of the classifier in different settings. As expected the classification accuracy is best on the original data. The figure shows that the performance of the classifier degrades in resolution transfer. The decrease in performance is expected because the classifier is learned on data in different resolution. The results obtained are promising because resolution transfer provides additional facilities to classify data in different resolution for which the class labels are not available. Furthermore, the negligible decrease in performance can also be attributed to curse of dimensionality [4]. In addition, the performance of multiclass classification is less than 40 % which is comparatively less but considering that there are 34 classes, the accuracy is plausible because random classifier would generate accuracy of less than 10 %. If all samples are classified as the cancer with the highest number of samples, i.e., Neuroepithelial tumors, the accuracy would be approximately 13 %. Therefore, the performance achieved by our methodology is plausible and provides a novel methodology to classify cancers across different resolutions.

3.1 Simulated Data

We also evaluated our methodology on a simulated data set. The simulated dataset was simple with 1000 data samples and 5 dimensions. We randomly sampled a number between 1 and 4, and generate row for the data sample where each element in the sample is equal to the randomly sampled number. For example, if we sample number 3, all the elements in the row are 3. We continue this process until 1000 data samples have been achieved. We then consider first variable as the class and remaining 4 variables as the features. We convert the four variables to binary using decimal to binary conversion system with 3 bits such that 4 dimensional data are transformed into 12 dimensional 0–1 vectors. We then randomly flip the bits of 1200 (10 %) data elements to add noise to the dataset. Similarly, we vertically concatenated the 12 dimensional data each dimension one by one to generate 24 dimensional data.

We then group each digit separately again into four groups to run the clustering experiments. Since, we know the number of clusters in the data, i.e., 4, we do not run model selection algorithm in this case. We then evaluate our methodology on this simulated multiresolution data in the similar vein as in Sect. 3. In this experiment, there was larger discrepancy in classification accuracy in multiclass classification. The original algorithm as well as the clusters labels used as class labels produced accuracy nearing 97 % while the resolution transfer produced classification accuracy nearing 75 %. Despite the addition of noise, the data is overly simple and in such simple datasets, classifiers often overfit. In one vs one and one vs the rest experiments all the methods produced plausible accuracy of 98.5 %.

4 Summary and Conclusions

In this contribution, we were interested in transferring the classifier learning across different resolutions. In our setting, we had access to class labels only in one resolution while the class labels were missing in other resolutions. We learn the classifier in the resolution with the class labels and transfer the learned classifier to classify the data in resolutions with missing class labels. Furthermore, our proposed methodology intrinsically reduces the data dimensionality to less than one–third in the coarse resolution and to less than one–eighth in the fine resolution as an added advantage of the proposed resolution transfer. We experimented our methodology on a simulated dataset, and chromosomal aberrations patterns to classify cancers with plausible results.

References

1. Adhikari, P.R., Hollmén, J.: Patterns from multiresolution 0–1 data. In: UP 2010: Proceedings of the ACM SIGKDD Workshop on Useful Patterns, pp. 8–16. ACM, New York (2010)
2. Adhikari, P.R., Hollmén, J.: Multiresolution mixture modeling using merging of mixture components. In: Hoi, S., Buntine, W. (eds.) Proceedings of 4th Asian Conference on Machine Learning, volume 25 of ACML 2012. JMLR Workshop and Conference Proceedings, pp. 17–32 (2012)
3. Baudis, M.: Genomic imbalances in 5918 malignant epithelial tumors: an explorative meta-analysis of chromosomal CGH data. BMC Cancer 7(1), 226 (2007)
4. Bellman, R.E.: Adaptive Control Processes - A Guided Tour. Princeton University Press, Princeton (1961)
5. Blum, A., Mitchell, T.: Combining labeled and unlabeled data with cotraining. In: Proceedings of 11th Annual Conference on Computational Learning Theory, COLT 1998, pp. 92–100. ACM, New York (1998)
6. Hsu, C.-W., Chang, C.-C., Lin, C.-J.: A practical guide to support vector classification. National Taiwan University, Technical report, Department of Computer Science (2003)
7. Kyriakopoulou, A., Kalamboukis, T.: Clustering as a prior step to classification: an empirical study. Int. J. Artif. Intel. Tools 20(03), 531–548 (2011)
8. Mardis, E.R.: A decade's perspective on DNA sequencing technology. Nature 470(7333), 198–203 (2011)
9. Myllykangas, S., Himberg, J., Böhling, T., Nagy, B., Hollmén, J., Knuutila, S.: DNA copy number amplification profiling of human neoplasms. Oncogene 25(55), 7324–7332 (2006)
10. Pan, S.J., Yang, Q.: A survey on transfer learning. IEEE Trans. Knowl. Data Eng. 22(10), 1345–1359 (2010)
11. Shaffer, L.G., Tommerup, N.: ISCN 2005: An International System for Human Cytogenetic Nomenclature (2005) Recommendations of the International Standing Committee on Human Cytogenetic Nomenclature. Karger (2005)
12. Smyth, P.: Model selection for probabilistic clustering using cross-validated likelihood. Stat. Comput. 10(1), 63–72 (2000)
13. Willsky, A.S.: Multiresolution markov models for signal and image processing. Proceed. IEEE 90(8), 1396–1458 (2002)

Very Short-Term Wind Speed Forecasting Using Spatio-Temporal Lazy Learning

Annalisa Appice[1]([✉]), Sonja Pravilovic[2], Antonietta Lanza[1],
and Donato Malerba[1]

[1] Dipartimento di Informatica, Università degli Studi di Bari Aldo Moro
via Orabona, 4, 70126 Bari, Italy
{annalisa.appice,antonietta.lanza,donato.malerba}@uniba.it
[2] Faculty of Information Technology, Mediterranean University
Vaka Djurovica b.b., 81000 Podgorica, Montenegro
sonja.pravilovic@unimediteran.net

Abstract. A wind speed forecast corresponds to an estimate of the upcoming production of a wind farm. The paper illustrates a variant of the Nearest Neighbor algorithm that yields wind speed forecasts, with a fast time resolution, for a (very) short time horizon. The proposed algorithm allows us to monitor a grid of wind farms, which collaborate by sharing information (i.e. wind speed measurements). It accounts for both spatial and temporal correlation of shared information. Experiments show that the presented algorithm is able to determine more accurate forecasts than a state-of-art statistical algorithm, namely auto. ARIMA.

1 Introduction

The growing integration of wind farms into the power grid requires precise forecasts of upcoming energy productions at different time scales, depending on the intended application. Very short-term forecasts (≤ 6 h) can be used for the turbine active control, short-term forecasts (48–72 h) may serve for wind power scheduling, as well as for economic dispatch, while longer time scales (up to 5–7 days ahead) may be considered for planning the maintenance of wind farms and transmission lines. Depending on the nature information processed, wind forecasting approaches can be classified into physical, statistical and hybrid approaches. A physical approach [1,6] uses weather forecast, while a statistical approach [2,8–10,13]) is based on vast amount of historical data (time series) without considering meteorological conditions. A hybrid approach [4,7]) uses both weather forecasts and time series analysis.

In this paper, we address the problem of very short-term wind speed forecasting by resorting to a statistical approach. We describe a spatio-temporal lazy learning-based algorithm, called WiNN (spatio-temporal WInd Nearest Neighbor algorithm), which is based on a variant of the Nearest Neighbor algorithm, which accounts for two forms of autocorrelation: the temporal autocorrelation,

N. Japkowicz and S. Matwin (Eds.): DS 2015, LNAI 9356, pp. 9–16, 2015.
DOI: 10.1007/978-3-319-24282-8_2

which may exist in wind speed time series data, and the spatial autocorrelation, which may exist between wind speed data measured at near wind farms. For every wind farm, lazy learning is performed to produce forecasts from wind speed values observed in a given spatio-temporal neighborhood. The main advantage of lazy learning is that the computation of a complex forecasting model of the historical data à la ARIMA is avoided. In addition, it can be easily extended to account for spatial autocorrelation as well.

The paper is organized as follows. In Sect. 2, we illustrate the data scenario and formulate the learning problem considered. In Sect. 3, we present the proposed algorithm, while in Sect. 4, we discuss the dataset and the relevant results. Finally, Sect. 5 draws some conclusions and outlines some future work.

2   Data Setting and Learning Problem

Data Scenario A wind farm grid is defined as a geophysical streaming system (K, Z, T), where: (1) K is the set of wind farms spanned on a bi-dimensional[1] XY representation of the geographic space, (2) Z is the wind speed variable and (3) T is the time line discretized in equally spaced time points denoted as $1, 2, \ldots, t, \ldots$. In this data scenario, $z(k, t)$ denotes a measure of Z collected from a certain wind farm $k \in K$, at a specific time point $t \in T$. A *wind speed stream* $z(k)$ is the stream of measures $z(k, t)$ collected at wind farm $k \in K$ for each time point $t \in T$, that is, $z(k) = z(k, 1), z(k, 2), \ldots, z(k, t), \ldots$. Following the *sliding window model* [3], $z(k)$ is decomposed into consecutive sliding windows of equal size w, namely $z(k) = z(k, 1 \rightarrow w), z(k, 2 \rightarrow w + 1), \ldots, z(k, t - w + 1 \rightarrow t), \ldots$, where $z(k, t - d - w + 1 \rightarrow t - d)$ denotes the *backward data window* of wind farm k at time t, with backward horizon w and temporal delay $d = 0, 1, \ldots, t - w$.

Forecasting Problem. Given a wind farm grid (K, Z, T) and a time horizon w, a forecasting service aims at producing, at each time point t and for every wind farm $k \in K$, the predictions of upcoming w values of Z, which are henceforth denoted as $\hat{z}(k, t + 1), \hat{z}(k, t + 2), \ldots, \hat{z}(k, t + w)$. In this study, the forecasting service for the upcoming w measurements of Z is based on the backward w measurements of Z collected over the grid. Backward data can be selected with a possible temporal delay that is at worst w-sized.

3   WiNN

WiNN is a lazy learning algorithm that allows us to yield (very-) short term forecasting of wind speed in a wind farm grid (K, Z, T). Input parameters of the algorithm are a spatial radius r, a window size w and a similarity threshold δ. The top-level description of the algorithm is reported in Algorithm 1.

First, for every target wind farm $k \in K$, WiNN applies a spatial filter, in order to determine a spatial neighborhood with center k and radius r (Algorithm 1, lines 1–3).

---

[1] Multi-dimensional representation of geographic space can be equally dealt.

Algorithm 1. WiNN(K, Z, T)

1: **for** $k \in K$ **do**
2:     compute $\sigma(k, r)$ {Definition 31}
3: **end for**
4: **for** $t \in T$ **do**
5:     **for** $k \in K$ **do**
6:         **for** $d \in 1, \ldots, w$ **do**
7:             compute $\tau(k, t, -d)$ {Definition 32}
8:             $\eta(k, t, -d) \leftarrow \sigma(k) \cap \tau(k, t, -d)$ {Definition 33}
9:         **end for** {Forecasting}
10:         **for** $f \in 1, \ldots, w$ **do**
11:             compute $\mathcal{L}(k, t + f)$ {Definition 34}
12:             $\hat{z}(k, t + f) \leftarrow \text{knn}(\mathcal{L}(k, t + f))$ {Definition 35}
13:         **end for**
14:     **end for**
15: **end for**

Definition 1 (Spatial Neighborhood). *The spatial neighborhood $\sigma(k, r)$ is the set of reference wind farms of K ($\sigma(k, r) \subseteq K$) such that:*

$$\sigma(k) = \{h \in K | h \neq k \text{ and } geoDistance(k, h) \leq r\}, \tag{1}$$

where $geoDistance(k, h)$ is the geographic distance computed between the spatial coordinates $xy(k)$ and $xy(h)$, respectively.

This phase is performed when the processing of the wind data streams produced by (K, Z, T) starts. It is repeated only when changes occur in the grid structure, i.e. a farm is either deleted from or added to the grid. After this initialization phase, WiNN processes data (wind speed measurements) as they are produced by the grid. At every streaming time point, it applies a temporal filter to every target farm (Algorithm 1, line 7), in order to identify neighbor farms whose backward data are correlated, with a temporal delay, to the backward data measured by the target farm at the present time (Algorithm 1, line 8). The temporal delay d ranges between 1 and w.

Definition 2 (Temporal Neighborhood). *Let d be a positive, integer-valued temporal delay with $d \leq w$. The temporal neighborhood $\tau(k, t, -d)$ is the set of reference wind farms of K ($\tau(k, t, -d) \subseteq K$) such that:*

$$\tau(k, t, -d) = \{h \in K | h \neq k \text{ and } dataDistance(k, h, t, -d, w) \leq \delta\}, \tag{2}$$

where $dataDistance(k, h, t, -d, w)$ is the Euclidean distance computed between the target backward data $z(k, t - w + 1 \rightarrow t)$ and the reference, d-delayed backward data $z(h, t - d + w + 1 \rightarrow t - d)$, that is

$$dataDistance(k, h, t, -d, w) = \sum_{i=1}^{w} (z(h, t - d - w + i) - z(k, t - w + i))^2.$$

In this study, the definition of this kind of neighborhood with a temporal delay is motivated by the characteristics of the physical variable (wind speed) that we are considering for the forecasting problem. Wind can be considered as a moving object over space, so it is reasonable that the wind speed measured from the target wind farm k at time point t is more similar (i.e. higher correlated) to the wind speed measured from a reference wind farm h at a time point before t (i.e. $t - d$) rather than to the wind speed measured from h at t. In this study, we construct neighborhoods with various temporal delay values, in order to be able to properly model wind as a moving object also without accounting for information concerning the wind direction.

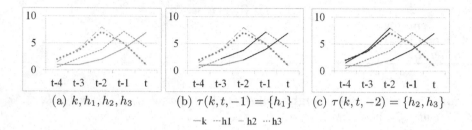

(a) k, h_1, h_2, h_3 (b) $\tau(k, t, -1) = \{h_1\}$ (c) $\tau(k, t, -2) = \{h_2, h_3\}$

—k ⋯h1 – h2 ⋯h3

Fig. 1. Example of temporal neighborhoods constructed with window size $w = 3$, time delay $d = 1$ (1(b)) and $d = 2$ (1(c)) for target wind farm k by considering reference wind farms h_1, h_2 and h_3, respectively (1(a)).

Example 1. *Let us consider one target wind farm k, as well as three reference wind farms, namely, h_1, h_2 and h_3 which measure wind speed data processed with window size $w = 3$(Fig. (1(a)). We can construct a temporal neighborhood of k with time delay $d = -1$ by comparing $z(k, t - w + 1 \rightarrow t)$ to $z(h_1, t - w \rightarrow t - 1)$, $z(h_2, t - w \rightarrow t - 1)$ and $z(h_3, t - w \rightarrow t - 1)$ (Fig. 1(b)), while we can construct the temporal data neighborhood of k with time delay $d = -2$ by comparing $z(k, t - w + 1 \rightarrow t)$ to $z(h_1, t - w - 1 \rightarrow t - 2)$, $z(h_2, t - w - 1 \rightarrow t - 2)$ and $z(h_3, t - w - 1 \rightarrow t - 2)$ (Fig. 1(c)).*

A spatial-temporal neighborhood is built by applying a spatial filter and a temporal filter in cascade. WiNN builds, for every wind farm $k \in K$, w spatial-temporal neighborhoods, one for every delay $d = 1, 2, \ldots, w$ (Algorithm 1, line 8).

Definition 3 (Spatio-Temporal Neighborhood). *The spatio-temporal neighborhood $\eta(k, t, -d)$ is the set of reference wind farms of K ($\eta(k, t, -d) \subseteq K$), which satisfy both the spatial neighborhood filter $\sigma(k)$ (Definition 1) and the temporal neighborhood filter $\tau(k, t, -d)$ (Definition 2) simultaneously, that is, $\eta(k, t, -d) = \sigma(k) \cap \tau(k, t, -d)$.*

Subsequently, spatio-temporal neighborhoods constructed with temporal delays $d = 1, 2, \ldots, w$ (Algorithm 1, line 11) are processed, in order to populate d learning datasets $\mathcal{L}(k, t + 1), \mathcal{L}(k, t + 2), \ldots, \mathcal{L}(k, t + w)$.

Definition 4 (Lazy Learning Data Set). *Let* $\{\eta(k,t,-1)\}_{d=1,2,\ldots,w}$ *be the set of spatio-temporal neighborhoods (Definition 3) associated with* $k \in K$ *at time t and constructed with the temporal delay* $d = 1, 2, \ldots w$, *respectively. The learning data set* $\mathcal{L}(k,t+d)$ *is the set of timestamped data points (reference farm, timestamp, measured wind speed), that is defined as follows:*

$$\mathcal{L}(k, t+d) = \bigcup_{f \geq d} \{(h, t - f + d, z(h, t - f + d)) | h \in \eta(k, t - f)\}.$$

Every learning set $\mathcal{L}(k, t + d)$, with $d = 1, 2, \ldots, w$, is constructed in order to forecast $\hat{z}(k, t + d)$. Lazy learning is performed by resorting to a spatio-temporal version of the Nearest Neighbour formula (Algorithm 1, lines 12). A spatio-temporal distance is computed, in order to estimate the weight according to any sampled backward data point can contribute to the forecast value.

Definition 5 (k-NN). *The forecast value* $\hat{z}(k, t + d)$ *is determined as follows:*

$$\hat{z}(k, t+d) = \frac{\displaystyle\sum_{(h,t',z)\in\mathcal{L}(k,t+d)} \omega((k, t+d), (h, t'))z(h, t')}{\displaystyle\sum_{(h,t',z)\in\mathcal{L}(k,t+d)} \omega((k, t+d), (h, t'))}, \tag{3}$$

where $\omega((k, t+d), (h, t')) = \frac{1}{st((k,t+d),(h,t'))^3}$ *and* $st((k, t+d), (h, t')) = \frac{1}{2} scaled_{01}$
$(d(k, h)) + \frac{1}{2} scaled_{01} d(t + d, t')).$

It is noteworthy that $st(\cdot, \cdot)$ is computed as the sum of the scaling in the interval $[0,1]$ of the distance $(d(k, h) = geoDistance(k, h))$ computed between the geographic coordinates of target wind farm k and neighbor reference farm h, as well as of the scaling in the interval $[0,1]$ of the distance $(d(t+d, t') = t+d-t')$ computed between the timestamps associated to the forecast value $(t + d)$ and to the sampled neighbor (t').

4   Experimental Study

The experiments have been carried out using real world data publicly provided by the DOE/NREL/ALLIANCE3 (http://www.nrel.gov/). The data consist of wind speed measurements from 1326 different locations at 80 m of height in the Eastern region of the US. The data were collected in 10 min intervals during the year of 2004 (time line). This wind farm grid is able to produce 580 GW, and each farm produces between 100 MW and 600 MW. Experiments are run on an Intel(R) Core(TM) i7 920 @2.67GHz running Windows 7 Professional. In this study, we have evaluated the sensitivity of both accuracy and efficiency of WiNN to the set-up of the spatial radius. In addition, we have analyzed the accuracy of WiNN compared to that of auto.ARIMA [5]. The selected competitor is a state-of-art statistical forecasting algorithm. Both algorithms have been evaluated in (very-) short forecasting setting. Forecasts have been produced by considering a time horizon of six hours with wind speed forecasting performed every 10 min ($w = 36$). δ is automatically determined for each considered temporal delay d, as a percentage ($\delta\% = 10\%$) of the maximum Euclidean distance computed between each pair of backward time series, selected for every farm of the grid, at time t and with delay d.

Evaluating WiNN. We have considered the entire wind farm grid, in order to analyze the size of constructed spatial neighborhoods, the average learning time spent, at each time point t, to forecast upcoming wind speed values (over six hours), as well as the accuracy of produced forecasts. Learning time is measured in milliseconds and averaged on the number of time points, as well as on the number of wind farms processed. WiNN is run repeatedly with a radius r varying between 100 km, 250 km and 400 km. Figure 2 (left side) reports the average learning time spent by every wind farm, in order to complete the considered forecasting task. We observe that by increasing the radius, the number of reference wind farms processed, as well as the average distance between neighbor farms increase. Learning times increase accordingly, but they are greatly lower than 10 min. Thus, the forecasting service deployed with WiNN can work in (near)-real time. Figure 2 reports the root mean squared error of the forecasts produced from the wind farm grid. Errors are calculated from the forecasts produced at each time point of the considered time line, for the forward 36 time points (6 h). We observe that the forecasting error decreases by increasing the number of reference neighbor farms processed. However, the reduction of forecasting error is negligible when the radius of spatial neighborhoods is enlarged from 250 Km to 400 km, while the average learning time doubles (from 8 msec to 17 msec) in the same case. Hence, we consider that the choice $r = 250$ km can guarantee an acceptable trade-off between accuracy and efficiency.

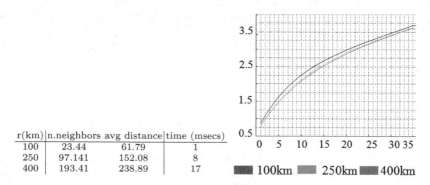

r(km)	n.neighbors	avg distance	time (msecs)
100	23.44	61.79	1
250	97.141	152.08	8
400	193.41	238.89	17

Fig. 2. WiNN: radius r varying among 100 km, 250 km and 400 km. Left side: size of constructed spatial neighborhoods (column 2), average geographic distance between spatial neighbors (column 3) and learning time spent in millisecs (column 4) to complete forecasting. Right side: average rmse (axis Y) of forecasts produced from the entire grid of wind farms at time $t+d$ with $t \in T$ and $d = 1, 2, \ldots, 36$ (axis X) (Color figure online).

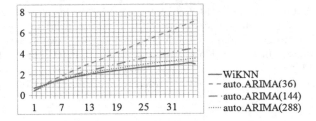

Fig. 3. WiNN vs auto.ARIMA: analysis of the forecasting accuracy measured per forecasting time point.

Comparing WiNN with auto.ARIMA. We have considered 15 wind farms randomly selected across the grid. For every selected wind farm, we have compared forecasting errors of WiNN with forecasting errors of auto.ARIMA. WiNN is run by setting radius r equal to 250 Km and window size w equal to 36. This means that it forecasts, at each time point of the considered time line, the upcoming 36 values of the wind speed and uses backward windows of size equal to 36, in order to determine these forecasts. Similarly to WiNN, the competitor auto.ARIMA forecasts, at each time point of the considered time line, upcoming 36 values of wind speed. These forecasts are determined by considering backward historical data with size ranging between 36 (six hours), 144 (24 h) and 288 (48 h). Figure 3 reports the root mean squared error of the forecasts produced from the fifteen selected wind farms for the upcoming 36 time points (6 h). Errors are calculated from the forecasts produced at each time point of the considered time line. Results show that auto.ARIMA outperforms WiNN if we consider the forecasts of the wind speed associated with the future time points close to t (i.e. $t+1, t+2, t+3, t+4$ and $t+5$), but WiNN outperforms auto.ARIMA if we consider the forecasts associated with the future time points distant from t ($t+6, t+7, \ldots, t+36$). Additionally, auto.ARIMA can produce lower errors, which are closer to the errors produced by WiNN, only by augmenting the amount of historical data to be learned. These results confirm the efficacy of dealing with spatial information when forecasting wind speed in a wind farm grid. Results also show the feasibility of the lazy learning approach that we have formulated for this forecasting task.

5   Conclusion

This paper presents a data mining algorithm that resorts to a spatio-temporal variation of Nearest Neighbor algorithm in order to produce accurate forecasting of very short term wind speed values in a wind farm grid. The efficacy of the proposed algorithm is compared to that of a state-of-art statistical model. As future work, we plan to investigate the combination of the neighborhood construction described in this paper with the spatial aware versions of the ARIMA model described in [11,12].

Acknowledgments. Authors thank Giuseppe Mumolo for his support in developing the algorithm presented. This work is carried out in partial fulfillment of the research objectives of both the Startup project "VIPOC: Virtual Power Operation Center" funded by the Italian Ministry of University and Research (MIUR) and the European project "MAESTRA - Learning from Massive, Incompletely annotated, and Structured Data (Grant number ICT-2013-612944)" funded by the European Commission.

References

1. Albert, W., Chi, S., J.H., C.: An improved grey-based approach for electricity. Elect. Power Syst. Res. **67**, 217–224 (2003)
2. Almeida, V., Gama, J.: Collaborative wind power forecast. In: Bouchachia, A. (ed.) ICAIS 2014. LNCS, vol. 8779, pp. 162–171. Springer, Heidelberg (2014)
3. Gaber, M.M., Zaslavsky, A., Krishnaswamy, S.: Mining data streams: a review. ACM SIGMOD Rec. **34**(2), 18–26 (2005)
4. Giebel, G., Badger, J., Perez, I.M., Louka, P., Kallos, G.: Short-term forecasting using advanced physical modelling-the results of the anemos project. Results from mesoscale, microscale and CFD modelling. In: EWEA 2006 (2006)
5. Hyndman, R., Khandakar, Y.: Automatic time series forecasting: the forecast package for R. J. Stat. Softw. **26**(3), 1–22 (2008)
6. Lange, M., Focken, U.: New developments in wind energy forecasting. Power Eng. Soc. IEEE Gen. Meet. **2008**, 1–8 (2008)
7. Negnevitsky, M., Johnson, P., Santoso, S.: Short term wind power forecasting using hybrid intelligent systems. Power Eng. Soc. IEEE Gen. Meet. **2007**, 1–4 (2007)
8. Ohashi, O., Torgo, L.: Wind speed forecasting using spatio-temporal indicators. In: ECAI 2012 - PAIS 2012 (System Demonstrations Track). Frontiers in Artificial Intelligence and Applications, vol. 242, pp. 975–980. IOS Press (2012)
9. Potter, C., Negnevitsky, M.: Very short-term wind forecasting for tasmanian power generation. IEEE Trans. Power Syst. **21**(2), 965–972 (2006)
10. Pravilovic, S., Appice, A., Lanza, A., Malerba, D.: Wind power forecasting using time series cluster analysis. In: Džeroski, S., Panov, P., Kocev, D., Todorovski, L. (eds.) DS 2014. LNCS, vol. 8777, pp. 276–287. Springer, Heidelberg (2014)
11. Pravilovic, S., Appice, A., Malerba, D.: An intelligent technique for forecasting spatially correlated time series. In: Baldoni, M., Baroglio, C., Boella, G., Micalizio, R. (eds.) AI*IA 2013. LNCS, vol. 8249, pp. 457–468. Springer, Heidelberg (2013)
12. Pravilovic, S., Appice, A., Malerba, D.: Integrating cluster analysis to the ARIMA model for forecasting geosensor data. In: Andreasen, T., Christiansen, H., Cubero, J.-C., Raś, Z.W. (eds.) ISMIS 2014. LNCS, vol. 8502, pp. 234–243. Springer, Heidelberg (2014)
13. Sideratos, G., Hatziargyriou, N.: An advanced statistical method for wind power forecasting. IEEE Trans. Power Syst. **22**(1), 258–265 (2007)

Discovery of Parameters for Animation of Midge Swarms

Judith Bjorndahl, Ashley Herman, Richard Hamilton,
Howard J. Hamilton[(✉)], and Mark Brigham

University of Regina, 3737 Wascana Pkwy, Regina, SK, Canada
Howard.Hamilton@uregina.ca

Abstract. We describe a method of discovering suitable parameters for simulating and animating the swarming behaviour of non-biting midges (*Diptera: Chironomidae*). A characteristic of animal aggregations that can be emulated by software is the emergence of complex behaviours from simple rules. Here the well-characterized swarming behaviour of non-biting midges is used to create a rule-based behaviour model for them. To test the effectiveness of this model in creating the emergent qualities of real swarms, success criteria are derived from quantitative swarm data. We propose using a genetic algorithm to automate the identification of parameter settings that optimize the effectiveness of the model.

Keywords: Midge · Swarm · Simulation · Animation · Genetic algorithm

1 Introduction

This paper describes a method of discovering suitable parameters for simulating and animating the swarming behaviour of Chironomidae (non-biting midges, a type of small insect). In this case, *swarming* describes the behaviour of individuals in which they aggregate but do not align their motions (adapted from [2]). This behaviour is different from *flocking*, in which the individuals have aligned movement and travel together [11], and it is also different from forming a torus, in which all of the individuals travel in a circle [2]. Male chironomus midges are well known for exhibiting swarming behaviour [4].

The problem of simulating midge swarms is typical of many problems in the biological simulation of animal behaviours: the goal is to find a model of behaviour for an individual animal that when simulated across time for many individuals gives a close approximation to the behaviour of groups of individuals observed in nature. The term *emergent behaviour* refers to an apparently complex behaviour which results from combining relatively simple behaviours of many individuals [1].

One approach to solving the problem of simulating midges is to (a) choose a combination of *tendencies* (influences on the behaviour of an individual midge), (b) formulate a behaviour model based on parameterized rules that encompasses the characteristics of the midges and the selected tendencies, (c) select appropriate values for all parameter settings, (d) run an animated simulation, (e) evaluate the degree of realism of the behaviours in the resulting animation, and (f) repeat until satisfactory results are obtained. If a human researcher is responsible for choosing the tendencies and parameter settings and evaluating the results, this process can be relatively inefficient.

© Springer International Publishing Switzerland 2015
N. Japkowicz and S. Matwin (Eds.): DS 2015, LNAI 9356, pp. 17–24, 2015.
DOI: 10.1007/978-3-319-24282-8_3

An alternative approach to the same problem, which is the approach we took, is to attempt to increase efficiency by applying a genetic algorithm to automatically select the tendencies and parameter values and to evaluate the results. For this alternative approach, we require a way of scoring the quality of a behaviour model based on the results of a simulation run or runs. Conveniently, in three recent studies, Puckett, Kelley, and Ouellette have described in detail the movement of individual midges (*Chironomus riparius*) in a swarm [6, 9, 10]. We suggest specific criteria that allow us to assess the fidelity of the model to nature, based on this quantitative data and separate qualitative observations of midge behaviour [4, 7]. From this research, we derived specific criteria to measure the fidelity to nature of a model of midge swarming. We show how to transform the criteria into real-valued functions and how to combine them into a scoring function for midge behaviour in a simulation run.

Our overall goal is to improve the appearance of simulated swarms of midges in animations. A key difference between simulation and animation is that in simulation one wants to have accurate physics, i.e., the equations should be consistent with known physical laws, while in animation, it is sufficient to have consistency with detailed observations. We are attempting first to provide physics-based simulation and secondly to animate the results. Our software can be run with 3D graphical display enabled or disabled. Of course, to produce animations in real time (60 frames per second), the complexity of calculations must be controlled in some fashion to allow the behaviours to be calculated in real time. The resulting animations will still need to be evaluated by a person, in a further step we do not attempt in this paper.

The remainder of the paper is organized as follows. The second section describes background information on midges and swarms of midges, with emphasis on previous models of their swarming behaviour. The third section describes our approach, including mention of the relevant tendencies, the range of possible values for parameters, and a scoring function for evaluating behaviours in simulation runs. We also explain how a genetic algorithm can be applied. The final section gives our conclusions and suggestions for future research.

2   Background

In this section, we present relevant background information on the characteristics of midges and swarms of midges and a summary of previous research on observing and simulating swarms of midges.

Midges and Swarms: Chironomidae is a family in the order *Diptera* (true flies). Chironomids are small (less than 2 cm long [3]) and relatively ubiquitous. Male chironomids form mating swarms. A typical swarm is an aggregation of males "dancing" back and forth in a certain location [10]. Swarms tend to form in certain preferred locations, often based on visual cues or air currents [4, 7]. For example, the midges may swarm directly above a light coloured area (such as a light-coloured straw hat) on a darker background [7]. For simplicity, for our simulation, we select a specific 3D position as the centre of the swarm.

Other than avoiding collisions, midges exhibit little interaction with each other in the swarm [9]. Midges tend to "dance" back and forth in a certain pattern, hovering around one spot and then darting to another. In still conditions (as simulated here), midges have complex paths that take them throughout the swarm [9, 10]. Puckett, Kelley, and Ouellette measured the paths of swarming midges in still conditions and calculated velocities, accelerations, and other statistics related to the swarm. For example, they found that a midge makes a high-curvature turn about 1–3 times per second and travel an average of 3.8 cm between such turns [9]. A *high-curvature turn* exists when the curvature of the midge's path, κ, exceeds a threshold value, κ_0. The value of κ is calculated as $\kappa = |v \times a| / |v|^3$ [10], where v is the velocity vector and a is the acceleration vector of the midge. The threshold value κ_0 is $1/3$ cm^{-1} [10].

Models of Midge Swarms: A relevant early model for swarms is the flocking model used for boids, i.e., creatures similar to schooling fish and flocking birds [11]. Flocking behaviour is typically simulated with three specific tendencies: (a) *cohesion* between the individuals, which attracts them to each other; (b) *separation* between individuals, which prevents them from getting too close together, and (c) *alignment,* which causes the individuals to travel in the same direction [2, 5, 11]. In swarms, however, there is no alignment because there is no uniform direction of travel [2]. As well, swarming midges do not interact with each other via long-distance cohesive influences [9]. Additionally, midges only exhibit separation behaviour when they are avoiding collisions, a behaviour which only occurs when they are within 1.2 cm of each other [9]. Therefore, different tendencies should be used to simulate swarming from those typically used for schooling and flocking simulations.

Couzin et al. proposed a collective behaviour model with separation, cohesion, and parallel alignment, which is able to simulate four collective behaviours: swarm, torus, dynamic parallel group, and highly parallel group [2]. The swarming behaviour model, implemented with long-range cohesion, was determined not to be biologically realistic by Puckett, Kelley, and Ouellette, who compared the acceleration statistics of the model with those of a real swarm [9]. Wang et al. created a swarming model in which the movements of the swarm members are governed by random noise and collision avoidance, without any cohesion [13]. This model appears to make realistic animations, but we were unable to compare their model to quantitative data. To improve efficiency, our approach uses simpler calculations than that of Wang et al.

3   Approach

Our approach to modelling the swarming behaviour of midges has four major facets. The first step is to collect some tendencies observed to affect the behaviour of individual midges and formalize these tendencies as rules to give a behaviour model. The relative weights attached to these tendencies are treated as parameters of the model. The second step is to examine reports by biologists describing the characteristics of midges (e.g., average length and maximum speed in still air) and determine reasonable ranges of values for these characteristics. These characteristics are also treated as parameters of the behaviour model. The third step is to develop a scoring function to rate the fidelity to nature of the behaviour in a simulation run, based on specific statistical measures suggested by biologists. The final step is to apply a genetic algorithm to discover a set of values for the parameters that yield simulation runs that obtain high scores. Each of these facets is now examined in turn.

3.1   Tendencies

We model the following tendencies in the movement of individual midges:

1. **Location Preference**: The midges have a preference to swarm in a certain location
    L, based on topographical features [4, 7]. In the simulation, the swarming location
    is selected arbitrarily.
2. **Central Preference**: The midges have a preference to be near to a large number of
    other midges, as swarming increases the chances of mating, and larger swarms offer
    more protection from predators [8]. In our approach, this tendency is interpreted as
    a tendency to fly toward the centre of the swarm. The center of the swarm is calcu-
    lated as the average position of all midges.
3. **Maximum Radius**: Midges in a swarm have been observed to stay within a radius
    defined based on the size of the swarm. The equation for the radius is:

$$r = \sqrt[3]{\frac{V}{\frac{4}{3}\pi}} \tag{1}$$

where V is calculated as the number of midges in the swarm times 200 cm^3/midge
[10], the mean volume per individual midge. If the midges fly farther from the center
of the swarm than the radius, this tendency causes them to turn and return to being
inside the specified radius.
4. **Dancing Preference**: The dancing movements are roughly figure-eight motions
    based on an imaginary vertical line through the center of the swarm [4].
5. **Random Changes**: The midges appear to change direction at random after a certain,
    presumably random time interval. When the direction is changed, there is a slight
    bias to go toward the preferred location L [9]. Although the direction and time
    interval may not actually be random, no other explanation has yet received wide-
    spread acceptance.
6. **Collision Avoidance**: If a midge is within a certain distance from another midge, it
    accelerates rapidly away from the other midge [9].

Given the above tendencies, we propose reformulating each tendency as an influence
on the *desired velocity D* of an individual midge. The actual velocity $V_{t+\Delta t}$ for the next
time step is computed by blending the previous velocity V_t and the desired velocity D
based on the idea of a steering vector, a maximum acceleration A_{max}, and a maximum
speed s_{max} (adapted from Reynolds [12]).

$$A = truncate(D/\Delta t, A_{max}) \tag{2}$$

$$V_{t+\Delta t} = truncate(V_t + A * \Delta t, s_{max}) \tag{3}$$

where *truncate* reduces the length of the vector given as the first argument to the length
(norm) specified by the second argument.

We assume that the relative importance of the tendencies can be described by
constant weights. As previously mentioned, these weights are parameters in the
behaviour model that are changed by the genetic algorithm.

Characteristics of midges were obtained from research papers and, where data were not available, educated guesses. These characteristics and their possible values are listed below:

- body length: [9, 3]
  - minimum: 0.2 to 1.0 cm
  - maximum = minimum + x, where x is 0.1 to 0.3 cm
- preferred speed in still air: 16.8 cm/s [10]
  - minimum: 10.0 to 14.0 cm/s
  - maximum: 17.5 to 21.0 cm/s
- maximum acceleration: 100 to 500 cm/s^2 [6]
- zone of repulsion: 1.2 cm [9]
- maximum turning rate: 2π rad/s

The mass of a midge is calculated using Eq. 4 [3]

$$y = -0.001929 + 0.0066637x \tag{4}$$

where y is the mass in grams and x is the body length in centimeters. When x is small, y can become negative, so we set a minimum mass of 0.0005 g. The minimum and maximum body length, minimum and maximum speed, and maximum acceleration characteristics are used as parameters by the genetic algorithm.

3.2 Criteria Used to Determine the Fidelity of a Model

As mentioned in Sect. 1, quantitative measures, where available, allow automated assessment of a behaviour model. Considerable progress has been made in deriving quantitative measures related to midge behaviour from detailed observations. The most detailed measures were derived by Puckett, Kelley, and Ouellette [6, 9, 10]. We selected nine of these measures to use as success criteria when testing models. We do not have access to the original researchers' raw data, but we were able to estimate the minimum and maximum observed values for some of the measures from graphs in their papers. In other cases, the relevant values were explicitly stated. Following the example of Puckett and Ouellette, we use a swarm size of 10 midges. They referred to this size as the *asymptotic swarm size*, i.e., the smallest size at which performance is consistent with that of larger swarms [10]. To assess the fidelity of a model to nature, we analyze the extent to which it satisfies the following nine criteria (each stated as a measure and a desired value or range):

1. Mean velocity in every coordinate direction = 0 [6] (C1)
2. Standard deviation of velocity [6]: (C2)
   (a) σ_{vx} = 15 to 19 cm/s (horizontal x)
   (b) σ_{vy} = 15 to 19 cm/s (horizontal y)
   (c) σ_{vz} = 7 to 13 cm/s (vertical z)
3. Mean acceleration in every coordinate direction = 0 [6] (C3)
4. Standard deviation of acceleration σ_a = 100 to 200 cm/s^2 [6] (C4)

5. Slope of the function between the mean acceleration (a) in a coordinate direction and the position (p) in that coordinate direction relative to the centre [6] (C5)
   (a) between $a_x = -2p_x + 13$ and $a_x = -11p_x + 5$ (horizontal x)
   (b) between $a_y = -2p_y + 13$ and $a_y = -11p_y + 5$ (horizontal y)
   (c) between $a_z = -0.5p_z - 5$ and $a_z = -4.0p_z - 5$ (vertical z)
6. Mean nearest neighbour distance = 3.2–4.0 cm [9] (C6)
7. Mean acceleration (a_{NN}) towards nearest neighbour as a function of distance from nearest neighbour: $a_{NN} = 3.75d_{NN} + 2.5$ [9] (C7)
8. Mean free path = 3.4 to 3.8 cm [9, 10] (C8)
9. Relative frequency of encounters < 0.01 [9] (C9)

The mean free path is calculated as the average distance travelled between consecutive turns for all midges [9]. The relative frequency of encounters is calculated as the number of times a midge is within 1.2 cm of another individual divided by the total number of high-curvature turns for that midge, then averaged over all individuals.

   To simplify the task of combining the nine criteria, we devised a set of nine functions, C_i for $0 \leq i \leq 9$, each yielding a real value in [0, 1], to represent them. The scoring function f is a weighted average of these functions, with equal weights ($w_i = \frac{1}{9}$) assigned to each.

$$f = \sum w_i C_i \tag{5}$$

   We assume that these functions should be continuous over the real numbers. For measures with acceptable ranges, a value of 1 is assigned to any result in the desired range and an exponential function is used to assign a value in [0, 1] to any value outside the range. Given an acceptable range [a, b] for criterion i, we first calculate the midpoint $m = \frac{(a+b)}{2}$ and the half width $h = \frac{(b-a)}{2}$. Then we specify function C_i as applied to a value u as:

$$C_i(u) = \min\left(1, e^{-(|u-m|+h)}\right) \tag{6}$$

This function is defined for u in $[-\infty, \infty]$. For criteria that are defined separately for each coordinate dimension, three results are first produced using Eq. 6 and then they are averaged (arithmetic mean).

   We use an extended version of the same technique for measures with slopes. The intuition is that in a sphere centred on the desired location L, the functions should yield values of 1, while outside the sphere, the function should yield values less than 1 and falling to 0 as distance increases. We assume that the area of interest with respect to each of the three coordinate directions is from $L_w - r$ to $L_w + r$, where L_w is the coordinate for dimension w (i.e., any of x, y, and z) of the preferred location L and r is the swarm maximum radius. Given g_{min} (the function derived by linear regression from the observed swarm with the lowest values), g_{max} (the function derived from the observed swarm with the highest values), and g_{sim} (the function derived from our simulation run), we examine the two edges of the swarm independently for a given coordinate dimension and average the results. For dimension w, the lower edge is evaluated using Eq. 6 with

$a = g_{min}(L_w - r)$, $b = g_{max}(L_w - r)$, and $u = g_{sim}(L_w - r)$. The higher edge is evaluated with $a = g_{min}(L_w + r)$, $b = g_{max}(L_w + r)$, and $u = g_{sim}(L_w + r)$. The two results are averaged to give a score for dimension w. As described above, the results for the three dimensions are then averaged.

The score reflects the magnitudes of the differences between the statistics derived from a simulation run and the desired values for these statistics, as listed above. For example, a model with a mean nearest neighbour distance of 3.5 cm gets a value of 1 for C_6, whereas one with a mean nearest neighbour distance of 2.0 cm gets a lower value. To obtain the relevant statistics, the program collects the position, velocity, and acceleration, all in relation to the swarm centre, and the nearest neighbour distance for each midge in every frame.

3.3 Genetic Algorithm

The genetic algorithm used to select parameters is a simple one using a population of behaviour models. Each behaviour model m is represented by a vector with entries for every parameter (treated as the chromosomes). The initial population is generated by choosing random values within the acceptable range for each parameter. Mutation is performed either (a) by adjusting the value of a parameter by a small increment or decrement while staying in the acceptable range or (b) by randomly choosing a different acceptable value for the parameter. Crossover is performed by choosing two models m_1 and m_2, randomly selecting a crossover point along the chromosome vector, and creating a new model m_3 with values from m_1 up to the crossover point and values from m_2 after the crossover point. The processing cycle is as follows: given a population of models, (a) evaluate each model in the population by performing an individual simulation run with the parameters specified for this model and then scoring the results of the simulation run, and (b) create a new population by keeping a fraction of the existing models, adding some new models produced by mutation, and adding some models produced by crossover. The cycle is repeated for a series of generations until an acceptable model is produced or a maximum count is reached. The model with the highest score in the final generation is selected as the best model.

Initially, for the genetic algorithm, we recommend creating a new population by keeping 50 % of the current members, creating 40 % by mutation, and creating 10 % by crossover. Of the mutations, we recommend that half be created by making a small change to one parameter and half by choosing a new random value for the parameter.

4 Conclusions and Further Research

We proposed an approach to the discovery of suitable parameters for a model of midge swarming behaviour. This approach formalizes tendencies into a rule-based model, determines ranges of acceptable values for the parameters to this model from the biological literature, develops a scoring function to rate the fidelity to nature of the behaviour in a simulation run, and finally applies a genetic algorithm to discover a set of values for the parameters that yield simulation runs that obtain high scores.

The novelty of this work is that we provide a method for automatically generating and validating a model of midge behaviour by comparison to quantitative experimental results. Previous researchers who proposed such models validated their results by qualitative assessments of simulations and animations.

Our basic model can be made more complex by considering different circumstances. For example, our current model assumes that the air is still, whereas, in nature, there is almost always at least a light breeze, which is known to affect how the swarm behaves [4]. Additionally, one could adjust the decision-making of the simulated midges to allow them able to judge swarm size and density, and then join or leave swarms based on their size. One could allow the form of the functions representing tendencies to be modified under control of the genetic algorithm. One could also investigate the utility of the midge swarming model in a simulation of the midge-hunting behaviour of bats, where the midges will have to react to a solid object (namely, a bat), passing near or through their swarm.

References

1. Ballerini, M., et al.: Empirical investigation of starling flocks: a benchmark study in collective animal behaviour. Anim. Behav. **76**, 201–215 (2008)
2. Couzin, I.D., Krause, J., James, R., Ruxton, G.D., Franks, N.R.: Collective memory and spatial sorting in animal groups. J. Theor. Biol. **218**(1), 1–11 (2002)
3. Encarnação, J., Dietz, M.: Estimation of food intake and ingested energy in Daubenton's bats during pregnancy and spermatogenesis. Eur. J. Wildl. Res. **52**, 221–227 (2006)
4. Gibson, N.H.E.: On the mating swarms of certain Chironomidae (Diptera). Trans. Roy. Entomol. Soc. London. **95**(6), 263–294 (1945)
5. Hala, S., Hamilton, H.J., Domenici, P., Bjorndahl, J.: Simulating the bubble net hunting behaviour of humpback whales to aid in evaluating biological theories. Unpublished manuscript (2014)
6. Kelley, D.H., Ouellette, N.T.: Emergent dynamics of laboratory insect swarms. Sci. Rep. **3**, 1073 (2013)
7. Lindeberg, B.: The swarm of males as a unit for taxonomic recognition in the Chironomids (Diptera). Ann. Zool. Fenn. **1**(1), 72–76 (1964)
8. Neems, R.M., Lazarus, J., Mclachlan, A.J.: Swarming behaviour in male chironomid midges: a cost-benefit analysis. Behav. Ecol. **3**, 285–290 (1992)
9. Puckett, J.G., Kelley, D.H., Ouellette, N.T.: Searching for effective forces in laboratory insect swarms. Sci. Rep. **4**, 4766 (2014)
10. Puckett, J.G., Ouellette, N.T.: Determining asymptotically large population sizes in insect swarms. J. R. Soc. Interface **11**, 99 (2014)
11. Reynolds, C.: Flocks, herds and schools: a distributed behavioral model. Comput. Graph. **21**, 25–34 (1987)
12. Reynolds, C.W.: Steering behaviors for autonomous characters. In: Proceeding Game Developers Conference 1999, San Jose, California, pp. 763–782 (1999)
13. Wang, X., Jin, W., Deng, Z., Zhou, L.: Inherent noise-aware insect swarm simulation. Comput. Graph. Forum. **33**(6), 51–62 (2014)

No Sentiment is an Island

Sentiment Classification on Medical Forums

Victoria Bobicev[1](✉) and Marina Sokolova[2,3]

[1] Technical University of Moldova, Chișinău, Moldova
vika@rol.md
[2] Institute for Big Data Analytics, Halifax, Canada
sokolova@uottawa.ca
[3] University of Ottawa, Ottawa, Canada

Abstract. In this study we propose a new method to classify sentiments in messages posted on online forums. Traditionally, sentiment classification relies on analysis of emotionally-charged words and discourse units found in the classified text. In coherent online discussions, however, messages' non-lexical meta-information can be sufficient to achieve reliable classification results. Our empirical evidence is obtained through multi-class classification of messages posted on a medical forum.

1 Motivation

A rapid growth in the Internet access from 70 % of the population in 2010 to 81 % in 2014 has caused an increase in online networking from 38 % of the population in 2011 to 46 % in 2014[1]. European Commission's strategy on Big Data (July, 2014) highlights that "Data is at the centre of the future knowledge economy and society"… and that to seize the opportunities of the large and complex resulting datasets, and be able to process such 'big data', initiative must be supported e.g. in the health sector (personalized medicine). Health-care of the future will be based on community, collaboration, self-caring, co-creation and co-production using technologies delivered via the Web (Cambria et al. 2012)."

Online medical forums are platforms on which interested parties (e.g., patients, family members) collaborate for better health. The best forums promote empowerment of patients and improve quality of life for individuals facing health-related problems. An online survey of 340 participants of HIV/AIDS-related Online Support Groups revealed four most important factors that contribute to the patient empowerment: receiving social support, receiving useful information, finding positive meaning and helping others (Mo and Coulson 2012). On surveyed medical forums, personal testimonials attract attention of up to 49 % of the participants, whereas 25 % of the participants are motivated by

---

[1] http://ec.europa.eu/eurostat/.

© Springer International Publishing Switzerland 2015
N. Japkowicz and S. Matwin (Eds.): DS 2015, LNAI 9356, pp. 25–32, 2015.
DOI: 10.1007/978-3-319-24282-8_4

scientific and practical content (Balicco and Paganelli 2011). In a survey of online infertility support groups, empathy and shared personal experience constituted 45.5 % of content, gratitude – 12.5 %, recognized friendship with other members – 9.9 %, whereas the provision of information and advice and requests for information or advice took 15.9 % and 6.8 % respectfully (Malik and Coulson 2010). In many testimonials, informative content intervenes with emotions, e.g. *For a very long time I've had a problem with feeling really awful when I try to get up in the morning* ties up the author's poor feeling and her daily routine.

Restricted communication environment of online support groups can amplify relations between communication competence and emotional well-being, especially for patients diagnosed with potentially life-threatening diseases (Shaw et al. 2008). A study of 236 breast-cancer patient posting online showed that quality of life and psychological concerns can be affected in both desired and undesired ways. Giving and receiving emotional support has positive effects on emotional well-being for breast cancer patients with higher emotional communication skills, while the same exchanges have detrimental impacts on emotional well-being for those with a lower emotional communication competence (Yoo et al. 2014). Challenges arise, however, when sentiments should be analyzed in a large data set: traditional tools, e.g. general-purpose emotional lexicons, are not efficient on medical forums, whereas domain-specific lexicons tend to over-fit the data (Bobicev et al. 2015a, 2015b).

Our current work proposes that coherent online discussions allow classification of sentiments by using information of the post's position in the discussion, sentiments of the neighboring posts and the author's activity level. Further, we test this approach in multi-class sentiment classification of data gathered from an online medical forum.

2   Related Work

Strong relationship exists between language of an individual and her health status (Rhodewalt 1984). Language expressions of negative and undesirable events can be predictors of cardio-vascular disease risks. This connection has led to the development of Linguistic Inquiry and Word Count (LIWC) (Pennebaker and Francis 2001). A software program calculates the degree to which people use different categories of words across a wide array of texts, including emails, speeches, poems, or transcribed daily speech. This tool automatically determines the degree to which any text uses positive and negative emotions, self-references, cognitive and social words.

Qiu et al. (2011) studied dynamics among positive and negative sentiments expressed on Cancer Survivors Network. They estimated that 75 %–85 % of the forum participants change their sentiment in a positive direction through online interactions with other community members.

The sheer volume of on-line messages commands the use of Sentiment Analysis to analyse emotions en masse. Taking advantage of Machine Learning technique, Sentiment Analysis has made considerable progress when applied population health (Chee et al. 2009) as well as on social networks (Zafarani et al. 2010). Empirical evidence shows a strong performance of Naive Bayes, K-Nearest Neighbor, Support Vector Machines, as well as scoring functions and sentiment-orientation methods that use Point-wise Mutual Information (Liu and Zhang 2012). Sentiment Analysis studies mostly identify text's sentiment through the text vocabulary (e.g., positive and negative adjective, positive and negative adverbs) and style (e.g., use of negations, modal verbs) (Taboada et al. 2011).

The In Vitro Fertilization (IVF) data set has been introduced in (Sokolova and Bobicev 2013). The data consists of 80 annotated discussions (1321 posts) gathered from the IVF Ages 35+ sub-forum[2]. Each post was annotated by two raters using three sentiment categories: 'confusion', 'encouragement' and 'gratitude', and one 'factual' category, a category transitional between 'factual' and 'encouragement' was named 'endorsement'. Each post was assigned with one of the labels: 'confusion' (117 posts), 'encouragement' (310 posts), 'gratitude' (124 posts), 'factual' (433), 'endorsement' (162 posts), and 176 'ambiguous' posts on which annotators disagreed. The annotators reached a strong agreement with Fleiss Kappa = 0.737. A detailed description of the manual annotation process can be found in (Sokolova and Bobicev 2013). Previously, sentiments transitions in this data had been studied by applying a domain-specific lexicon HealthAffect and a general-purpose emotional lexicon SentiWordNet (Bobicev et al. 2015a; Bobicev et al. 2015b).

In the current work, we, however, hypothesize that texts related in their content and context can be efficiently classified into sentiment categories without invoking vocabulary of these texts.

3 Problem Statement

We observed a certain pattern of sentiments transactions within discussions: in the first message, the author who started the discussion usually requested help with finding information or emotional help (*confusion* accounted for 56 % of the initial posts). The following posts were either with encouragement (24 %) or provided the factual information requested by the first author (30 %). In many cases, the discussion initiator either updated the interlocutors on the factual progress (39 %) or expressed gratitude for their helpful comments (33 %) (Bobicev et al. 2015b).

The following discussion exemplifies the discussion flow:

---

[2] http://ivf.ca/forums/forum/166-ivf-ages-35/.

post_id_140964 Hi Everyone,I am only five days past my ET and I'm tempted to do a pregnancy home test by next Wednesday. How many of you did home tests or were you patient enough to complete the 2WW without testing at home? I know most of you have mentioned how long the wait seemed or was. ...

post_id_140968 I did do a HPT, first one was 6dp3dt and it was negative, then I did another one 8dp3dt and it was very faintly positive....and here we are with a happy healthy 2 month old! it's totally up to you whether you test, everyone has their opinions! good luck! let us know what you decide!

post_id_140971 I always test too! I just can't help it. I get too anxious and I would much rather know before beta day. ...

post_id_140995 Its really hard to patient and not poas. I just wanted to mention that if you test too early and its negative it can really devasting! The wait is brutal but so is a negative. I tested early also and it was so hard to deal with. You have to decide what you can handle. Best of luck with your cycle.

post_id_141010 I have to admit that I was a chicken when it came to the poas. I never did. I was not sure I could handle it. I just waited it out. You will have to go with your gut. Good luck with whatever you decide.

post_id_141032 I believe that you should do what is right for you. I know some people like to wait and they have the patience of a saint! I know as long as you can keep yourself from being too bummed out if it is a BFN then go for it. There are many people who can and some who can't. I do know that no matter what you will eventually do what your mind wants to do.   It is just human nature. The urge is SO strong. If you do test early, GOOD LUCK!

post_id_141036 Wow! Thanks to all of you for the info and encouragement. Now I don't have to worry about that anymore and I'll certainly buy First Response and do my testing. Again, thanks and all the best to all of you too in every step of this fertility journey.

Working on sentiment analysis *sans* vocabulary content, we pursued the following goals and connected them with feature sets representing the messages:

(A) Our first goal was to demonstrate that there are patterns of sentiments in forum's discussions and they mutually influence each other. Hence, we built a representation which reflected sentiment transitions in discussions. Having two annotation labels for each post we decided to use them both as features rather than merge them. This allowed us to disambiguate the *ambiguous* label, which appeared when two annotators selected different sentiment labels for the post. We then represented each post through the two labels assigned by each annotator to the previous post and two labels assigned by each annotator to the following post; posts lacking this information (e.g., the first post in discussion) were assigned a label "none" (Set I - 4 categorical features).

(B) We then concentrated on the position of the posts within the discussion, as its position can affect the expressed sentiments.

We built three binary features showing whether the previous, current and next messages are first, middle, or last ones. We used these features to enhance the previous representation (Set II – 4 categorical features + 3 binary features = 7 features).

(C) We were interested in the impact of the longer sequences of sentiment transitions on the post's sentiment. To assess this impact, we represented the post by four labels assigned by each annotator to the two previous messages and by four labels assigned by each annotator to the two following messages (Set III - 8 categorical features). To investigate whether this information can be enhanced by the post's position, we expanded Set III with the three position features of the post (Set IV – 8 categorical + 3 binary features = 11 features).

(D) Next, we aimed to represent the influence of author's activity on the post sentiments. We built three features to present the post author's activity: a binary feature pr showing whether the author belongs to the most active authors of this forum (aka a prolific author); a binary feature i indicating whether the author of the post is the one who started this discussion; a binary feature f indicating whether the author posted in this discussion for the first time. Note that these features are independent and can simultaneously be true.

To investigate the mitigating impact of the author's activity, we enhanced the post representation through Set IV by all the three features (Set V – 11 features + 3 features = 14 features) and by each feature separately (Set VI – 11 features + pr = 12 features, Set VII – 11 features + f = 12 features, Set VIII – 11 features + n = 12 features).

Note that all the 12 feature sets omit references to the content of the post they represent.

For multi-class classification, we apply Support Vector Machines (SVM, the logistic model, normalized poly kernel, WEKA toolkit) and Conditional Random Fields (CRF, the default model, Mallet toolkit). SVM has shown a reliable performance in sentiment analysis of social networks. At the same time, we expect CRF to benefit from the feature sets that are sequences of mutually dependent random variables.

4   Empirical Evidence

We worked on four multi-class classification tasks:

6-class classification where 1322 posts are classified into confusion, encouragement, endorsement, gratitude, facts, ambiguous; the majority class F-score = 0.162;

5-class classification where the ambiguous class is removed and remaining 1146 posts are classified into the other 5 classes; the majority class F-score = 0.207;

4-class classification where 1322 posts are grouped as following: facts and endorsement classes make up a (factual) class, encouragement and gratitude classes become a positive class, and confusion and ambiguous classes remain; the majority class F-score = 0.280;

3-class classification where 176 ambiguous messages are removed and the remaining 1146 messages are classified in positive, confusion and factual as in 4-class classification; the majority class F-score = 0.355.

The best classifiers were found by 10-fold cross-validation. We calculated the macro-average F-score. Table 1 reports SVM's performance for each task, and Table 2 – on CRF. The feature sets are the same as in Sect. 3.

Analyzing the results of SVM, we notice that the best F-score is consistently obtained when the feature set conveys all the three aspects of the author's activity (Set V). The impact of the activity attributes is especially noticeable when we compare the results with those obtained on Set IV for 5-, 4-, and 3-class tasks: F-score = 0.448, F-score = 0.495, F-score = 0.594 respectively.

Table 1. Classification results for SVM. For each task, the highest F-score is in **bold**, the lowest F-score – in *italics*.

Sets	Feat.	6-class			5-class			4-class			3-class		
		P	R	F	P	R	F	P	R	F	P	R	F
I	4	0.621	0.605	**0.613**	0.619	0.614	0.616	0.681	0.652	**0.665**	0.732	0.704	**0.717**
II	7	0.613	0.602	0.607	0.616	0.610	0.613	0.654	0.635	0.644	0.711	0.711	0.711
III	8	0.516	0.490	0.502	0.580	0.540	0.558	0.570	0.524	0.545	0.669	0.609	0.637
IV	11	0.516	0.487	0.500	0.566	0.535	0.548	0.568	0.528	0.546	0.671	0.612	0.640
V	14	0.438	0.436	0.436	0.507	0.500	*0.503*	0.489	0.480	*0.483*	0.640	0.625	0.631
VI	12	0.437	0.423	0.429	0.511	0.516	0.513	0.500	0.482	0.489	0.627	0.610	*0.617*
VII	12	0.569	0.544	**0.555**	0.637	0.624	**0.629**	0.559	0.524	0.541	0.657	0.610	0.631
VIII	12	0.489	0.467	*0.477*	0.561	0.522	0.540	0.536	0.506	0.520	0.661	0.611	0.634

The situation changes when we consider the classification results obtained by CRF. For 6-, 4-, 3- class classification, the most predictive feature set is the one that shows sentiment labels of the preceding and following posts, i.e. Set I. Enhancement of the four labels with indicators of the post position in the discussion outputs slightly lower results (Set II). However, these results are still higher than those obtained on other sets. 5-class classification is the only task where CRF benefited from a full spectrum of information available to it. Recall that in this task we removed the ambiguous posts, i.e., the ones labeled with two different labels, and kept original labels assigned by annotators.

Table 2. Classification results for CRF. For each task, the highest F-score is in **bold**, the lowest F-score – in *italics*.

Sets	Feat.	6-class			5-class			4-class			3-class		
		P	R	F	P	R	F	P	R	F	P	R	F
I	4	0.390	0.410	*0.355*	0.45	0.463	0.418	0.494	0.518	*0.487*	0.609	0.611	0.600
II	7	0.395	0.408	0.356	0.441	0.458	*0.414*	0.475	0.528	0.489	0.620	0.620	0.605
III	8	0.365	0.400	0.369	0.445	0.468	0.444	0.478	0.500	*0.484*	0.587	0.593	*0.587*
IV	11	0.371	0.407	0.373	0.449	0.471	0.448	0.493	0.513	0.495	0.595	0.601	0.594
V	14	0.431	0.460	**0.431**	0.524	0.528	**0.517**	0.507	0.538	**0.515**	0.632	0.630	**0.624**
VI	12	0.377	0.412	0.379	0.457	0.479	0.457	0.509	0.534	0.513	0.618	0.617	0.610
VII	12	0.404	0.437	0.408	0.506	0.514	0.501	0.498	0.527	0.504	0.620	0.618	0.612
VIII	12	0.387	0.416	0.387	0.470	0.479	0.463	0.503	0.530	0.508	0.631	0.629	0.623

If compared with the previous work on the same data (Bobicev et al. 2015a, 2015b). The best F-score = 0.613 for 6 class classification improved on the previously reported F-score = 0.491. Note, that we obtained this result based on the neighboring posts' sentiment labels, whereas the classification in (Bobicev et al. 2015a) was done on representing messages through emotional lexicons.

5 Discussion

In this work, we proposed a method that eschews the use of a lexical content in sentiment classification of online discussions. Using a data set gathered from a medical forum, we have shown that sentiments can be reliably classified when posts are represented through sentiment labels of the previous and following posts, enhanced by information about the author activity and the post position in the discussion. We solved 6-,5-,4-, and 3-class classification problems. On the most difficult 6-class classification task, the best F-score = 0.613 improves on the previously obtained F-score = 0.491.

SVM's performance improved when we added information about the post's author (i.e., prolificness, the initiator of the discussion, the discussion's newcomer). CFR performance, however, demonstrated that relationship between sentiments in the consecutive posts provide for a higher classification F-score than longer sentiment sequences. Overall, CRF outperformed SVM due to its ability to gauge information from a sequence of elements.

In this work, we applied a supervised learning approach which relies on manually annotated data. To reduce dependency on manual annotation, we plan a transition to semi-supervised learning. We have shown that sentiment transitions help to predict the sentiment of the current post. A vast volume of messages posted on social media makes the use of fully annotated data unrealistic. Thus, we plan to combine a lexicon-based sentiment classification with the features discussion in this work. A possible approach would be to use Markov chains to disambiguate ambiguous sentiment labels.

References

Balicco, L., Paganelli, C.: Access to health information: going from professional to public practices. In: Information Systems and Economic Intelligence: 4th International Conference – SIIE 2011 (2011)

Bobicev, V., Sokolova, M., Oakes, M.: What goes around comes around: learning sentiments in online medical forums. J. Cogn. Comput. (2015a). http://link.springer.com/article/10.1007/s12559-015-9327-y

Bobicev, V., Sokolova, M., Oakes, M.: Sentiment and factual transitions in online medical forums. In: Barbosa, D., Milios, E. (eds.) Canadian AI 2015. LNCS, vol. 9091, pp. 204–211. Springer, Heidelberg (2015b)

Cambria, E., Benson, T., Eckl, C., Hussain, A.: Sentic PROMs: application of sentic computing to the development of a novel unified framework for measuring health-care quality. Expert Syst. Appl. **39**, 10533–10543 (2012)

Chee, B., Berlin, R., Schatz, B.: Measuring population health using personal health messages. In: Proceedings of AMIA Symposium, pp. 92–96 (2009)

Liu, B., Zhang, L.: A survey of opinion mining and sentiment analysis. In: Aggarwal, C.C., Zhai, C.X. (eds.) Mining Text Data, pp. 415–463. Springer, US, New York (2012)

Malik, S.H., Coulson, N.S.: Coping with infertility online: an examination of self-help mechanisms in an online infertility support group. Patient Educ. Couns. **81**(2), 315–318 (2010). Elsevier

Mo, P.K.H., Coulson, N.S.: Developing a model for online support group use, empowering processes and psychosocial outcomes for individuals living with HIV/AIDS. Psychol Health. **27**(4), 445–459 (2012)

Pennebaker, J.W., Francis, M., Booth, R.: Linguistic Inquiry and Word Count Manual (2001)

Qiu, B., Zhao, K., Mitra, P., Wu, D., Caragea, C., Yen, J., Greer, G., Portier, K.: Get online support, feel better–sentiment analysis and dynamics in an online cancer survivor community. In: Privacy, Security, Risk and Trust (PASSAT) and 2011 IEEE Third International Conference on Social Computing (SocialCom), pp. 274–281 (2011)

Rhodewalt, F.: Self-involvement, self-attribution, and the type a coronary-prone behavior pattern. J. Pers. Soc. Psychol. **47**(3), 662–670 (1984)

Shaw, B., Han, J., Hawkins, R., McTavish, F., Gustafson, D.: Communicating about self and others within an online support group for women with breast cancer and subsequent outcomes. J. Health Psychol. **13**(7), 930–939 (2008)

Sokolova, M., Bobicev, V.: What sentiments can be found on medical forums? In: Proceedings of RANLP 2013, pp. 633–639

Taboada, M., Brooke, J., Tofiloski, M., Voll, K., Stede, M.: Lexicon-based methods for sentiment analysis. Comput. Linguist. **37**(2), 267–307 (2011)

Yoo, W., Namkoong, K., Choi, M., Shah, D.V., Tsang, S., Hong, Y., Gustafson, D.H.: Giving and receiving emotional support online: communication competence as a moderator of psychosocial benefits for women with breast cancer. Comput. Hum. Behav. **30**, 13–22 (2014)

Zafarani, R., Cole, W.D., Liu, H.: Sentiment propagation in social networks: a case study in LiveJournal. In: Chai, S.-K., Salerno, J.J., Mabry, P.L. (eds.) SBP 2010. LNCS, vol. 6007, pp. 413–420. Springer, Heidelberg (2010)

Active Learning for Classifying Template Matches in Historical Maps

Benedikt Budig[✉] and Thomas C. van Dijk

Chair of Computer Science I, Universität Würzburg, Würzburg, Germany
{benedikt.budig,thomas.van.dijk}@uni-wuerzburg.de

Abstract. Historical maps are important sources of information for scholars of various disciplines. Many libraries are digitising their map collections as bitmap images, but for these collections to be most useful, there is a need for searchable metadata. Due to the heterogeneity of the images, metadata are mostly extracted by hand—if at all: many collections are so large that anything more than the most rudimentary metadata would require an infeasible amount of manual effort. We propose an active-learning approach to one of the practical problems in automatic metadata extraction from historical maps: locating occurrences of image elements such as text or place markers. For that, we combine template matching (to locate possible occurrences) with active learning (to efficiently determine a classification). Using this approach, we design a human computer interaction in which large numbers of elements on a map can be located reliably using little user effort. We experimentally demonstrate the effectiveness of this approach on real-world data.

Keywords: Active learning · Threshold detection · Human computer interaction · Template matching · Historical maps · Knowledge discovery

1 Introduction

In this paper we apply proper data mining techniques to a problem in the digital humanities. Many (university) libraries and archives have an extensive collection of historical maps. Besides being valuable historical objects, these maps are an important source of information for researchers in various scientific disciplines. This ranges from the actual history of cartography to general history, as well as the geographic and social sciences. To give a non-trivial example: onomastics, the study of the origin and history of proper names, makes extensive use of historical maps.

With the progressing digitisation of libraries and archives, these maps become more easily available to a larger number of scholars. A basic level of digitisation consists of scanned bitmap images, tagged with some basic bibliographic information such as title, author and year of production. In order to make the maps searchable in more useful ways, further metadata describing the contained information is desirable. A particularly useful class of metadata is a *georeferenced*

© Springer International Publishing Switzerland 2015
N. Japkowicz and S. Matwin (Eds.): DS 2015, LNAI 9356, pp. 33–47, 2015.
DOI: 10.1007/978-3-319-24282-8_5

Fig. 1. Place markers and text on several historical maps from the Franconica collection. Note the variety of visual styles, both in the pictographs and the lettering.

index of the contained geographical features (such as labeled cities and rivers) and geopolitical features (such as political or administrative borders). In this context, a *georeferenced* map element is one that is associated with a real-world, geographical location (in some coordinate reference system). This enables queries that are useful for actual research practice, such as "all 17th century maps that include the surroundings of modern-day Würzburg," or comparing the evolution of place-name orthography in different regions. It also enables analyses of the geographic/geodetic accuracy or distortion of the maps, which is of historical and cartographic interest.

Unfortunately, analysing the contents of historical maps is a complex and time-consuming process. For the most part, this information extraction task is performed manually by experts—if at all. For example, it currently takes the Würzburg University Library between 15 and 30 hours to georeference just the labeled settlements in a typical map from their collection.[1] To see why it takes so long, consider that the number of labeled place markers in a map can be in the order of several thousand.

Automated tools for this task are scarce, for a variety of reasons. For one, there is a large variety of drawing styles in historical maps. This makes it hard for a single algorithm or software tool to automatically perform well on a large set of maps: see Fig. 1 for some examples of the range of styles that occur in the Franconica collection.[2] Secondly, there is the question of input. When an historian georeferences a map, he or she brings a wealth of background information and the ability to do additional research when required. Finally, there is the issue of correctness: in general, algorithms for extracting semantic information from bitmap images are far from perfect. This is to be expected since these problems are truly difficult for computers. To the curators of historical map collections, however, the correctness of metadata can be of paramount importance (not to mention: a matter of pride).

In light of the above difficulties, we have developed an active-learning system for a generally-applicable subproblem in this area. In this paper we demonstrate

---

[1] Personal communication with Dr. H.-G. Schmidt, head of the Manuscripts and Early Prints department, Würzburg University Library.

[2] Würzburg University Library, http://www.franconica-online.de/.

that active learning is suitable for this real-world task. As a first step, a user indicates a rectangular crop around the map element he or she is looking for, such as ⚔ or 🏛. We use standard techniques from image processing to find a set of *candidate matches*, but the problem remains to determine which of these candidate matches are in fact semantically correct. We model this as a classification problem and use pool-based batch-mode active learning. Experiments show that the resulting human-computer interaction is efficient.

2   Related Work

Since the digitisation and analysis of historical maps is of increasing interest to libraries, several systems simplifying this complex process have been developed. Most of these systems provide convenient graphical interfaces, but still rely heavily on users to manually annotate or even georeference the input maps. See for example Fleet et al.'s *Georeferencer* [7] and the system by Simon et al. [23]. For the postprocessing of georeferenced maps, Jenny and Hurni [13] introduced a tool that is able to analyse the geometric and geodetic accuracy of historical maps and then visualise the identified distortions.

Some research has gone into image segmentation specifically for bitmap images of (historical) maps. Höhn [9] introduced a method to detect arbitrarily rotated labels in historical maps; Mello et al. [15] dealt with the similar topic of identifying text in historical maps and floor plans. These systems are rather sensitive to their parameters, requiring careful tweaking in order to perform well. In a further paper, Höhn et al. [10] specifically raise this as an area for improvement: their experiments work well, but do not necessarily generalise to a large variety of maps. The system of Mello et al. was developed for a large set of rather homogeneous maps, which means that it was merited to spend significant effort to find good parameter values. In contrast, we aim to handle diverse maps, each with relatively small user effort. We therefore specifically address finding model parameters.

There is not much research available on fully-algorithmic information retrieval specifically from historical maps. Automatic approaches exist, but only for restricted inputs—that is, developed specifically to digitise a particular corpus. For example, Leyk et al. [14] describe a method to find forest cover in a specific set of 19th century topographic maps. Arteaga [1] extracts building footprints from a set of historical maps from the New York Public Library (NYPL). The effectiveness of these approaches is in part due to the homogeneity of these relatively recent maps. The tests in this paper are performed on much older maps (16th and 18th century).

We approach the above problem using active learning (see Settles [20] for a survey). In particular, we use batch-mode learning [4,8,11]. Our approach is pool based, that is, we have a discrete set of items that we wish to classify and we can only query the oracle on those items. In effect, we learn a threshold [3], based on logistic regression. See Schein and Ungar [18] for a general discussion of active learning for logistic regression.

The design of our system takes into account the human factors involved in using a human as oracle. This combines aspects of human-computer interaction (HCI) and knowledge discovery, as advocated for example by Holzinger [12]. Such factors can be incorporated in the algorithms used, as in *proactive learning* [6]. For our purposes we found that standard active learning suffices.

3   Design Rationale

Our general goal is to georeference bitmap scans of historical maps. We focus on a specific subtask of this larger goal in order to get a manageable problem. This modular approach—with subgoals more modest than "understand this map"— allows for rigorous problem statements and, thereby, reproducible experiments and comparability; this is in contrast to monolithic software systems, where it can be unclear how any specific detail influences the outcome. Competing systems for a certain step can then be proposed and evaluated. Such a "separation of concerns" in systems for processing historical maps is also advocated, for example, by Shaw and Bajcsy [22] and Schöneberg et al. [19]. The latter propose a pipeline with separate tasks operating independently; our (interactive) system could serve as a module in such a system.

The task we discuss in this paper is finding pictographs and textual elements. This is an information extraction step that lifts from the unstructured level of a bitmap image to data that is combinatorial in nature: a list of locations of map elements. Finding approximate matches of an example image is a classic problem in image processing (see for example Brunelli [2] for an overview). This approach can be used for a variety of map elements, from settlement pictographs, to forests, to text labels: we find approximate repeat occurrences of an example image. However, standard techniques yield only a list of candidates along with "matching scores:" this still needs to be converted into into a yes/no classification. In this paper we focus on efficiently learning a classifier in this setting.

Specifically in our application, the user provides a *template* by indicating the bounding box for an interesting map element. This could be a prototypical pictograph on the map, such as a house (🏠), a tree (🌳) or even individual characters (*a*, *e*, *n*). See Fig. 3 for an example: here the user wants to find all occurrences of the character 'a' and inputs the red rectangle in the leftmost image. The template matching algorithm comes up with—among thousands of others—the three matches indicated in the other images. The remaining problem is to decide which of these matches are in fact semantically correct.

The usefulness of recognising individual characters should not be underestimated, since standard optical character recognition (OCR) does not perform well when applied directly to an entire historical map: consider for example Fig. 1, particularly the middle image where the text is not clearly separated from the other map elements. Even in such messy maps, there are usually several characters that are particularly recognisable (which ones might depend on the handwriting). Given one typical example of a character, our method can be used to find most of the other occurrences of the character with high precision.

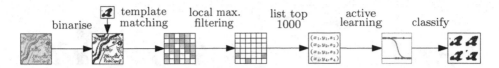

Fig. 2. Overview of the consecutive steps in our method. The input is a bitmap generated by scanning a historical map, and a template to search for. The output is a list of positive matches and their location in the image.

If we do this for a small number of different characters, a later pipeline step can cluster these results to find out where the text elements are (for example: labels). This can be used as a preprocessing step for OCR, in case the OCR algorithm would otherwise get confused by overlapping map elements or is computationally too expensive to be run on the entire map. Because of this application, we prefer our system to have a tendency to side with precision over recall: false negatives are not a disaster if we use a suitable set of characters, since it is likely that at least some character occurrences within each label will be found. This approach based on finding a small set of specific characters as preprocessing is also used by Leyk et al. [14].

In our experiments we have used a basic template matching algorithm, which we briefly sketch here. Since all our maps are effectively black and white, we first binarise to a 1-bit-per-pixel bitmap. Then we consider a sliding window, which calculates a matching *score* for every possible position, to pixel precision: when the template is shifted to a certain position, how many pixels are equal between the template and the image, and how many are different? Following standard procedure, we take the percentage of equal pixels as our matching score.[3] If the score is high for a certain pixel (that is, for a certain position of the template), it is likely that a slight shift of the template still results in a good score; we therefore throw out all pixels that do not have maximal score in their 8-neighbourhood. Of the remaining pixels, we select the 1000 highest-scoring ones: this parameter is chosen generously such that all true positive matches survive this step. In this way, the template matching algorithm is used as a data reduction and projection step that takes place before the classification happens. See Fig. 2 for an overview of the different steps in this process.

This leaves the classifier. We choose to classify based on a score threshold, or equivalently: a rank threshold. A threshold that more-or-less cleanly separates the true positive matches from the true negative matches does indeed exist in our experiments: we have manually created ground truth for the templates in Table 1 and find ROC curves with area under curve of around 0.9.

Because the maps and the templates vary wildly, picking a single threshold value will not work. Some literature in fact handwaves this issue (for example [9]) by hand picking the value for their experiments. This is valid when the objective

---

[3] Note that this basic approach is not invariant to scale and rotation. It is naturally robust against *small* variations, but some historical maps would require a more advanced template matching algorithm.

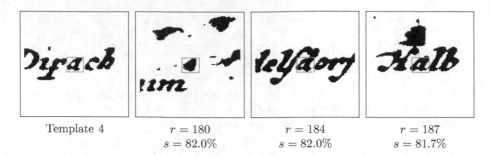

Template 4	$r = 180$	$r = 184$	$r = 187$
	$s = 82.0\%$	$s = 82.0\%$	$s = 81.7\%$

Fig. 3. Various crops from the same historical map. The red rectangle in the leftmost image indicates the crop used as template; the other three are computed candidate matches. Note that these three matches have similar rank and score, but do not all represent semantic matches of the template. In the ground truth we reject the rank-180 match (probably a hill) and accept the rank-187 match ('a'). The ground truth of Experiment 4 accepts the rank-184 match ('d'): see Fig. 4 for the reasoning.

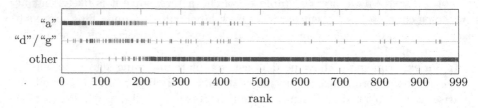

Fig. 4. Distribution of the contents of the first thousand matches for Template 4, ordered by rank. Matches containing either "a", "d" or "g" can be separated fairly well from the remaining matches using a threshold (e.g. rank $\leq$ 200). In contrast, a discrimination of strictly the matches showing "a" will not have high accuracy.

is to show that a certain algorithm *can* achieve high accuracy, but does not show usefulness of the method in practice. In order to efficiently classify the potential matches given to us by the template matching algorithm, we will employ pool-based active learning with a human user as oracle.

Since a given candidate match either contains the desired element (correct) or does not contain it (incorrect), we describe it with a dichotomous variable. We then use logistic regression as a model to discriminate between correct and incorrect matches. In the experiments section we show that logistic regression is a suitable classifier when trained on complete ground truth (all labels).

Since acquiring labels is the most time-consuming step in our system—it involves a human—we use active learning. Following standard practice, we use the following batch-mode query strategy. As input it takes the list of candidate matches, ordered by rank, and a parameter k, the size of a batch. (We examine the choice of k in the experiments section.) The algorithm starts by assuming the best-scoring match is correct and the worst-scoring match is incorrect and (trivially) fits an initial model. Then, in each iteration it picks the k unlabeled matches that are most uncertain (according to the current model) and asks the user to label this batch; the results are stored and the model is retrained. After any number of iterations, this gives the following classifier: return the user-provided label if available, and give the most likely answer according to the logistic regression model otherwise.

4   Experiments

In order to evaluate the efficacy of our method, we have implemented the proposed system and applied it to several real-world datasets. This section describes our findings.

4.1   Evaluation Settings

We implemented our method primarily in Python, using the *Scikit-learn* library[4] for logistic regression. The template matching is implemented in C++. All experiments presented in this section have been run on a desktop PC. Neither runtime nor memory were an issue; template matching takes up to a couple of second on practical maps and batch selection occurs in realtime.

To evaluate our active learning approach, we created nine real-world data sets. These were created by analysing template matching results from actual historical maps, using various templates: the combination of a map and a template identifies a data set. For every data set, we considered the thousand highest-ranking matches and manually determined if they are correct. This gives us a ground truth containing nine times 1,000 samples; Table 1 gives an overview of these datasets.[5] Note that for some templates we have accepted several characters, not just the exact character in the template. This improves classification performance for reasons illustrated in Fig. 4; see also Deseilligny [5]. Choosing which characters to accept for a certain template currently involves some user judgment, but the sets shown in the table seem widely applicable.

The samples in these data sets have only one feature: their score according to template-matching algorithm. These scores also imply a ranking of the samples. In each of the following experiments, there was no clear difference between using the actual scores and using the implied rank. For the rest of the paper, we report the results of using the sample's rank as its (only) feature.

In order to assess how difficult the classification for a particular template is, and if learning is even feasible, we use ROC analysis for a threshold classifier. Figure 5 shows an area under curve of over 0.85 for all experiments, showing that the approach is feasible for a wide range of templates. As an additional measure of difficulty, we trained the logistic regression model on a full ground truth of each data set. This allows us to calculate the self information (or: surprisal) for every sample, relative to this model. Table 1 shows the sum of self information

---

[4] See [17] and http://scikit-learn.org/.

[5] Available at http://www1.pub.informatik.uni-wuerzburg.de/pub/data/ds15/.

Table 1. Data sets used in our experiments. Each line describes one data set: the name of the map, a thumbnail of the template, characters that were considered positive matches, the area under curve according to Fig. 5 and the self-information relative to the logistic regression model trained on all samples.

	Historical Map	Template	Accepted	AUC	Self-Info
1	*Carte Topo. D'Allemagne* (1787)	*b*	b, h	0.85	462.91 bit
2	*Franciae Orientalis* (1570)	*a*	a, g, d	0.90	566.95 bit
3	*Franciae Orientalis* (1570)	*e*	e	0.87	642.02 bit
4	*Circulus Franconicus*, De Wit (1706)	*a*	a, g, d	0.92	444.48 bit
5	*Das Franckenlandt* (1533)	*a*	a, g	0.87	590.50 bit
6	*SRI Comitatus Henneberg* (1743)	*n*	n, m, h	0.92	524.85 bit
7	*SRI Comitatus Henneberg* (1743)	*e*	e	0.87	524.01 bit
8	*Circulus Franconicus*, De Wit (1706)	*i*		0.88	560.29 bit
9	*Circulus Franconicus*, Seutter (1731)	o		0.99	146.16 bit

All maps in this table are taken from the Franconica collection (http://www.franconica-online.de/) of the Würzburg University Library. Identifiers: **1**: 36/A 1.16-41; **2, 3**: 36/A 20.39; **4, 8**: 36/A 1.17; **5**: 36/G.f.m.9-14,136; **6, 7**: 36/A 1.13; **9**: 36/A 1.18.

over all matches of each template. This can also be regarded as a measure of the classification difficulty for the particular template: high self information hints at a larger number of outliers and/or a wider interval of rank overlap between the positive and negative samples. This interpretation is confirmed by the fact that the data sets collected on maps from the 16th century have higher self information than those on maps from the 18th century. On many of the older maps, elements indeed seem harder for humans to distinguish due to the heterogeneous style of handwriting and the suboptimal state of preservation.

We measured the classification performance of our algorithm using accuracy and F1 score. (See for example Parker [16] for definitions of these standard evaluation criteria.) Values for precision and recall of our classifier will also be discussed. Recall that for our application, precision is more important than recall: a missed character or text label might still be located later using another template, whereas false positives could potentially disturb subsequent pipeline steps (such as OCR) significantly.

4.2   Evaluation Results

Classification Performance. We have run our algorithm on the nine real-world data sets introduced above. Following Settles [21], we use *learning curves* to show the performance of our method. We use batch size $k = 3$ unless stated otherwise. A discussion of the choice of this parameter value follows later. As a baseline, we have used a random strategy, where the batch of samples to be labeled is picked uniformly at random from the pool of unlabeled samples.

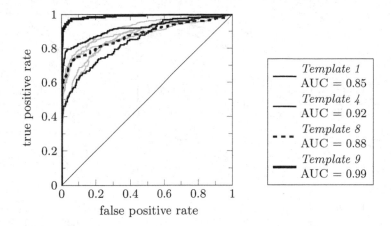

Fig. 5. ROC curves for the data sets in Table 1. Highlighted are the lowest and highest area under curve values for templates containing characters (Templates 1 and 4). Of the two templates for place markers, one shows typical performance (Template 8) and one performs exceptionally well (Template 9).

We now show that our active strategy outperforms this baseline strategy in almost every situation.

Figure 6 shows the learning curves of our active learning approach in comparison to the random strategy. The plots describe the accuracy of both classifiers against the number of iterations; the number of labeled samples is three times this number, as we set $k = 3$. For the random strategy, we performed 100 runs and show mean, 10th, and 90th order statistic of the achieved accuracy. As we can see in the figure, the accuracy of the active learning strategy dominates the accuracy of the random strategy at almost every iteration. Only in the very beginning (number of iterations below approximately 15), this is not consistently true. However, the active learning approach is near the 90th percentile performance of the random strategy in these situations as well.

Now we look at additional performance measures. The results in this experiment refer to Template 6, as a typical example. Figure 7 shows the performance of the active learning approach in comparison to three runs of the random sampling strategy. Note that after 15 iterations, the active learning classifier dominates the three random classifiers in accuracy, precision and F1 score. The random approach does better only in terms of recall, which we find acceptable, as discussed before. The same observations hold for a larger number of random runs and for the remaining data sets; plots are omitted for space.

It can additionally be noted that, in contrast to the random approach, all four scores increase monotonically after the first few iterations when using the active learning method. Thus, when adding additional labels, the classifier's performance is highly likely to improve. This property is especially valuable for the design of proper user interaction when using active learning: from the users'

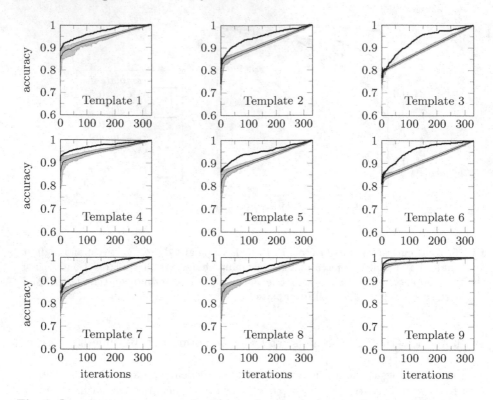

Fig. 6. Learning curves comparing the performance of our active learning strategy ($k = 3$) to the random baseline strategy. The bold black line indicates the accuracy of our method over the iterations. The thin black line shows the mean accuracy of 100 runs of the random strategy; the grey area indicates 10th to 90th percentile.

point of view, it is hard to accept that additional effort in labeling leads to a decrease in quality.

In the next experiment, we consider the self information of the samples that our system selects, in comparison to those chosen by the random baseline strategy. We calculate the self information as before (see Evaluation Settings). For almost any number of iterations, the total self information in the samples from the active learning strategy is considerably higher than in those from the random baseline strategy. Figure 8 illustrates this for four templates; the same holds for the remaining five data sets. This behaviour of the active learning strategy is desirable, because higher self information means that the labeled samples were *indeed* hard to classify for the logistic regression model and therefore having them labeled by the user is valuable. In contrast, the random strategy presents a substantial number of samples whose labels are comparatively clear (for example because they have a very high rank), thereby wasting the user's time.

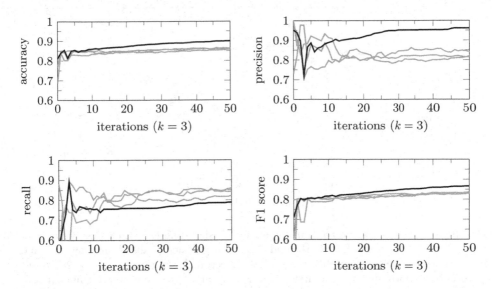

Fig. 7. Statistics for our active learning strategy (black) and three runs of the random baseline strategy (grey) on Template 6. Note that after 15 iterations, this strategy outperforms the random baseline strategy in all measures except recall.

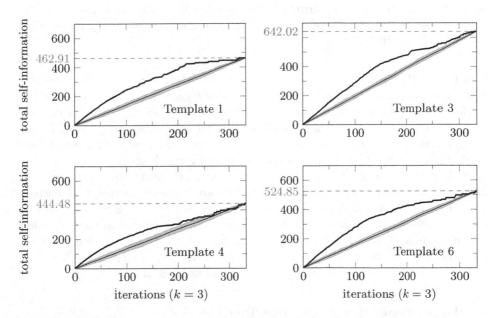

Fig. 8. Self-information of all samples that have been labeled up to a given iteration. The labels picked by the active learning strategy (bold black) are considerably more informative than those selected by the random baseline strategy (black: mean, grey area: 10th to 90th percentile). Note that in the end, each strategy has labeled all samples and achieves the self information of the ground truth as listed in Table 1.

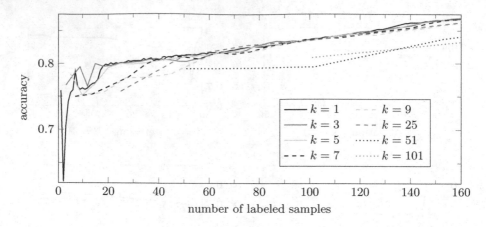

Fig. 9. Accuracy of our active learning strategy using different batch sizes k on Template 3. For values of k between 3 and 7, accuracy is acceptable from the start and increases for increasing number of samples. Exceedingly large values ($k \geq 25$) result in inferior performance for the first few iterations and these represent significant user effort due to the batch size.

Runtime. In our decidedly unoptimised implementation, it takes a total of approximately one second of runtime to calculate 100 batches of size $k = 3$. As this represents 100 batches of user interaction, the system is clearly suitable for realtime applications.

Choice of Parameters. Our system depends on the batch size k. We have run a set of experiments to evaluate the influence of k on our system's classification performance. Figure 9 shows that the performance of our system does not depend very strongly on the choice of k, as long as no extreme values are chosen. Based on this data set, we might recommend values between 3 and 7. This conclusion holds for the remaining templates (plots omitted for lack of space).

When choosing the parameter k, human factors should also be taken into account. The time taken to decide if a displayed candidate match is correct (that is, to label a sample) varies with the batch size. Since selecting and delivering a new set of samples to the user requires a perceptible amount of time (both technologically and cognitively), a larger batch size may cause less user disturbance. For this reason—and aesthetic reasons—we currently use $k = 9$ in our web-based prototype implementation of the user interface.

5   Prototype of a User Interface

In addition to the experimental setup described above, we have implemented a prototype user interface for our system. This allows us to assess not only the abstract, but also the practical suitability of our approach. Figure 10 shows two screenshots of our implementation. The interface on the left allows users

(a) An overview of a map, the corresponding templates and the candidate matches.

(b) Classification interface with batch size $k = 9$.

Fig. 10. Screenshots showing two user interfaces from our web-based prototype. Note that (a) is intended to be used on large screens, while (b) can be used conveniently on smartphones as well.

to browse a historical map, crop templates and start the template matching process. With the interface on the right, users can classify samples selected by the active learning system (in the screenshot $k = 9$). By clicking or touching any of the nine tiles, the tile turns around and shows a green check mark to indicate that the sample was classified as positive. Once the user is finished inspecting the nine samples, he or she presses the "Next" button. The samples that remain unchecked will be considered negative and a new batch of samples chosen by the active learning algorithm will be presented. Our implementation of the user interface is web-based (using HTML5 and JavaScript), so it can be used seamlessly on any device that runs a modern browser. In particular, the classification interface can be conveniently used on smartphones, which enables crowdsourcing of this task. (See also Arteaga [1].)

Using our prototype, it takes a user with some experience approximately 25 seconds to do 4 iterations (that is, to classify 36 samples, since $k = 9$). This includes the runtime of our active learning algorithm and client-server overhead. According to our experimental results in the preceding section, the stated number of labels is already enough to achieve good classification results for a typical template. Projecting these numbers, our approach allows the effective classification of 10 templates within 5 min, assuming the templates have been selected beforehand. In contrast, even with significant experience it takes about 10 to 15 min to generate the full ground truth for a single template. This leaves quite some time to select the templates and still achieve a factor-10 improvement in template throughput. (Recall that the user is probably looking for many templates on the same map.) This shows that our system, and the proposed user interaction, is well-suited for this application.

6   Conclusion

In this paper, we have tackled a real-world problem from a knowledge-discovery perspective: the extraction of information from historical maps. We have focused on the detection of occurrences of certain elements in bitmap images, and introduce a practical approach that solves this problem. Our proposed system uses template matching for feature extraction from the image, and batch-mode active learning to detect appropriate thresholds. Particularly this active-learning step addresses an open problem in the literature on metadata extraction from historical maps. We implemented this approach and experimentally demonstrate that it performs well on real data sets from practice. In combination with the user interface we propose, our system is able to save users a significant amount of time when georeferencing historical maps. Directions for future work include the following.

In a practical setting, our system clearly extends to other (historical) documents besides maps. Early experimentation shows, for instance, that the system also works well for locating specific glyphs in medieval manuscripts. Our prototype is currently being integrated into the existing workflow at Würzburg University Library, which will enable user studies on a proper scale.

On a more abstract level, our active-learning approach with human-computer interaction is not limited specifically to historical documents and template matching. We expect that many other computer-vision methods that depend sensitively on parameter selection can benefit from this strategy.

Acknowledgments. We thank Wouter Duivesteijn for fruitful discussion and helpful comments. We thank Hans-Günter Schmidt of the Würzburg University Library for providing real data and practical use cases.

References

1. Arteaga, M.G.: Historical map polygon and feature extractor. In: Proceedings of the 1st ACM SIGSPATIAL International Workshop on MapInteraction, pp. 66–71 (2013)
2. Brunelli, R.: Template Matching Techniques in Computer Vision: Theory and Practice. Wiley, New York (2009)
3. Bryan, B., Nichol, R.C., Genovese, C.R., Schneider, J., Miller, C.J., Wasserman, L.: Active learning for identifying function threshold boundaries. Adv. Neural Inf. Process. Syst. **18**, 163–170 (2006)
4. Chen, Y., Krause, A.: Near-optimal batch mode active learning and adaptive submodular optimization. In: Proceedings of the 30th International Conference on Machine Learning, pp. 160–168 (2013)
5. Deseilligny, M.P., Le Men, H., Stamon, G.: Character string recognition on maps, a rotation-invariant recognition method. Pattern Recogn. Lett. **16**(12), 1297–1310 (1995)
6. Donmez, P., Carbonell, J.G.: Proactive learning: cost-sensitive active learning with multiple imperfect oracles. In: Proceedings of the 17th ACM Conference on Information and Knowledge Management, pp. 619–628 (2008)

7. Fleet, C., Kowal, K.C., Pridal, P.: Georeferencer: crowdsourced georeferencing for map library collections. D-Lib Mag. **18**(11/12) (2012)

8. Guo, Y., Schuurmans, D.: Discriminative batch mode active learning. In: Advances in Neural Information Processing Systems 20, Proceedings of the 21st Annual Conference on Neural Information Processing Systems, pp. 593–600 (2007)

9. Höhn, W.: Detecting arbitrarily oriented text labels in early maps. In: Sanches, J.M., Micó, L., Cardoso, J.S. (eds.) IbPRIA 2013. LNCS, vol. 7887, pp. 424–432. Springer, Heidelberg (2013)

10. Höhn, W., Schmidt, H.G., Schöneberg, H.: Semiautomatic recognition and georeferencing of places in early maps. In: Proceedings of the 13th ACM/IEEE-CS Joint Conference on Digital Libraries, pp. 335–338 (2013)

11. Hoi, S., Jin, R., Zhu, J., Lyu, M.: Batch mode active learning and its application to medical image classification. In: Proceedings of the 23rd International Conference on Machine Learning, pp. 417–424 (2006)

12. Holzinger, A.: Human-computer interaction and knowledge discovery (HCI-KDD): what is the benefit of bringing those two fields to work together? In: Cuzzocrea, A., Kittl, C., Simos, D.E., Weippl, E., Xu, L. (eds.) CD-ARES 2013. LNCS, vol. 8127, pp. 319–328. Springer, Heidelberg (2013)

13. Jenny, B., Hurni, L.: Cultural heritage: studying cartographic heritage: analysis and visualization of geometric distortions. Comput. Graph. **35**(2), 402–411 (2011)

14. Leyk, S., Boesch, R., Weibel, R.: Saliency and semantic processing: extracting forest cover from historical topographic maps. Pattern Recogn. **39**(5), 953–968 (2006)

15. Mello, C.A.B., Costa, D.C., dos Santos, T.J.: Automatic image segmentation of old topographic maps and floor plans. In: Proceedings of the 2012 IEEE International Conference on Systems, Man, and Cybernetics, pp. 132–137 (2012)

16. Parker, C.: An analysis of performance measures for binary classifiers. In: Proceedings of the 11th International Conference on Data Mining, pp. 517–526 (2011)

17. Pedregosa, F., Varoquaux, G., Gramfort, A., Michel, V., Thirion, B., Grisel, O., Blondel, M., Prettenhofer, P., Weiss, R., Dubourg, V., Vanderplas, J., Passos, A., Cournapeau, D., Brucher, M., Perrot, M., Duchesnay, E.: Scikit-learn: machine learning in python. J. Mach. Learn. Res. **12**, 2825–2830 (2011)

18. Schein, A.I., Ungar, L.H.: Active learning for logistic regression: an evaluation. Mach. Learn. **68**(3), 235–265 (2007)

19. Schöneberg, H., Schmidt, H.G., Höhn, W.: A scalable, distributed and dynamic workflow system for digitization processes. In: Proceedings of the 13th ACM/IEEE-CS Joint Conference on Digital Libraries, pp. 359–362 (2013)

20. Settles, B.: Active learning literature survey. Computer Sciences Technical report 1648, University of Wisconsin-Madison (2010)

21. Settles, B.: Active Learning. Synthesis Lectures on Artificial Intelligence and Machine Learning. Morgan and Claypool Publishers, San Rafael (2012)

22. Shaw, T., Bajcsy, P.: Automation of digital historical map analyses. In: Proceedings of the IS&T/SPIE Electronic Imaging 2011, vol. 7869 (2011)

23. Simon, R., Haslhofer, B., Robitza, W., Momeni, E.: Semantically augmented annotations in digitized map collections. In: Proceedings of the 11th Annual International ACM/IEEE Joint Conference on Digital Libraries, pp. 199–202 (2011)

An Evaluation of Score Descriptors Combined with Non-linear Models of Expressive Dynamics in Music

Carlos Eduardo Cancino Chacón[✉] and Maarten Grachten

Austrian Research Institute for Artificial Intelligence, Vienna, Austria
{carlos.cancino,maarten.grachten}@ofai.at
http://www.ofai.at/research/impml/

Abstract. Expressive interpretation forms an important but complex aspect of music, in particular in certain forms of classical music. Modeling the relation between musical expression and structural aspects of the score being performed, is an ongoing line of research. Prior work has shown that some simple numerical descriptors of the score (capturing dynamics annotations and pitch) are effective for predicting expressive dynamics in classical piano performances. Nevertheless, the features have only been tested in a very simple linear regression model. In this work, we explore the potential of a non-linear model for predicting expressive dynamics. Using a set of descriptors that capture different types of structure in the musical score, we compare the predictive accuracies of linear and non-linear models. We show that, in addition to being (slightly) more accurate, non-linear models can better describe certain interactions between numerical descriptors than linear models.

Keywords: Musical expression · Non-linear basis models · Artificial neural networks · Computational models of music performance

1 Introduction

Performances of written music by humans are hardly ever precise acoustical renderings of the notes in the score, as a computer would produce —nor are they expected to be. A natural human performance involves an interpretation of the music, in terms of structure, but also in terms of affective content [5,22], which is conveyed to the listener by local variations in tempo and loudness, and (depending on the expressive possibilities of the instrument) the timing, articulation, and timbre of individual notes.

Musical expression is a complex phenomenon. Becoming an expert musician takes many years of training and practice, and rather than adhering to explicit rules, achieved performance skills are to a large degree the effect of implicit, procedural knowledge. That is not to say that regularities cannot be found in the way musicians perform music. Decades of empirical research have identified a number of factors that jointly determine the way a musical piece

© Springer International Publishing Switzerland 2015
N. Japkowicz and S. Matwin (Eds.): DS 2015, LNAI 9356, pp. 48–62, 2015.
DOI: 10.1007/978-3-319-24282-8_6

is rendered [11,21]. For example, aspects such as phrasing [29], meter [25], but also intended emotions [20], all have an effect on expressive variations in music performances.

A better understanding of musical expression is not only desirable in its own right. The potential role of computers in music creation will also depend on accurate computational models of musical expression. For example, music software such as MIDI sequencers and music notation editors may benefit from such models in that they enable automatic or semi-automatic expressive renderings of musical scores.

Several methodologies have been used to study musical expression. Complementary to controlled experiments that investigate a single aspect of performance, data mining and machine learning paradigms set out to discover regularities in musical expression using data sets comprising musical performances [23,31]. Given the implicit nature of expressive performance skills, the benefit of the latter approach is that it may reveal patterns that have gone as of yet unnoticed, because perhaps they do not relate in any obvious ways to existing scholarly knowledge about expressive performance.

A computational framework has been proposed in [13], to model the effect of structural aspects of a musical score on expressive performances of that score, in particular expressive dynamics (the relative intensity with which the notes are performed). This framework, referred to as the Linear Basis Model (LBM), follows the machine learning paradigm in that it estimates the parameters of a model from a set of recorded music performances, for which expressive parameters such as local loudness, tempo, or articulation, can be measured or computed.

An important characteristic of the LBM is its use of *basis functions* as a way to describe structural properties of a musical score, ranging from the metrical position of the notes, to the presence and scope of certain performance directives. For instance, a basis function for the performance directive *forte* (f), may assign a value of 1 to notes that lie within the scope of the directive, and 0 to notes outside the scope. Another basis function may assign a value of 1 to all notes that fall on the first beat of a measure, and 0 to all other notes. But basis functions are not restricted to act as indicator functions; They can be any function that maps notes in a score to real values. For example, a useful basis function proves to be the function that maps notes to (powers of) their MIDI pitch values. Given a set of such basis functions, each representing a different aspect of the score, the intensity of notes in an expressive performance is modeled simply as a linear combination of the basis functions. The resulting model has been used for both predictive and analytical purposes [13,15].

The original formulation of the LBM used a least squares (LS) regression to compute the optimal model parameters. A probabilistic LBM using the Bayesian linear regression assuming zero mean Gaussian priors with isotropic covariance was presented in [15], and then expanded to Gaussian priors with arbitrary mean and covariance in [4].

Although the linear model produces surprisingly good results given its simplicity, a question that has not been answered until now is whether the same basis function framework can benefit from a more powerful, non-linear model.

It is conceivable that *interactions* of score properties produce an effect on performance, rather than each of the properties in isolation. Moreover, it may be that certain properties covary with musical expression, but not in a linear fashion. Therefore, in this paper, we propose a Non-Linear Basis Model (NLBM), that enables non-linear combinations of basis functions through the use of supervised Feedforward Neural Networks (FFNN). These models have been successful in a variety of tasks, ranging from handwritten digit recognition to robot control. FFNNs are powerful models for learning non-linear transformations: with enough hidden units they can represent arbitrarily complex but smooth functions.

Thus, the purpose of this paper is to investigate whether the basis-function modeling approach to expressive dynamics benefits from non-linear connections between the basis-functions and the targets to be modeled. To this end, we run a comparison of the LBM and the NLBM approaches on a data set of professional concert performances of Chopin's piano works. Apart from the predictive accuracy of both models, we present a (preliminary) qualitative interpretation of the results, by way of a sensitivity analysis of the models.

The outline of this paper is as follows: In Sect. 2, we discuss prior work on computational models of musical expression. In Sect. 3, the basis-function modeling approach for musical expression is presented in some more detail. A mathematical formulation of the presented non-linear model is provided in Sect. 4. In Sect. 5, we describe the experimental comparison mentioned above. The results of this experimentation are presented and discussed in Sect. 6. Conclusions are presented in Sect. 7.

2   Related Work

Musical performance represents an ongoing research subject that involves a wide diversity of scientific and artistic disciplines. On the one hand, there is an interest in understanding the cognitive principles that determine the way a musical piece is performed [5,22] such as the effects of musical imagery in the anticipation and monitoring of the performance of musical dynamics [2]. On the other hand, computational models of expressive music performance attempt to investigate the relationships between certain properties of the musical score and performance context with the actual performance of the score [32]. These models can serve mainly analytical purposes [30,33], by showing the relation between structural properties of the music and its effect in the performance of such music, mainly predictive purposes [28], i.e. the models are used to render expressive performances, or both [7,13,17]. Computational models of music performance tend to follow two basic paradigms: *rule based* approaches, where the models are defined through music-theoretically informed rules that intend to map structural aspects of a music score to quantitative parameters that describe the performance of a musical piece, and *data-driven* (or *machine learning*) approaches, where the models try to infer the rules of performance from analyzing patterns obtained from (large) datasets of observed (expert) perfomances [14,31].

One of the most well-known rule-based systems for musical music performance was developed at the Royal Institute of Technology in Stockholm (referred to as the KTH model) [10]. This system is top-down approach that describes expressive performances using a set of (music theoretically sound/cognitively plausible) performance rules that predict aspects of timing, dynamics and articulation, based on a local musical context. On the other hand, the model proposed in this paper represents a bottom-up approach that uses a lower level encoding of a musical score in order to learn how different aspects of the score contribute to generate an expressive performance of a musical piece.

Among the machine learning methods for musical expression is the model proposed by Bresin [3]. This model uses artificial neural networks (NNs) in a supervised fashion in two different contexts: 1) to learn and predict the rules proposed by the KTH model and 2) to learn the performing style of a professional pianist using an encoding of the KTH rules as inputs. As in the case of the KTH model, the NLBM proposed in this paper uses a lower level representation of the score, and makes less assumptions on how the different score descriptors contribute to the expressive dynamics.

On the other hand, Van Herwaarden et al. [18] present an unsupervised approach to modeling musical dynamics using restricted Boltzmann machines. This approach uses a piano roll representation of musical scores to explain the musical dynamics of performed piano music. In order to predict expressive dynamics of a score, the features learned by this model are trained in a supervised fashion using LS regression. The choice of a note-centered representation of a musical score makes this system able to model harmonic context based on relative pitch, but insensitive to absolute pitch. Furthermore, this encoding of a score does not include performance directives written by the composer, such as dynamics or articulation markings (such as *piano*, staccato, etc.). Both the KTH system and previous work on LBMs have shown that the encoding of pitch and dynamics/articulation markings plays an important role in the rendering of expressive performances.

A broader overview of computational models of expressive music performance can be found in [14, 32].

3   The Basis-Function Model of Expressive Dynamics

In this section, we describe the basis-function modeling (BM) approach, independent of the linear/non-linear nature of the connections to the expressive parameters. We consider a *musical score* a sequence of elements [13]. These elements include note elements (e.g. pitch, duration) and non-note elements (e.g. dynamics and articulation markings). The set of all note elements in a score is denoted by $\mathcal{X}$. Musical scores can be described in terms of *basis functions*, i.e. numeric descriptors that represent aspects of the score. Formally, we can define a basis function φ as a real valued mapping $\varphi \colon \mathcal{X} \mapsto \mathbb{R}$. In a similar way, musical expression is characterized in a quantitative way by a number of *expressive parameters*. In particular, expressive dynamics is conveyed by the MIDI velocities of the performed notes. Further expressive parameters capture aspects of note timing and local tempo (e.g. inter-onset intervals between consecutive notes), and articulation (the proportion of the duration of a note with respect to its inter-onset interval). Although the basis-function approach can be applied without any alteration to model all of these expressive parameters, the focus in this study will be on expressive dynamics. By defining basis functions as functions of notes, instead of functions of time, the BM framework allows for modeling forms of music expression related to simultaneity of musical events, like the microtiming deviations of note onsets in a chord, or the melody lead [12], i.e. the accentuation of the melody voice with respect to the accompanying voices by playing it louder and slightly earlier.

The BM framework relies on the simplifying assumption that given all score information, the expressive parameters for each note are independent from those of other notes. This assumption implies that temporal dependencies within parameters are not explicitly modeled. One advantage of non-linear models over previous work is that this framework allows for modeling of mutual dependencies between expressive parameters.

Figure 1 illustrates the idea of modeling expressive dynamics using basis functions schematically. Although basis functions can be used to represent arbitrary properties of the musical score (see Sect. 3.1), the BM framework was proposed with the specific aim of modeling the effect of *dynamics markings*. Such markings are hints in the musical score, to play a passage with a particular dynamical character. For example, a *p* (for *piano*) tells the performer to play a particular passage softly, whereas a passage marked *f* (for *forte*) should be performed loudly. Such markings, which specify a constant loudness that lasts until another such directive occurs, are modeled using a step-like function, as shown in the figure. A gradual increase/decrease of loudness (*crescendo*/*diminuendo*) is indicated by right/left-oriented wedges, respectively. Such markings are encoded by ramp-like functions. A third class of dynamics markings, such as *marcato* (i.e. the "hat" sign over a note), or textual markings like *sforzato* (*sfz*), or *forte piano* (*fp*), indicate the accentuation that note (or chord). This class of markings is represented through (translated) unit impulse functions. In the BM approach, the expressive dynamics (i.e. the MIDI velocities of performed notes) are modeled as a combination of the basis functions, as displayed in the figure.

3.1 Groups of Basis Functions

As stated above, the BM approach encodes a musical score into a set of numeric descriptors. In the following, we describe various groups of basis functions, each group representing a different aspect of the score. This list should by no means be taken as an exhaustive (or accurate) set of features for modeling musical expression. It is a tentative list that encodes basic information, either directly available, or easily computable from a symbolic representation of the musical piece (such as MusicXML).

I **Dynamics Markings.** Bases that encode dynamics markings, such as shown in Fig. 1. For each of the constant loudness markings (*p*, *pp*, *f* etc.), two additional ramp-function are included that allows for a gradual change

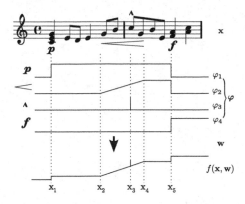

Fig. 1. Schematic view of expressive dynamics as a function $f(\mathbf{x}, \mathbf{w})$ of basis functions φ, representing dynamic annotations

towards the loudness level indicated by the marking. Such bases are referred to as *anticipation* functions, and we distinguish between *long* and *short* anticipations, according to how gradual is the change towards the target dynamics marking. Additionally, basis functions that describe gradual changes in loudness, such as *crescendo* and *diminuendo*, are represented through a combination of a ramp function, followed by a constant (step) function, that continues until a new constant dynamics marking (e.g. f) appears, as illustrated by φ_2 in Fig. 1.

II **Polynomial Pitch Model.** Grachten et al. [13] proposed a third order polynomial model to describe the dependency of dynamics on pitch. This model can be integrated in the BM approach by defining each term in the polynomial as a separate basis function, i.e. "pitch", "pitch^2", and "pitch^3".

III **Vertical Neighbors.** Two basis functions that evaluate to the number of simultaneous notes with lower and higher pitches, respectively.

IV **IOI.** The inter-onset-interval (IOI) is the time between the onsets successive notes; For note i, three basis functions encode the IOIs between $(i, i-1)$, $(i-1, i-2)$, and $(i-2, i-3)$, respectively.

V **Ritardando.** Encoding of markings that indicate gradual changes in the tempo of the music; Includes functions for *rallentando*, *ritardando*, *accelerando*.

VI **Slur.** Description of *legato* articulations, which indicate that musical notes are performed smoothly and connected, i.e. without silence between each note. The encoding of this bases functions is through parabolic functions that act locally where such a slur is present on the score.

VI **Duration.** A basis function that encodes the duration of a note.

VIII **Rest.** Indicates whether notes precede a rest.

IX **Metrical.** Representation of the time signature of a piece, and the position of each note in the bar. For example, the basis function labeled *4/4 beat 0* evaluates to 1 for all notes that start on the first beat in a 4/4 time signature, and to 0 otherwise.

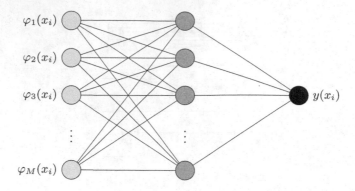

Fig. 2. The architecture of the used NLBM for modeling expressive dynamics

X **Repeat.** Takes into account repeat and ending bars, i.e. explicit markings of that indicate the structure of a piece by indicating the end of a particular section (which can be repeated), or the ending of a piece.

XI **Accent.** Accents of individual notes or chords, such as the *marcato* in Fig. 1.

XII **Staccato.** Encodes *staccato* markings on a note, an articulation indicating that a note should be temporally isolated from its successor, by shortening its duration.

XIII **Grace Notes.** Encoding of musical ornaments that are melodically and or harmonically nonessential, but have an embellishment purpose.

XIV **Fermata.** A basis function that encodes markings that indicate that a note should be prolonged beyond its normal duration.

4 Non-linear Basis Model

In this section we provide a mathematical formulation of the Non-Linear Basis Model (NLBM) model for modeling expressive dynamics. Let $\mathbf{x} = (x_1, \ldots, x_N)^T \in \mathbb{R}^N$ be a vector representing a set of N notes in a musical score and $\mathbf{y} = (y_1, \ldots, y_N)^T \in \mathbb{R}^N$ be a vector representing of an expressive parameter for each note. In this paper, we focus on expressive dynamics, but this framework can be used for other parameters. Let $\boldsymbol{\varphi}(x_i) = (\varphi_1(x_i), \ldots, \varphi_M(x_i))^T \in \mathbb{R}^M$ be a vector whose elements are the values of the basis functions for note x_i. The influence of these basis functions in the expressive parameter can be modeled in a non-linear way using the framework of Feed Forward Neural Networks (FFNNs). These neural networks can be described as a series of functional transformations [1], i.e. a series of non-linear activations of linear combinations of the inputs. Using this formalism, we can write the parameter y as the output of a fully-connected FFNN with L *hidden layers* as

$$y(x_i, \mathbf{w}) = f^{(L)} \left(\sum_{j=1}^{D_L} w_j^{(L)} h_j^{(L-1)}(x_i) + w_0^{(L)} \right), \tag{1}$$

where $\mathbf{h}^{(l)}(x_i) \in \mathbb{R}^{D_l}$ is the activation of the l-th hidden layer, whose k-th component is given by

$$h_k^{(l)}(x_i) = f^{(l)} \left(\sum_{j=1}^{D_l} w_{kj}^{(l)} h_j^{(l-1)}(x_i) + w_{k0}^{(l)} \right), \tag{2}$$

and activation of the first hidden layer is then given as a function of the basis functions as

$$h_k^{(1)}(x_i) = f^{(1)} \left(\sum_{j=1}^{M} w_{kj}^{(1)} \varphi_j(x_i) + w_{k0}^{(1)} \right). \tag{3}$$

The set of all parameters is denoted by $\mathbf{w}$, where $\mathbf{w}^{(l)} = \{w_0^{(l)}, w_1, \ldots, w_{D_l}^{(l)}\}$ are the parameters of the l-th hidden layer[1], and $f^{(l)}$ represent the activation function of the l-th layer. Common (non-linear) activation functions are sigmoid, hyperbolic tangent, softmax and rectifier $(ReLU(x) = \max(0, x))$. Since we are using the FFNN in a regression scenario, the activation function of the last hidden layer is set to the identity function, i.e. $f(x) = x$ [1]. Figure 2 shows the scheme of an FFNN with one hidden layer.

Given a set of training data consisting of input $\mathbf{x}$ and target data $\mathbf{t}$, the model parameters can be estimated in a supervised way by minimizing a loss function, as

$$\hat{\mathbf{w}} = \operatorname*{argmin}_{\mathbf{w}} \mathcal{L}(\mathbf{y}(\mathbf{x}, \mathbf{w}), \mathbf{t}). \tag{4}$$

A usual loss function for supervised regression problems is the *mean squared error (MSE)*, i.e.

$$\mathcal{L}_{MSE}(\mathbf{y}, \mathbf{t}) = \frac{1}{N} \sum_i (y_i(\mathbf{x}, \mathbf{w}) - t_i)^2. \tag{5}$$

As previously stated, the NLBM is able to model mutual dependencies between the basis functions. The output of the model can be written as a linear combination of the last hidden layer, i.e.

$$y(\mathbf{h}^L, \mathbf{w}^{(l)}) = \sum_{j=1}^{D_L} w_j^{(L)} h_j^{(L-1)} + w_0^{(L)} = \mathbf{w}^{(L)^T} \tilde{\mathbf{h}}^{(L-1)}, \tag{6}$$

where $\tilde{\mathbf{h}}^{(L-1)} = \left(1, h_1^{(L-1)}, \ldots, h_{D_L}^{(L-1)} \right)^T$. Since $\tilde{\mathbf{h}}^{(L-1)}$ is a non-linear activation of linear combinations of the input units, it can model the dependencies and interactions of the basis functions. Therefore, we can understand the training of the NLBM as finding Least Squares solution of a non-linear encoding of the input basis functions.

---

[1] In the machine learning literature $\{w_1, \ldots, w_{D_l}^{(l)}\}$ and $w_0^{(l)}$ are respectively referred to as the set of *weights* and the *bias* of the l-th layer.

5   Experiments

To determine to what degree the model is able to account for expressive dynamics, encoded as MIDI velocities of performed notes (see Sect. 5.1), the accuracy of
the predictions of the trained model was tested using a 10-fold cross validation.
We report several measures to characterize the accuracy of the learned models.
Firstly, we report MSE, the mean squared error of the predictions, which is the
most direct measure of how close the model predictions are to their targets. Secondly the Pearson correlation coefficient (r), expresses how strongly predictions
and target are correlated. Lastly, the coefficient of determination R^2, expresses
the proportion of variance explained by the model.

5.1   Data Set

The Magaloff corpus [9] consists of the complete Chopin piano solo works performed by the renown Russian-Georgian pianist Nikita Magaloff (1912–1992)
during a series of concerts in Vienna, Austria in 1989. These performances were
recorded using a Bösendorfer SE computer-controlled grand piano, and then
converted into standard MIDI format. These performances have been aligned to
their corresponding musical scores. One of the unique properties of this corpus
is that the hammer velocities of each performed note have been recorded in a
precise way, and converted to MIDI velocities. This dataset comprises more than
150 pieces and over 300,000 performed notes, adding up to almost 10 hours of
music.

5.2   Model Training

We trained several NLBM models with different configurations. Of the training
data in each fold, 70 % was used for updating the parameters, and 30 % was used
as validation set. The model was trained using RMSProp [6]. This method is a
mini batch variant of stochastic gradient descent that adaptively updates the
learning rate by dividing the gradient by an average of its recent magnitude. In
order to avoid overfitting, dropout and early stopping were used. Dropout prevents overfitting and provides a way of approximately combining different neural
networks efficiently by randomly removing units in the network, along with all its
incoming and outgoing connections. These methods have been effectively used
to improve the results in several applications including image processing [19, 26].
    The number of hidden units, activation function of the hidden layers and the
hyper-parameters (learning rate, batch size and probability of dropout $p_{dropout}$)
were empirically selected using a grid search. The results presented below are
those of the best model on the test set. This network has one hidden layer
model with 100 $ReLU$ hidden units, and a linear output layer with a single unit,
$p_{dropout} = 0.5$, a learning rate of 0.0001 a batch size of 16000 and was trained
for an average of 1037 epochs. It is interesting to notice that with the current
training methods, the accuracy of the model was not benefitted by the addition
of more hidden layers.

Table 1. Predictive results for MIDI Velocity, averaged over a 10-fold cross-validation on the Magaloff piano performance corpus. A smaller value of MSE is better, while larger r and R^2 means better performance.

Model	MSE	r	R^2
LBM	0.780	0.472	0.223
LBM (Bayesian)	0.774	0.475	0.226
LBM (best regularized)	0.771	0.477	0.228
NLBM	**0.757**	**0.492**	**0.242**

The LBM models were trained using the original LS solution, a regularized LS that imposes a constraint in the l_2 norm on the model parameters [1] and the Bayesian LBM reported in [15]. The damping coefficient for the regularized LS was selected empirically through a grid search, and the reported results correspond to those with the lowest MSE on the test set (denoted as "best regularized" in Table 1).

6   Results and Discussion

In this section, we present and discuss the results of the cross-validation experiment. We first present the predictive accuracies, and continue with a more qualitative analysis of the results.

Table 1 shows the accuracy the LBM and the NLBM Models in the 10-fold cross-validation scenario. All three accuracy measures show that the NLBM model gives a small but consistent improvement over all LBM models. A t-test was performed over the MSE, showing that the difference between the LBM with lowest MSE (the regularized LBM, from now on referred to as the best LBM), and NLBM is statistically significant ($t(316344) = 4.64$ at $p = 3.5 \times 10^{-6}$). This may not seem surprising, since FFNNs are known to be *universal approximators*, i.e. they can uniformly approximate any continuous function on a compact input domain to arbitrary accuracy, given that the model has enough hidden units [1]. However, the limited amount of training data, and the approximate nature of the parameter optimization techniques may well limit the improvement in accuracy in practice.

Prior work has revealed that a major part of the variance explained by the LBM is accounted for by the basis functions that represent dynamic markings and pitch, respectively, whereas other basis functions had very little effect on the predictive accuracy of the model [13]. To gain a better insight into the role that different basis functions play in each of the models, the learned models must be studied in more detail. For the LBM this is straight-forward: Each of the basis-functions is linearly related to the target using a single weight, so that the magnitude of a weight is a direct measure of the impact of the corresponding basis-function on the target. In a non-linear model such as the NLBM, the weights of the model cannot be interpreted in such a straight-forward way. To accommodate for this, we use a more generic method to analyze the behavior of computational models, referred to as *sensitivity analysis*.

6.1 Sensitivity Analysis

In order to account for the effects of the different basis functions, a *variance based sensitivity analysis* was performed on the trained LBM and NLBM models [24]. In this way, the sensitivity of the model as a function of the input basis functions φ given the parameters $\mathbf{w}$, i.e. $y = f(\varphi \mid \mathbf{w})$ is explained through a decomposition of the variance of y into terms depending on the input basis functions and their interactions. The *first order sensitivity coefficient* S_{1_i} measures the additive effect of the basis function φ_i in the model output, while S_{T_i}, the *total effect index*, accounts for all higher order effects (interactions) of a factor φ_i. These sensitivity measures are given respectively by

$$S_{1_i} = \frac{V_{\varphi_i}(\mathbb{E}_{\varphi \setminus \varphi_i}(y \mid \varphi_i))}{V(y)} \qquad \text{and} \qquad S_{T_i} = \frac{\mathbb{E}_{\varphi \setminus \varphi_i}(V_{\varphi_i}(y \mid \varphi_i))}{V(y)}, \qquad (7)$$

where V_{φ_i} is the variance with respect to the i-th basis function, $\mathbb{E}_{\varphi \setminus \varphi_i}$ is the expected value with respect to all basis functions but φ_i and $V(y)$ is the total variance of y. It can be shown that $\sum_i S_{T_i} \geq 1$, with the equality occurring if the model is linear (as is the case with LBM), and $S_{1_i} = S_{T_i}$. Both quantities are estimated using a quasi-Monte Carlo method proposed by Saltelli et al. [24], that generates a pseudo random sequence of samples using low-discrepancy (Sobol sequences) to estimate the expected values and variances in the above equations.

Table 2 lists the basis functions that contribute the most to the variance of the model, ordered according to S_{T_i} for the best LBM and the NLBM models. These results show that the polynomial model (the basis functions pitch, pitch^2, and pitch^3) and the dynamics annotations (the basis-functions for *f*, *ff*, *ff* and their anticipations, *pp* anticipation, and *sotto voce*) have the strongest impact on the predicted MIDI velocities in the LBM models. This is consistent with findings reported in [13]. The other basis functions in the LBM list pertain to time signatures that occur relatively rarely: 12/8 time signature occurs in 4 pieces; the high S_T values for those bases may well be due to an overfitting of the model to the particularities of those pieces.

The list of bases to which the NLBM model is most sensitive (Table 2, right half) shows a similar pattern, i.e. the strongest effect on the predicted dynamics come from the dynamics annotations, with a smaller contribution from the polynomial pitch model. Comparing the total effect index and the first order sensitivity coefficient shows that the non-linear effects in the NLBM model capture interactions between the certain basis functions, e.g. *diminuendo* (*dim.*) with $S_T = 0.173$ and $S_1 = 0.087$ and *crescendo* (*cresc.*) with $S_T = 0.133$ and $S_1 = 0.051$. These results also suggest an increased total effect index for gradual basis functions (like *cresc.* or *dim.*).

Figure 3 illustrates how the NLBM model can account for interactions between the *cresc.* and *dim.* These bases interact in ca. 28 % of the Magaloff corpus. In this context, interaction should be understood as those instances where the value of both basis functions is non-zero at the same time, i.e. when *dim* appears after a *cresc.*, before a new constant loudness dynamics markings appear on the score (see Fig. 1 and Sect. 3.1). The lower half of the

Table 2. Basis functions with the largest sensitivity coefficients for the best LBM and NLBM models; Averages are reported over the 10 runs of the cross-validation.

LBM			NLBM		
Basis function	S_T	S_1	Basis function	S_T	S_1
pitch^3	0.187	0.187	ff	0.182	0.160
ff	0.112	0.112	$diminuendo$	0.173	0.087
duration	0.085	0.085	$crescendo$	0.133	0.051
pitch	0.081	0.081	fff	0.115	0.095
fff	0.080	0.080	f	0.095	0.082
f	0.044	0.044	duration	0.082	0.052
pitch^2	0.022	0.022	pitch^3	0.046	0.041
pp	0.021	0.021	pp	0.032	0.020
f anticipation long	0.016	0.016	pitch^2	0.017	0.015
ff anticipation long	0.015	0.015	4/4 weak beat	0.016	0.014
12/8 beat 1	0.013	0.013	p	0.015	0.008
4/4 weak beat	0.013	0.013	p anticipation short	0.014	0.013
fz	0.013	0.013	f anticipation long	0.013	0.010
12/8 beat 2	0.012	0.012	ff anticipation long	0.013	0.012
accent	0.011	0.011	mp	0.012	0.008
12/8 beat 7	0.011	0.011	p anticipation long	0.012	0.010
3/4 beat 1	0.011	0.011	pitch	0.012	0.009
12/8 beat 8	0.010	0.010	accent	0.010	0.009
p	0.010	0.010	fz	0.010	0.008
6/8 beat 1	0.009	0.009	mf	0.010	0.005

figure shows the *cresc.* and *dim.* basis functions in two different contexts: *cresc.* alone and the effects of *cresc.* after *dim.* The upper leftmost figure represents the case of the dynamics predicted by the best LBM using the crescendo basis function alone. The upper center figure shows the predicted dynamics by the NLBM using only *cresc.*, while the upper rightmost figure shows the interaction of a *cresc.* after a *dim.* for both NLBM and the best LBM models. Here it is possible to see a diminished effect of the *cresc.* on predicted dynamics by the NLBM when it appears after a *dim.* On the other hand, these results also illustrate the inability of the LBM to model interactions between basis functions. These results also suggest that the NLBM model might be able to capture a more "natural" dynamics curve for basis function that represent gradual changes, like *cresc.*, and polynomial pitch model. The interaction between *cresc.* and *dim.* illustrates how the NLBM model can capture interactions between basis functions that the (simpler) LBM model is not able to describe.

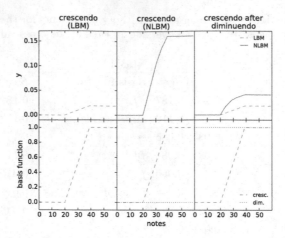

Fig. 3. Example of the effect of the interaction of *crescendo* after a *diminuendo* for both LBM and NLBM models.

The results in Table 2 suggest that some of the most important basis functions for both the LBM and NLBM correspond to certain rules in the KTH model, as is the case of the polynomial pitch model and the *High Loud* phrasing rule[2].

7   Conclusions

In this paper, a neural-network based model for musical expression was presented. This model is shown to perform better than previous work based on linear basis models. A sensitivity analysis performed on the two models suggests that the new non-linear approach is able to capture certain interactions of basis functions that cannot be captured in linear models.

In this work, we used simple music-theoretically informed numerical descriptors to capture certain aspects of the score. The results presented above suggest that new basis functions could improve the performance of the presented model.

Additionally, the presented results suggest that the LBM model benefits from bases that contain redundant information (such as long and short anticipation and the polynomial pitch model). It would be interesting to determine wether the NLBM model can capture the similar effects, without recurring to the use of such basis functions, e.g. by using only pitch instead of pitch, $pitch^2$ and $pitch^3$. Another interesting question would be to investigate to what degree the nonlinear mappings from basis functions to targets improves the accuracy of the model for non-binary basis functions.

An interesting approach from the music-theoretic side would be the use of basis functions that encode structural (i.e. form) and harmonic information of the piece. Among these basis functions could be the use of key identification algorithms and pattern identification techniques [27].

---

[2] See Table 1 in [10] for an overview of the rules of the KTH model.

Furthermore, it would be interesting to use a combination of unsupervised learned features (using Deep Learning) and music-theoretic-informed features for analyzing and predicting expressive music performance, expanding previous work by van Herwaarden et al. [18]. Following previous work on Bayesian LBMs [15], the presented framework can also be expanded into a fully probabilistic approach using the framework of Bayesian neural networks [1].

As stated in Sect. 3, neither the NLBM, nor the LBM (in both its deterministic and Bayesian formulations) allow for modeling of temporal dependencies within parameters. This issue can be addressed by using a temporal model, such as recurrent neural networks (RNNs) [16], conditional random fields (CRFs) or considering the temporal autocorrelation [8].

Acknowledgment. This work is supported by European Union Seventh Framework Programme, through the Lrn2Cre8 (FET grant agreement no. 610859) and the PHENICX (grant agreement no. 601166) projects.

References

1. Bishop, C.M.: Pattern Recognition and Machine Learning. Springer, New York (2006). Microsoft Research Ltd
2. Bishop, L., Bailes, F., Dean, R.T.: Performing musical dynamics. Music Percept. **32**(1), 51–66 (2014)
3. Bresin, R.: Artificial neural networks based models for automatic performance of musical scores. J. New Music Res. **27**(3), 239–270 (1998)
4. Cancino Chacón, C.E., Grachten, M., Widmer, G.: Bayesian linear basis models with gaussian priors for musical expression. Technical report, October 2014
5. Clarke, E.F.: Generative principles in music. In: Sloboda, J. (ed.) Generative Processes in Music: The Psychology of Performance, Improvisation, and Composition. Oxford University Press, New York (1988)
6. Dauphin, Y.N., de Vries, H., Chung, J., Bengio, Y.: RMSProp and equilibrated adaptive learning rates for non-convex optimization. arXiv 1502, 4390 (2015)
7. De Poli, G.S.C., Rodà, A., Vidolin, A., Zanon, P.: Analysis and modeling of expressive intentions in music performance. In: Proceedings of the International Workshop on Human Supervision and Control in Engineering and Music, Kassel, Germany, 21–24, September 2001
8. Eck, D.: Beat tracking using an autocorrelation phase matrix. In: IEEE International Conference on Acoustics, Speech and Signal Processing - ICASSP 2007, pp. 1313–1316. IEEE, January 2007
9. Flossmann, S., Goebl, W., Grachten, M., Niedermayer, B., Widmer, G.: The magaloff project: an interim report. J. new music Res. **39**(4), 363–377 (2010)
10. Friberg, A., Bresin, R., Sundberg, J.: Overview of the kth rule system for musical performance. Adv. Cogn. Psychol. **2**(2–3), 145–161 (2006)
11. Gabrielsson, A.: Music performance research at the millennium. Psychol. Music **31**(3), 221–272 (2003)
12. Goebl, W.: Melody lead in piano performance: expressive device or artifact? J. Acoust. Soc. Am. **110**(1), 563–572 (2001)
13. Grachten, M., Widmer, G.: Linear basis models for prediction and analysis of musical expression. J. New Music Res. **41**(4), 311–322 (2012)

14. Grachten, M.: Summary of the music performance panel, MOSART workshop 2001, Barcelona. In: MOSART Workshop, pp. 1–17, March 2002
15. Grachten, M., Cancino Chacón, C.E., Widmer, G.: Analysis and prediction of expressive dynamics using Bayesian linear models. In: Proceedings of the 1st International Workshop on Computer and Robotic Systems for Automatic Music Performance, pp. 545–552, July 2014
16. Graves, A.: Generating Sequences With Recurrent Neural Networks. arXiv 1308, 850 (2013)
17. Grindlay, G., Helmbold, D.: Modeling, analyzing, and synthesizing expressive piano performance with graphical models. Mach. Learn. **65**(2–3), 361–387 (2006)
18. van Herwaarden, S., Grachten, M., de Haas, W.B.: Predicting expressive dynamics using neural networks. In: Proceedings of the 15th Conference of the International Society for Music Information Retrieval, pp. 47–52, July 2014
19. Hinton, G.E., Srivastava, N., Krizhevsky, A., Sutskever, I., Salakhutdinov, R.: Improving neural networks by preventing co-adaptation of feature detectors. arXiv 1207, 580 (2012)
20. Juslin, P.: Communicating emotion in music performance: a review and a theoretical framework. In: Juslin, P., Sloboda, J. (eds.) Music and Emotion: Theory and Research, pp. 309–337. Oxford University Press, New York (2001)
21. Palmer, C.: Anatomy of a performance: sources of musical expression. Music Percept. **13**(3), 433–453 (1996)
22. Palmer, C.: Music performance. Annu. Rev. Psychol. **48**, 115–138 (1997)
23. Ramirez, R., Hazan, A.: Rule induction for expressive music performance modeling. In: ECML Workshop Advances in Inductive Rule Learning, September 2004
24. Saltelli, A., Annoni, P., Azzini, I., Campolongo, F., Ratto, M., Tarantola, S.: Variance based sensitivity analysis of model output. Design and estimator for the total sensitivity index. Comput. Phys. Commun. **181**(2), 259–270 (2010)
25. Sloboda, J.A.: The communication of musical metre in piano performance. Q. J. Exp. Psychol. **35A**, 377–396 (1983)
26. Srivastava, N., Hinton, G., Krizhevsky, A., Sutskever, I., Salakhutdinov, R.: Dropout: a simple way to prevent neural networks from overfitting. J. Mach. Learn. Res. **2014**(15), 1929–1958 (2014)
27. Temperley, D.: Music and Probability. Mit Press, Cambridge (2007)
28. Teramura, K., Okuma, H.: Gaussian process regression for rendering music performance. In: Proceedings of the 10th International Conference on Music Perception and Cognition (ICMPC), Sapporo, Japan (2008)
29. Todd, N.: The dynamics of dynamics: a model of musical expression. J. Acoust. Soc. Am. **91**, 3540–3550 (1992)
30. Widmer, G.: Machine discoveries: a few simple, robust local expression principles. J. New Music Res. **31**(1), 37–50 (2002)
31. Widmer, G.: Discovering simple rules in complex data: a meta-learning algorithm and some surprising musical discoveries. Artif. Intell. **146**(2), 129–148 (2003)
32. Widmer, G., Goebl, W.: Computational models of expressive music performance: the state of the art. J. New Music Res. **33**(3), 203–216 (2004)
33. Windsor, W.L., Clarke, E.F.: Expressive timing and dynamics in real and artificial musical performances: using an algorithm as an analytical tool. Music Percept. **15**(2), 127–152 (1997)

Geo-Coordinated Parallel Coordinates (GCPC): A Case Study of Environmental Data Analysis

Maha El Meseery and Orland Hoeber(✉)

Department of Computer Science, University of Regina, Regina, Canada
{elmeseem,orland.hoeber}@uregina.ca

Abstract. Knowledge discovery in scientific and research datasets is an extremely challenging problem due to the high dimensionality, heterogeneity, and complex relationships within the data. When these datasets also includes temporal and geospatial components, the challenges in analyzing the data become even more difficult. A number of visualization approaches have been developed and studied to support the exploration and analysis among such datasets, including parallel coordinate plots, dimensional subsetting, geovisualization, and multiple coordinated views. In this research, we combine and enhance these approaches in a system called Geo-Coordinated Parallel Coordinates (GCPC), with the goal of supporting interactive exploration, analytical reasoning, and knowledge discovery.

1 Introduction

With advances in data collection and storage technology, the volume and complexity of scientific and research datasets are becoming increasingly overwhelming. Analyzing and understanding these datasets is an essential step in hypothesis development and scientific discovery. Discovering new and unexpected knowledge requires making sense of large amounts of high-dimensional and interrelated data. The need to derive insights from data collected for a particular domain or problem is driving researchers to design, develop, and study new tools and techniques to support data analysis and knowledge discovery.

Knowledge discovery is the process of identifying and understanding new meaningful patterns and trends contained within datasets [14]. It is a complex process that requires multiple iterations of data processing and transformation, hypotheses generation, and finally interpretation and reasoning about what has been discovered [14]. Such a process is an extremely challenging problem when the data is high-dimensional and heterogeneous, contains complex relationships among attributes, and has important temporal and spatial aspects.

Modern knowledge discovery systems utilize automated data analysis methods based on research from various fields including data mining, statistics, artificial intelligence, and machine learning. Even though these automated methods may be used to identify previously unknown aspects of the data, they provide researchers with few explanations about how or why the knowledge has been acquired, and provide little aid to the researchers in interpreting what has been

© Springer International Publishing Switzerland 2015
N. Japkowicz and S. Matwin (Eds.): DS 2015, LNAI 9356, pp. 63–77, 2015.
DOI: 10.1007/978-3-319-24282-8_7

discovered. Exploratory data analysis takes a different approach, with the aim of keeping humans involved in the discovery process. This often includes iterative investigation of the data, with the support of automated data processing, leading to the understanding of the patterns and the acquisition of new knowledge.

Visual analytics is an emerging approach that is increasingly being employed to support exploratory analysis of data [7,23]. By combining information visualization, data processing, data mining, and interactive interfaces, analysts are able to explore, analyze, reason, and make sense of highly complex data [23]. Merging multiple visual and interactive representations helps analysts to generate hypotheses, identify new lines of inquiry, understand patterns, and derive new insight from what is being shown.

Our goal in this research is to develop a method to support the exploration and understanding of complex patterns and trends within high dimensional, heterogeneous, and geotemporal data. Geo-Coordinated Parallel Coordinates (GCPC) takes a visual analytics approach to the problem domain, using multiple coordinated views to simultaneously show, filter, and examine the data using parallel coordinates, micro-visualizations of the statistical features of the data, geovisualization, and investigative scatter plotting. Interactive and automated features support the knowledge discovery process, as well as the necessary task of hiding the complexity of the data to reveal the patterns.

The remainder of this paper is organized as follows. Section 2 provides a review of the key literature that has informed this research, including overviews of high dimensional data visualization, geovisual analytics, and multiple coordinated views. Section 3 outlines the key features of GCPC, followed by a case study in Sect. 4 that illustrates how these features can be used for data exploration and analysis activities, leading to new knowledge generation. The paper concludes with Sect. 5, which outlines the key contributions of this work, the limitations of the approach, and future work.

2   Literature Review

The use of visual analytics to support exploration, reasoning, and knowledge discovery within high dimensional geotemporal data is an active research domain with applications in many scientific fields [13,22]. The following literature review focuses on three main topics that are relevant to our work: high dimensional data visualization, geotemporal data visualization, and multiple coordinated views.

2.1   High Dimensional Data Visualization

While a multitude of approaches have been developed over the years to visualize high dimensional data, each has its limitations [30]. Dimensional reduction methods use computational techniques such as principle component analysis [2] or multidimensional scaling [30] to transform the data to a lower dimensional space while preserving the relative proximity between data points [11]. The end result is a visual representation of the data in a coordinate space that has no obvious correlation to the actual dimensions, introducing complexity while exploring the data [11]. Dimensional subsetting methods use algorithmic techniques or user preferences to select a small subset of the dimensions to visualize. The success of such an approach is dependant on choosing which dimensions contain the most relevant and useful information [30]. This approach can be extended by displaying several dimensionally subsetted views of the data in a small multiples configuration. However, as more views are added to explore the relationships between the different dimensions, it becomes increasingly difficult to detect and interpret patterns within the data [20]. Instead of using simple shapes to represent each data point within these plots, glyphs can be used to encode additional dimensions of the data with shape, size, colour, orientation, or other visual attributes. However, there is a limit to the number of dimensions that can be visualized using glyphs before they become incomprehensible [12].

A fundamentally different approach to the problem is the use of parallel coordinate plots, where data are represented within a structure that maps each of the dimensions to a parallel axis [21]. Individual data points are represented using polylines, intersecting each axis at the appropriate location for the value on the specific dimension. This approach is very flexible and scalable with respect to the dimensionality of the data; adding a new dimension can be achieved by adding a new parallel axis and extending the polylines to their appropriate values [17]. The primary value of this approach over other high dimensional data visualization techniques is that all aspects of the data are shown, and the relationship between pairs of attributes can be investigated by interactively placing their axes beside one another.

However, there are also a number of important limitations. When multiple data points intersect an axis at the same location, ambiguity is introduced. This can be addressed by interactively highlighting the data points, or replacing the polylines with curves or density functions, both of which make it easier to perceive and follow the data points through the parallel coordinate structure [17]. When a large number of data points are shown using parallel coordinates, overplotting may occur, resulting in visual clutter. Some have explored clustering and outlier detection algorithms in order to reduce the amount of data that is shown, and to highlight those data points that are different from the norm [16,32]. Since overplotting may also make it difficult to grasp the distribution of the data, adding statistical information to the parallel axes may be useful [18], but may also further contribute to the complexity of the display. When a particular dimension of the data represents discrete qualitative values, the ambiguity and overplotting problems become even more acute. Parallel sets provide an alternative for visualizing such categorical data, using ribbons between the coordinates to represent the frequency of each category [24]. However, combining this with traditional parallel coordinates when the data includes a combination of qualitative and quantitative data is not feasible.

2.2   Geotemporal Data Visualization

Geotemporal data visualization is a challenging problem due to the complexity of representing different scales and relations between the geospatial and temporal aspects of the data [7]. Simply mapping the data with traditional GIS tools, using different layers for each temporal range, limits the ability to dynamically analyze the features of the data. Several approaches have been proposed to support interactive visual analysis of geotemporal data. One of the earliest methods is the space-time cube, where location is represented in the map dimensions, and time represented in the third dimension [5]. Another straightforward method is using two coordinated visualizations: one to represent the temporal aspect of the data and the second to represent the geospatial aspect of the data. In this approach, a thematic dot map [22] or a choropleth map [16] may be used to visualize the location of the data; the temporal representation can then be used to interactively filter what is shown in the map. Others have studied methods for directly visualizing the temporal aspect of the data overlaid on the map (e.g., using a glyph to representing an aggregation of monthly temporal data [4]). However, by using visual attributes to represent the temporal aspect of the data, these attributes cannot be used to represent other multivariate aspects. Pre-processing the data to calculate changes over specified timeframes, and then visualizing these differences, can allow interesting features to be identified that would be difficult to discern otherwise [19]. Such an approach can be beneficial when the goal is to analyze how the data are changing over space and time.

2.3   Multiple Coordinated Views

Considering data that consists of both high dimensional attributes and geotemporal aspects, exploring and analyzing this type of data becomes an extremely challenging problem. A single visualization method is not adequate to support exploration, comparison, analysis, and knowledge discovery across the different aspects of such complex data. A common approach is to provide multiple visualizations of the data, which are linked together such that interactive manipulation in one (e.g., zooming, filtering, and focusing) results in a corresponding change in all others [27]. Views of the data that are customized to the specific meanings of the attributes have been used in this manner, such as the combination of a scatter plot matrix, time series visualizations, and word clouds [13,16]. From a geospatial analysis perspective, the coordinated combination of parallel coordinates and geovisualization approaches have been studied for many years [3,15]. However, there remains a shortage of geovisual analytics systems that support interactive analysis of high dimensional geotemporal data [6]. Furthermore, linking these approaches within an integrated analysis of the qualitative non-spatial attributes remains an open problem that we wish to address in this research.

3   Geo-Coordinated Parallel Coordinates (GCPC)

Discovering knowledge and testing hypothesis in environmental studies is a challenging problem due to the complexity of environmental data. Such data often

consists of multiple heterogeneous factors with complex interrelations and important spatial and temporal aspects. Motivated by the challenges of exploring
among such data, we have developed a geovisual analytics system to support
analysis and reasoning about high dimensional heterogeneous geotemporal data.
Geo-Coordinated Parallel Coordinates (GCPC) has been designed to enable
the interactive analysis activities described in Keim's visual analytics mantra:
"analyze first; show the important; zoom, filter, and analyze further; details on
demand" [23].

The core of the system is comprised of two tightly coordinated features: a
parallel coordinate plot and a geovisualization. These two views allow the system to represent the high dimensional, heterogeneous, temporal, and geospatial
aspects of the data simultaneously. An optional scatterplot view allows analysts
to interactively investigate correlations between pairs of factors. To further support exploration among the data, these visual components are linked through
coordinated interactions: filtering, zooming, and highlighting the data to focus
on interesting features in one view results in similar actions in the other views.
Selections for visual encoding (e.g., colour, size) are replicated across all views,
reinforcing the interpretation of the coordination across the views. A screenshot
of the system is provided in Fig. 1.

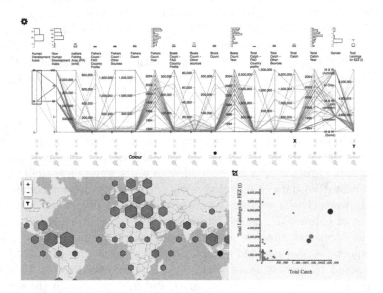

Fig. 1. The main view of the GCPC consists of parallel coordinates, micro-
visualizations of the statistical properties of each dimension, interactive controls for
configuring the visualizations, a geovisualization of the geospatial distribution of the
data, and the investigative scatterplot. Here, the data is filtered for specific values on
the first dimension and coloured based on the values in the sixth dimension (Colour
figure online).

Since environmental data analytics is seldom performed on just a single data set, the system was designed with data flexibility in mind. In order to load an arbitrary dataset, an automatic pre-processing step converts raw data into a tabular format with a single geospatial location per data point. The data types are automatically detected, allowing appropriate visual encodings within the GCPC interface.

The software was developed within a web-based interface, using the Data Driven Documents (D3) library [8] as the core. Existing parallel coordinates [10] and geovisualization [1] plugins were used and extended to add the additional visual and interactive features. In the remainder of this section, the specific features and design considerations of GCPC will be explained.

3.1   Parallel Coordinate Plot

The parallel coordinate plot in GCPC allows multiple interactions to support exploration within the high dimensional data. Since the direct relationship between a pair of dimensions can only be seen if the dimensions are placed adjacent to one another, the system supports interactive reordering of the coordinates. Dimensions of the data that are not relevant for the current analysis activity can be interactively hidden. Investigating interesting subsets of the data is supported by filtering the data using brushing operations on the coordinates. Such filtering is immediately applied throughout the system to support further investigation in the other views. Detailed analysis is also supported by allowing the user to zoom-in on the brushed data, resulting in a rescaling of the data displayed on the coordinate. This can allow for the study of a subset of data that is tightly clustered within a narrow range of values.

The analyst may choose to colour the data according to the value on a chosen dimension, which is also reflected in the other visualizations. In order to ensure proper interpretation of the colour encoding, the encoding scheme is different for each data type. Quantitative data is encoded with a continuous, perceptually ordered scale; ordinal data is encoded with a discrete perceptually ordered scale; and qualitative data is encoded with perpetually distinct colour scales. Colour scales were chosen with an awareness of colour theory and the human interpretation of colour [31], using ColorBrewer [9] as the starting point for the specific scale selections. Since a given dataset may contain multiple different temporal aspects, it was decided to not provide a single timeline to filter the data, but instead to include these within the parallel coordinates. This allows the temporal features to be studied, filtered, and manipulated in the same way as other aspects of the data.

Statistical Descriptors. One of the criticisms of using parallel coordinate plots is the difficulty in identifying the distribution of the data on a given dimension when there are a large number of data points. The compact nature of the parallel coordinates may result in overplotting and visual clutter, making it difficult to identify the precise data points going through a given value on a specific dimension. To address this limitation, micro-visualizations have been added to each parallel coordinate to illustrate the statistical properties of the data. While others have explored similar solutions by overlaying the statistical descriptors on the coordinates [18], GCPC provides these on top of each coordinate, allowing the information to be observed as needed, without interfering with the interpretation of the data shown in parallel coordinate plots.

The format of these statistical descriptors depends on the type of data they describe. Quantitative data is visualized using Tukey box plots [29], providing a compact representation of the median, quartile, and fifth/ninety-fifth percentiles. For qualitative and ordinal data, such measures are meaningless; instead histograms of the distributions are provided. Both formats allow the analyst to quickly observe and interpret the different types of data, providing an overview of the features of the dataset. Any filtering of the data automatically results in a recalculation of the statistical properties of what remains, and an update in these micro-visualizations.

Outlier Detection. A second criticism of parallel coordinate plots is the difficulty in identifying outliers, due to the significant visual weight that is given to the dominant pattern within the data. As a result, it is difficult to visually isolate data points that are different from the norm. In some cases, such outliers may be uninteresting, and there may be a desire to remove these to reduce the additional visual clutter they cause. In other cases, the outliers may be important for the analysis at hand, and there may be a desire to highlight them. In order to support outlier analysis, GCPC includes an automatic outlier detection algorithm.

The approach employed is designed specifically for high dimensional data, based on the comparison of angles between multi-dimensional vectors [25]. This is based on the observation that for an anomalous data point, the angle to other pairs of data points in the collection will be small because of its distance from the other data. Conversely, for data points that are not anomalous, they will be surrounded by other points, resulting in large angles to other pairs of data points. The statistical variance of the angle is computed for each point to all other pairwise points in the dataset, which is then used as an outlier score to rank the data points. A data point is labeled as outlier if the score is lower than an empirically set threshold. Although not entirely accurate, qualitative and ordinal data are mapped and normalized to numerical values in this process.

More specifically, the angle based outlier detection score ($ABOD$) of point A is computed as:

$$ABOD(A) = VAR_{B,C \epsilon D} \left(\frac{\langle \bar{A}B, \bar{A}C \rangle}{\| \bar{A}B \|^2 \| \bar{A}C \|^2} \right)$$

where B, and C represent all pairs of data points in the dataset D, and VAR is the statistical variance over these data. What is being calculated are the angles between point A and all pairs B and C, noramlized by the length of the vectors

$\| \ \overline{AB} \ \|, \| \ \overline{AC} \ \|$, which gives more weight to the score if points B, C are nearer to point A.

Because this algorithm must compare each data point to all other pairs, it is computationally expensive $(O(n^3))$. While classifying the data using an algorithm such as k-nearest neighbours can speed up the approach [25], for our purposes it is not necessary to calculate these outliers in real-time. Instead, the outlier ranking scores can be calculated during the pre-processing step and stored as part of the data, but keeping the cut-off threshold for outlier classification as an interactive parameter. The analyst may then choose how sensitive to make the outlier detection, and whether to use this to filter out the outliers or highlight them for detailed investigation.

3.2   Geovisualization

The purpose of the geovisualization is to allow the analyst to observe and interpret the spatial distribution of the data. This is an essential part of GCPC, allowing for the exploration among the relations between multiple factors and the geospatial aspects of the data. GCPC contains two main modes of displaying geospatial features on the map: a dot map that represents each point as a circle at the appropriate location, and hexagonal binning that represents aggregated spatial data on a hexagonal grid. While the process for producing the dot map is straightforward, there is some complexity in the creation of the hexagonal bin map. A grid of hexagonal polygons are layered over the map, and the data is aggregated based on which bin it falls into [26]. The default is to simply count the number of data points in each bin, but more complex aggregation such as total or average calculations are also possible. The size of the hexagons are then used to encode the data aggregated within the bins.

Settings below the parallel coordinate plot allow for the manipulation of two visual variables within the geovisualization: colour and size. When the dot map is shown, the size and colour of the dots are encoded based on the dimensions of the data chosen for these values. When the data is aggregated in the hexagonal bins, this colour and size encoding cannot be used directly. Instead, the size of the hexagon continues to be calculated as normal, but the colour is determined by the average value for quantitative data or the most frequent value for qualitative data.

The normal pan and zoom operations on the map allow the analyst to view more closely the geospatial relationships among the data. In order to further understand and explore among the data, the analyst may activate a geographical filter. The system allows the user to draw polygons to create arbitrary shapes that overlap a region of interest. The filter will remove all data points outside of the drawn shape, both within the geovisualization and also from the other visual representations. Coordinated highlighting allows the analyst to select specific data points within the map in order to isolate their attributes in the other dimensions using the other visual representations. These features enable the co-exploration of the data within both the high-dimensional elements and the geospatial elements.

3.3   Investigative Scatterplot and Correlation Analysis

Investigating the correlation between different factors and dimension is essential to understanding the complex relations within the data. While the order of the dimensions in the parallel coordinates may be manipulated to observe the pattern of the relationship, an analyst may wish to investigate such relationships in more detail and in a more fluid and interactive way. Selecting dimensions of the data to plot on the x- and y-axis using the controls under the parallel coordinates results in the creation of a scatterplot of the data. This enables a direct and intuitive analysis of the correlation between the selected attributes. Any selections of the colour and size encoding will also be present in this scatterplot, allowing for the interactive visualization of four dimensions of the data.

Because the analyst can easily change the dimension of the data to use for the axes, correlations can quickly be investigated and examined. Following the same coordinated interaction within the other views of the data, brushing over a region of this scatterplot will filter the other views, and selecting individual points will highlight their counterparts within the parallel coordinates and the geovisualization.

3.4   Data Inspection

During the exploration of the data, it is important to maintain the ability to drill down to the raw data in order to allow the analyst to inspect the actual values. This inspection may be used by the analyst to confirm what has been shown visually. When an individual data point is selected in any of the other views, a details window is populated with the complete set of data for this point. In addition, as the analyst filters the data using the parallel coordinates, geovisualization, and investigative scatterplot, they may wish to extract this specific subset of the data for detailed inspection and export into other software. A table view of the data supports this process, which only shows the data that matches the current filter settings.

4   Case Study

To demonstrate the features and utility of GCPC in the analysis of environmental data, we describe below three exploration scenarios of a dataset from the fisheries domain. The dataset was provided by the Too Big To Ignore (TBTI) research project, whose goal is to document and study the impact and importance of small scale fisheries around the world [28]. It consists of 127 data points over nineteen dimensions that include quantitative, qualitative, ordinal, temporal, and geospatial attributes that describe the small-scale fishing industry around the world. While the size of this dataset is relatively small, its high dimensionality, heterogeneity, and geotemporal attributes made it difficult to analyze using traditional means.

Initial Exploration. The analyst in this case study is an environmental researcher trying to explore and compare the impact of small scale fisheries across the broad range of attributes collected. Loading the dataset into GCPC will automatically identify the type of each dimension and calculate initial statistical distributions. As shown in Fig. 2, the system will default to showing all of the data in the parallel coordinates plot, and the locations at which this data was collected in the geovisualization. From this overview of the entire dataset, global patterns in the data may be observed, including the distribution of the data over the dimensions, the correlation between adjacent coordinates, and the geodistribution of the data. This initial assessment of the data can then be used as the basis for confirming what is known (e.g., the extent of small-scale fishing in Central America), and developing and evaluating new hypotheses about the data.

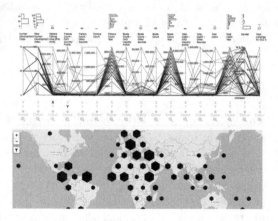

Fig. 2. The default view of the data loaded in GCPC.

One aspect of the data that can be readily observed from the overview is that it is highly irregular, with a small number of extreme values that extend the range of some coordinates (e.g., *Inshore Fishing Area* (third parameter), and *Fishers Count* (sixth parameter)). This pattern in the data causes the remaining data points to be clustered at the other end of the scale, making it difficult to discern their pattern. There are two mechanisms built into GCPC that can address this problem: using the automatic outlier detection to hide these data points that are substantially different than the norm, or using the interactive focusing, filtering, and zooming features on the coordinates of interest. Supposing that the analyst wants to retain interactive control over the analysis process, the first step in exploring these coordinates is to inspect the extreme values. By clicking on each, the researcher will observe that they correspond to countries with large fishing regions (i.e., Canada, Indonesia, and Australia). The data on these dimensions can be filtered easily, by interactively dragging a bounding box over the coordinates. Clicking on the zoom icon will cause the selected range to fill the available space for the coordinate in question. The results of this filtering and zooming operation can be seen in Fig. 3.

Knowing that there are outliers in the data, the analyst may wish to have the system automatically find these so that they can be evaluated, and then perhaps hidden when conducting future analyses of the data. Figure 4 shows these anomalies, and dims the remaining data so that the overall pattern can still be observed. The algorithm identified countries that are considerably different than the normal pattern of the data across multiple dimensions (e.g., Chile and Mexico). Another set of outliers detected were data points with multiple missing

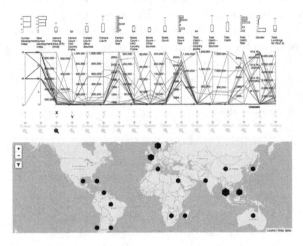

Fig. 3. Zooming the scale on the *Inshore Fishing Area* dimension from over 1,000,000 to 100,000 reveals the pattern at the lower values on this dimension.

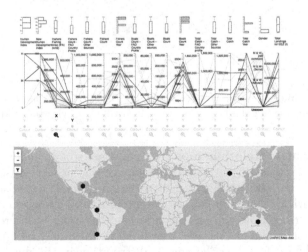

Fig. 4. Highlighting the outliers data allows the analyst to easily inspect these, and the subsequently hide them.

information, which render them substantially different than normal data (e.g., Australia). Isolating these anomalies from the rest of the data would be tedious and cognitively taxing had it been done manually.

Analysis of Attribute Relationships. After this initial observation and exploration, the researcher may be interested in the development and testing of an hypothesis that relates *Total Catch*, *Boats Count*, *Fishers Count*, and *Total Landings*. A first step in such an examination is to re-order the coordinates such that they are adjacent to one another. Doing so, allows the analyst to observe direct or inverse correlations easily. More complex relationships can be observed using the investigative scatterplot, mapping these attributes to the x-axis, y-axis, colour, and size options (see Fig. 5). Since the colour and size parameters are also represented on the map, the locations where the *Total Landings* and *Fishers Count* are large can be observed. This analysis shows a pattern of the correlation between these parameters, as well as the instances of data points that are counter to the pattern (e.g., Japan, with low *Fishers Count*, but high *Boats Count* and *Total Catch*).

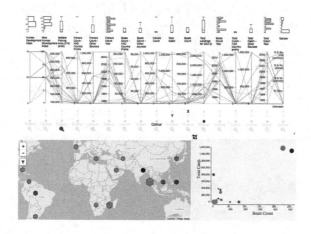

Fig. 5. Investigating the relations between *Total Catch*, *Boats Count*, *Fishers Count*, and *Total Landings* using the investigative scatter plot and visual encodings.

Analysis of Spatial Relationships. An important step in analyzing data such as this is to make comparisons across different geographical regions. Suppose the researcher wishes to study the gender distribution of small-scale fishers between Europe and Africa. The map can be zoomed to these regions independently, and then free-form shapes can be drawn around the areas of interest. Doing so filters the data shown in the parallel coordinates, which may be further filtered, perhaps in order to focus on the gender distribution in the most recent data. Screenshots showing these two analyses are provided in Fig. 6. From this, we can readily observe that in France, Italy, and Greece, it is not common for both men

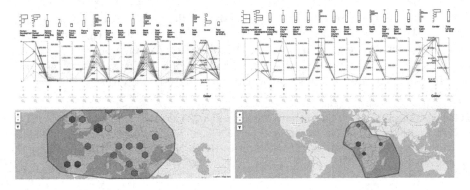

Fig. 6. Zooming the map and filtering the data to specific geographic regions allows for the isolation of this data within the parallel coordinates, enabling the comparison of parameters such as the differences on the *Gender* dimension.

and women to fish. However, in sub-Saharan Africa, recent data shows that both men and women are actively involved in the small-scale fisheries.

5   Conclusion and Future Work

In this paper, we presented Geo-Coordinated Parallel Coordinates, a visual analytics system designed to support the exploration and analysis of high dimensional, heterogeneous, geotemporal data. The main contribution of the system are: (1) the integration and coordination of multiple visualization and interaction techniques; (2) the micro-visualizations of the statistical information for each dimension added to the parallel coordinates; (3) the use of automatic outlier detection to allow highlighting and filtering of outlier data; and (4) the flexible design that allows arbitrary high-dimensional data to be loaded into the system.

Even though this paper demonstrated through a case study the benefits of GCPC in facilitating analytic reasoning through interactive exploration of the data, there are some limitations. The current implementation assumes the geospatial data are point data, and does not support data that represents geospatial regions. Complex data types such as hierarchical data and missing data cannot be represented within the parallel coordinate structure. GCPC does not currently support the analysis of complex temporal patterns such as temporal ranges and movement data. Even though we endeavoured to address some of the fundamental limitations of parallel coordinates in GCPC, it remains difficult to detect patterns over more than four dimensions of the data.

Future work may include addressing some of these limitations. Supporting different geospatial features (e.g., regions, trajectories) will allow the system to more readily support the analysis of such data. Modifying the angle-based outlier detection approach to more accurately detect differences in qualitative and ordinal data will improve the anomaly detection. Clustering the data may address the overplotting issues within the parallel coordinates, allowing high-level patterns within the data to be identified. Adding a separate view for temporal data (e.g., a timeline), will enable users to analyze different types of temporal data and identify complex geotemporal patterns and trends. Since this approach is highly interactive, we are currently in the planning stage for an empirical evaluation with expert data analysts, which will provide evidence of the value and usefulness of the approach for real-world data analysis activities.

Acknowledgements. The authors wish to thank the Too Big To Ignore (TBTI) project for their support and dataset they provided. This work was supported in part by grant from Social Sciences and Humanities Research Council (SSHRC) held by the second author.

References

1. Agafonkin, V.: Leaflet. http://leafletjs.com/ (2010). Accessed December 2014
2. Aggarwal, C.C.: Data Mining, The Textbook. Springer International Publishing, Switzerland (2015)
3. Andrienko, G., Andrienko, N.: Exploring spatial data with dominant attribute map and parallel coordinates. Comput. Environ. Urban Syst. **25**(1), 5–15 (2001)
4. Andrienko, G., Andrienko, N.: Spatio-temporal aggregation for visual analysis of movements. In: Proceedings of the IEEE Symposium on Visual Analytics Science and Technology, pp. 51–58 (2008)
5. Andrienko, G., Andrienko, N., Demsar, U., Dransch, D., Dykes, J., Fabrikant, S.I., Jern, M., Kraak, M.J., Schumann, H., Tominski, C.: Space, time and visual analytics. Int. J. Geogr. Inf. Sci. **24**(10), 1577–1600 (2010)
6. Andrienko, G., Andrienko, N., Keim, D., MacEachren, A.M., Wrobel, S.: Challenging problems of geospatial visual analytics. J. Vis. Lang. Comput. **22**(4), 251–256 (2011)
7. Andrienko, N., Andrienko, G., Gatalsky, P.: Exploratory spatio-temporal visualization: an analytical review. J. Vis. Lang. Comput. **14**(6), 503–541 (2003)
8. Bostock, M., Ogievetsky, V., Heer, J.: D3 data-driven documents. IEEE Trans. Vis. Comput. Graph. **17**(12), 2301–2309 (2011)
9. Brewer, C.A., Hatchard, G.W., Harrower, M.A.: Colorbrewer in print: a catalog of color schemes for maps. Cartography Geogr. Inf. Sci. **30**(1), 5–32 (2003)
10. Chang, K.: Parallel coordinates toolkit. https://syntagmatic.github.io/parallel-coordinates/ (2012). Accessed December 2014
11. Choi, J.Y., Bae, S.H., Qiu, X., Fox, G.: High performance dimension reduction and visualization for large high-dimensional data analysis. In: Proceedings of the IEEE/ACM International Conference on Cluster, Cloud and Grid Computing, pp. 331–340 (2010)
12. Chung, D.H., Legg, P., Parry, M.L., Bown, R., Griffiths, I.W., Laramee, R.S., Chen, M.: Glyph sorting: interactive visualization for multi-dimensional data. Inf. Vis. **14**(1), 76–90 (2015)
13. Dörk, M., Carpendale, S., Collins, C., Williamson, C.: VisGets: coordinated visualizations for web-based information exploration and discovery. IEEE Trans. Vis. Comput. Graph. **14**(6), 1205–1212 (2008)
14. Fayyad, U.M., Piatetsky-Shapiro, G., Smyth, P.: From data mining to knowledge discovery: an overview. Advances in knowledge discovery and data mining. AI Mag. **17**(3), 37–54 (1996)

15. Gahegan, M., Takatsuka, M., Wheeler, M., Hardisty, F.: Introducing geo VISTA studio: an integrated suite of visualization and computational methods for exploration and knowledge construction in geography. Comput. Environ. Urban Syst. **26**(4), 267–292 (2002)
16. Guo, D., Chen, J., MacEachren, A.M., Liao, K.: A visualization system for space-time and multivariate patterns (VIS-STAMP). IEEE Trans. Vis. Comput. Graph. **12**(6), 1461–1474 (2006)
17. Heinrich, J., Weiskopf, D.: State of the art of parallel coordinates. In: Proceedings of Eurographics - State of the Art Reports, pp. 95–116 (2013)
18. Ho, Q.V., Lundblad, P., Åström, T., Jern, M.: A web-enabled visualization toolkit for geovisual analytics. Inf. Vis. **11**(1), 22–42 (2012)
19. Hoeber, O., Wilson, G., Harding, S., Enguehard, R., Devillers, R.: Exploring geo-temporal differences using GTdiff. In: Proceedings of the IEEE Pacific Visualization Symposium, pp. 139–146 (2011)
20. Im, J.F., McGuffin, M.J., Leung, R.: GPLOM: the generalized plot matrix for visualizing multidimensional multivariate data. IEEE Trans. Vis. Comput. Graph. **19**(12), 2606–2614 (2013)
21. Inselberg, R.: The plane with parallel coordinates. Vis. Comput. **1**(2), 69–91 (1985)
22. Jänicke, S., Heine, C., Scheuermann, G.: GeoTemCo: comparative visualization of geospatial-temporal data with clutter removal based on dynamic delaunay triangulations. In: Csurka, G., Kraus, M., Laramee, R.S., Richard, P., Braz, J. (eds.) VISIGRAPP 2012. CCIS, vol. 359, pp. 160–175. Springer, Heidelberg (2013)
23. Keim, D.A., Andrienko, G., Fekete, J.-D., Görg, C., Kohlhammer, J., Melançon, G.: Visual analytics: definition, process, and challenges. In: Kerren, A., Stasko, J.T., Fekete, J.-D., North, C. (eds.) Information Visualization: Human-Centered Issues and Perspectives. LNCS, vol. 4950, pp. 154–175. Springer, Heidelberg (2008)
24. Kosara, R., Bendix, F., Hauser, H.: Parallel sets: interactive exploration and visual analysis of categorical data. IEEE Trans. Visual. Comput. Graph. **12**(4), 558–568 (2006)
25. Kriegel, H.P., Schubert, M., Zimek, A.: Angle-based outlier detection in high-dimensional data. In: Proceedings of the ACM SIGKDD International Conference on Knowledge Discovery and Data Mining, pp. 444–452 (2008)
26. Ramakrishna, A., Chang, Y.-H., Maheswaran, R.: An interactive web based spatio-temporal visualization system. In: Bebis, G., et al. (eds.) ISVC 2013, Part II. LNCS, vol. 8034, pp. 673–680. Springer, Heidelberg (2013)
27. Roberts, J.: State of the art: coordinated & multiple views in exploratory visualization. In: Proceedings of the International Conference on Coordinated and Multiple Views in Exploratory Visualization, pp. 61–71 (2007)
28. TBTI: Report on enhancing stewardship in small-scale fisheries through ecosystem approaches and other means. Technical report, Too Big to Ignore Latin America and the Caribbean Joint Workshop with Working Group 4, Curitiba, Brazil. TBTI, Canada, August 2013
29. Tufte, E.R.: The Visual Display of Quantitative Information, 2nd edn. Graphics Press, Cheshire (2001)
30. Ward, M., Grinstein, G., Keim, D.: Interactive Data Visualization: Foundations, Techniques, and Applications. A. K. Peters Ltd., Natick (2010)
31. Ware, C.: Information Visualization: Perception for Design, 3rd edn. Morgan Kaufmann Publishers Inc., Waltham (2013)
32. Zhou, H., Yuan, X., Qu, H., Cui, W., Chen, B.: Visual clustering in parallel coordinates. Comput. Graph. Forum **27**(3), 1047–1054 (2008)

Generalized Shortest Path Kernel on Graphs

Linus Hermansson[1]([⊠]), Fredrik D. Johansson[2], and Osamu Watanabe[1]

[1] Tokyo Institute of Technology, Tokyo, Japan
{linus3,watanabe}@is.titech.ac.jp
[2] Chalmers University of Technology, Gothenburg, Sweden
frejohk@chalmers.se

Abstract. We consider the problem of classifying graphs using graph kernels. We define a new graph kernel, called the generalized shortest path kernel, based on the number and length of shortest paths between nodes. For our example classification problem, we consider the task of classifying random graphs from two well-known families, by the number of clusters they contain. We verify empirically that the generalized shortest path kernel outperforms the original shortest path kernel on a number of datasets. We give a theoretical analysis for explaining our experimental results. In particular, we estimate distributions of the expected feature vectors for the shortest path kernel and the generalized shortest path kernel, and we show some evidence explaining why our graph kernel outperforms the shortest path kernel for our graph classification problem.

Keywords: Graph kernel · SVM · Machine learning · Shortest path

1 Introduction

Classifying graphs into different classes depending on their structure is a problem that has been studied for a long time and that has many useful applications [1,4,11]. It is generally regarded that the number of self-loop-avoiding paths between all pairs of nodes of a given graph is useful for understanding the structure of the graph [7,12]. Computing the number of such paths between all nodes is however a computationally hard task (usually #P-hard). Counting only the number of shortest paths between node pairs is however possible in polynomial time and such paths at least avoid cycles, which is why some researchers have considered shortest paths a reasonable substitute. When using standard algorithms to compute the shortest paths between node pairs in a graph we also get, as a by-product, the *number* of such shortest paths between all node pairs. Our approach for classifying graphs is based on taking this number of shortest paths into account.

One popular technique for classifying graphs is by using a *support vector machine* (SVM) classifier with graph kernels. This approach has proven successful for classifying several types of graphs [3,4,8]. Graph kernels that consider many different properties have been proposed. Such as graph kernels considering all walks [6], shortest paths [3], small subgraphs [14], global graph properties [9],

© Springer International Publishing Switzerland 2015
N. Japkowicz and S. Matwin (Eds.): DS 2015, LNAI 9356, pp. 78–85, 2015.
DOI: 10.1007/978-3-319-24282-8_8

etc. Different graph kernels can however give vastly different results depending on the types of graphs that are being classified. Analyzing how these graph kernels perform on particular datasets, gives us the ability of choosing graph kernels appropriate for the particular types of graphs that we are trying to classify.

One particular type of graphs, that appears in many applications, are graphs with a cluster structure. Such graphs appear for instance when considering graphs representing social networks. In this paper, in order to test how well our approach works, we test its performance on the problem of classifying graphs by the number of clusters that they contain. More specifically, we consider two types of models for generating random graphs, the Erdős-Rényi model [2] and the planted partition model [10], where we use the Erdős-Rényi model to generate graphs with one cluster and the planted partition model to generate graphs with two clusters (See Sect. 4 for details). The example task considered in this paper is to classify whether a given random graph is generated by the Erdős-Rényi model or by the planted partition model. Where we consider the standard supervised machine learning approach. For this classification problem, we use the standard SVM and compare experimentally the performance of the SVM classifier, with the *shortest path* (SP) kernel, and with our new *generalized shortest path* (GSP) kernel. We show that the SVM classifier that uses our GSP kernel outperforms the SVM classifier that uses the SP kernel, on several datasets.

We also give a theoretical analysis of the random feature vectors of the SP kernel and the GSP kernel, for the random graph models from the experiments. We give an estimation of expected feature vectors for the SP kernel and show that the they are relatively close between graphs with one cluster and graphs with two clusters. We then analyze the expected feature vectors for the GSP kernel, and we show some evidence that the expected feature vectors have a different structure between graphs with one cluster and graphs with two clusters.

2   Preliminaries

Here we introduce necessary notions and notation for our technical discussion. Throughout this paper we use symbols G, V, E (with a subscript or a superscript) to denote graphs, sets of nodes, and sets of edges respectively. We fix n and m to denote the number of nodes and edges of considered graphs. By $|S|$ we mean the number of elements of the set S.

We are interested in the length and number of shortest paths. In relation to the kernels we use for classifying graphs, we use *feature vectors* for expressing such information. For any graph G, for any $d \geq 1$, let n_d denote the number of pairs of nodes of G with a shortest path of length d. We call a vector $\boldsymbol{v}_{\mathrm{sp}} = [n_1, n_2, \ldots]$ a *SPI feature vector*. On the other hand, for any $d, x \geq 1$, we use $n_{d,x}$ to denote the number of pairs of nodes of G that has x number of shortest paths of length d, and we call a vector $\boldsymbol{v}_{\mathrm{gsp}} = [n_{1,1}, n_{1,2}, \ldots, n_{2,1} \ldots]$ a *GSPI feature vector*. Note that $n_d = \sum_x n_{d,x}$. Thus, a GSPI feature vector is a more detailed version of a SPI feature vector. In order to simplify our discussion we often use feature vectors by considering shortest paths from any fixed node of G.

We will clarify which version we use in each context. By $E[\boldsymbol{v}_{\mathrm{sp}}]$ and $E[\boldsymbol{v}_{\mathrm{gsp}}]$ we mean the expected SPI feature vector and the expected GSPI feature vector, for some specified random distribution. Note that the expected feature vectors are equal to $[E[n_d]]_{d\geq 1}$ and $[E[n_{d,x}]]_{d\geq 1, x\geq 1}$.

It should be noted that the SPI and the GSPI feature vectors are computable efficiently. We can use Dijkstra's algorithm for each node in a given graph, which gives all node pairs' shortest path length (i.e. a SPI feature vector) in time $\mathcal{O}(nm+n^2\log n)$. Note that by using Dijkstra's algorithm to compute the shortest path from a fixed node to any other node, the algorithm actually needs to compute *all* shortest paths between the two nodes, to verify that it really has found a shortest path. Thus it is possible to store the number of shortest paths between all node pairs, **without increasing the running time of the algorithm**, meaning that we can compute the GSPI feature vector in the same time as the SPI feature vector.

3   Shortest Path Kernel and Generalized Shortest Path Kernel

A graph kernel is a function $k(G_1, G_2)$ on pairs of graphs, which can be represented as an inner product $k(G_1, G_2) = \langle \phi(G_1), \phi(G_2) \rangle_{\mathcal{H}}$ for some mapping $\phi(G)$ to a Hilbert space $\mathcal{H}$, of possibly infinite dimension. In many cases, graph kernels can be thought of as similarity functions on graphs. Graph kernels have been used as tools for using SVM classifiers for graph classification problems [3,4,8].

The kernel that we build upon in this paper is the *shortest path* (SP) kernel, which compares graphs based on the shortest path length of all pairs of nodes [3]. By $D(G)$ we denote the multi set of shortest distances between all node pairs in the graph G. For two given graphs G_1 and G_2, the SP kernel, where we use the indicator function, is defined as:

$$K_{\mathrm{SPI}}(G_1, G_2) = \sum_{d_1 \in D(G_1)} \sum_{d_2 \in D(G_2)} \mathbb{1}\left[d_1 = d_2\right]. \tag{1}$$

Which is one of the most common versions of the SP kernel, used in for instance Borgwardt and Kriegel [3]. We call this version of the SP kernel the *shortest path index* (SPI) kernel. It is easy to check that $K_{\mathrm{SPI}}(G_1, G_2)$ is simply the inner product of the SPI feature vectors of G_1 and G_2.

We now introduce our new kernel, the *generalized shortest path* (GSP) kernel, which is defined by using *also* the number of shortest paths. For a given graph G, by $ND(G)$ we denote the multi set of numbers of shortest paths between all node pairs of G. Then the GSP kernel, where we use the indicator function, is defined as:

$$K_{\mathrm{GSPI}}(G_1, G_2) = \sum_{d_1 \in D(G_1)} \sum_{d_2 \in D(G_2)} \sum_{t_1 \in ND(G_1)} \sum_{t_2 \in ND(G_2)} \mathbb{1}\left[d_1 = d_2\right] \mathbb{1}\left[t_1 = t_2\right]$$

$$\tag{2}$$

Which we call the *generalized shortest path index* (GSPI) kernel. It is easy to see that this is equivalent to the inner product of the GSPI feature vectors of G_1 and G_2. It should be noted that we may consider other functions than the indicator function for the definitions of the SP and the GSP kernels.

4 Random Graph Models

We investigate the advantage of our GSPI kernel over the SPI kernel for a synthetic random graph classification problem. Our target problem is to distinguish random graphs having two relatively "dense parts", from simple graphs generated by the Erdős-Rényi model. Here by "dense part" we mean a subgraph that has more edges in its inside compared with its outside.

For any edge density parameter p, $0 < p < 1$, the Erdős-Rényi model (with parameter p) denoted by $G(n, p)$ is to generate a graph G (of n nodes) by putting an edge between each pair of nodes with probability p independently at random. On the other hand, for any p and q, $0 < q < p < 1$, the *planted partition model* [10], denoted by $G(n/2, n/2, p, q)$ is to generate a graph $G = (V^+ \cup V^-, E)$ (with $|V^+| = |V^-| = n/2$) by putting an edge between each pair of nodes u and v again independently at random with probability p if both u and v are in V^+ (or both u and v are in V^-) and with probability q otherwise.

In the following, we use the symbol p_1 to denote the edge density parameter of the Erdős-Rényi model and p_2 and q_2 to denote the edge density parameters of the planted partition model. We want to have $q_2 < p_2$ while keeping the expected number of edges the same for both random graph models (so that one cannot distinguish random graphs by just counting the number of edges). It is easy to check that this requirement is satisfied by setting $p_2 = (1 + \alpha_0)p_1$, and $q_2 = 2p_1 - p_2 - 2(p_1 - p_2)/n$, for some constant α_0, $0 < \alpha_0 < 1$. We consider the "sparse" situation for our experiments and analysis, and assume that $p_1 = c_0/n$ for sufficiently large constant c_0. Note that we may expect with high probability, that when c_0 is large enough, a random graph generated by both models have a large connected component but might not be fully connected [2]. In the rest of the paper, a random graph generated by $G(n, p_1)$ is called a *one-cluster graph* and a random graph generated by $G(n/2, n/2, p_2, q_2)$ is called a *two-cluster graph*.

5 Experiments

Here we compare the performance of the GSPI kernel with the SPI kernel on datasets where the goal is to classify if a graph is a one-cluster graph or a two-cluster graph. All datasets are generated using the models $G(n, p_1)$ and $G(n/2, n/2, p_2, q_2)$, described above. We generate 100 graphs each from the two different classes in each dataset. q_2 is chosen in such a way that the expected number of edges is the same for both classes of graphs. Note that the bigger difference there is between p_1 and p_2, the more different the one-cluster graphs are compared to the two-cluster graphs. In our experiments we generate graphs where $n \in \{200, 400, 600, 800, 1000\}$, $np_1 = c_0 = 40$ and $p_2 \in$

$\{1.2p_1, 1.3p_1, 1.4p_1, 1.5p_1\}$. Hence $p_1 = 0.2$ for $n = 200$, $p_1 = 0.1$ for $n = 400$ etc. In all experiments we calculate the normalized feature vectors for all graphs. By normalized we mean that each feature vector $\boldsymbol{v}_{\mathrm{sp}}$ and $\boldsymbol{v}_{\mathrm{gsp}}$ is normalized by its Euclidean norm. This means that the inner product between two feature vectors always is in $[0, 1]$. We then train an SVM using 10-fold cross validation and evaluate the accuracy of the kernels. We use Pegasos [13] for solving the SVM.

Table 1 shows the accuracy of both kernels on the different datasets. As can be seen neither of the kernels perform very well on the datasets where $p_2 = 1.2p_1$. This is because the two-cluster graphs generated in this dataset are very similar to the one-cluster graphs. As p_2 increases compared to p_1, the task of classifying the graphs becomes easier. As can be seen in the table the GSPI kernel outperforms the SPI kernel on nearly all datasets. In particular, on datasets where $p_2 = 1.4p_1$, the GSPI kernel has an increase in accuracy of over 20 % on several datasets. When $n = 200$ the increase in accuracy is over 40 %! Although the shown results are only for datasets where $c_0 = 40$, experiments using other values for c_0 gave similar results.

One reason that our GSPI kernel is able to classify graphs correctly when the SPI kernel is not, is because the feature vectors of the GSPI kernel, for the two classes, are a lot more different than for the SPI kernel. Due to space constraints however, visualizations of such feature vectors are not included in this paper.

Table 1. The accuracy of the SPI kernel and the GSPI kernel using 10-fold cross validation. The datasets where $p_2 = 1.2p_1$ are the hardest and the datasets where $p_2 = 1.5p_1$ are the easiest. Very big increases in accuracy are marked in bold.

Kernel	n	p_2	Accuracy
SPI	200	$\{1.2p_1, 1.3p_1, 1.4p_1, 1.5p_1\}$	$\{52.5\%, 55.5\%, \mathbf{54.5\%} \mathbf{56.5\%}\}$
GSPI	200	$\{1.2p_1, 1.3p_1, 1.4p_1, 1.5p_1\}$	$\{52.5\%, 64.0\%, \mathbf{99.0\%}, \mathbf{100.0\%}\}$
SPI	400	$\{1.2p_1, 1.3p_1, 1.4p_1, 1.5p_1\}$	$\{55.5\%, 63.5\%, \mathbf{75.5\%}, 95.5\%\}$
GSPI	400	$\{1.2p_1, 1.3p_1, 1.4p_1, 1.5p_1\}$	$\{54.0\%, 62.0\%, \mathbf{96.5\%}, 100.0\%\}$
SPI	600	$\{1.2p_1, 1.3p_1, 1.4p_1, 1.5p_1\}$	$\{58.0\%, 60.5\%, \mathbf{75.5\%}, 93.5\%\}$
GSPI	600	$\{1.2p_1, 1.3p_1, 1.4p_1, 1.5p_1\}$	$\{58.0\%, 67.0\%, \mathbf{94.0\%}, 100.0\%\}$
SPI	800	$\{1.2p_1, 1.3p_1, 1.4p_1, 1.5p_1\}$	$\{57.5\%, 59.0\%, 72.0\%, 98.0\%\}$
GSPI	800	$\{1.2p_1, 1.3p_1, 1.4p_1, 1.5p_1\}$	$\{57.5\%, 58.0\%, 82.0\%, 100.0\%\}$
SPI	1000	$\{1.2p_1, 1.3p_1, 1.4p_1, 1.5p_1\}$	$\{53.5\%, 55.0\%, \mathbf{66.0\%}, 98.5\%\}$
GSPI	1000	$\{1.2p_1, 1.3p_1, 1.4p_1, 1.5p_1\}$	$\{55.0\%, 62.0\%, \mathbf{87.5\%}, 100.0\%\}$

6 Analysis

In this section we give some approximated analysis of random feature vectors in order to give theoretical support for our experimental observations. We first show that one-cluster and two-cluster graphs have quite similar SPI feature vectors (as their expectations). Next we show some evidence that there is a non-negligible difference in their GSPI feature vectors. Throughout this section, we consider feature vectors defined by considering only paths from any fixed source node s and use a superscript $^{(z)}$ to denote the number of clusters in the graph. Thus, for example, $n_d^{(1)}$ is the number of nodes at distance d from s in a one-cluster graph, and $n_{d,x}^{(2)}$ is the number of nodes that have x shortest paths of length d to s in a two-cluster graph.

Here we introduce a way to state an approximation. For any functions a and b depending on n, we write $a \approx_{\text{rel}} b$ by which we mean $b\left(1 - \frac{c}{n}\right) < a < b\left(1 + \frac{c}{n}\right)$ holds for some constant $c > 0$ and sufficiently large n. Note that this closeness notion is closed under constant number of additions/subtractions and multiplications. For example, if $a \approx_{\text{rel}} b$ holds, then we also have $a^k \approx_{\text{rel}} b^k$ for any $k \geq 1$ that can be regarded as a constant w.r.t. n.

First we compare the SPI feature vectors. We consider relatively small[1] distances d so that d can be considered as a small constant w.r.t. n. We show that $\text{E}[n_d^{(1)}]$ and $\text{E}[n_d^{(2)}]$ are similar in the following sense.

Theorem 1. *For any constant d, we have $\text{E}[n_d^{(1)}] \in \text{E}[n_d^{(2)}](1 \pm \frac{2}{c_0-1})$, holds within our $\approx_{\text{rel}}$ approximation when $c_0 \geq 2 + \sqrt{3}$.*

Remark. For deriving this relation we assume that all paths in G exists independently, following the analysis of Fronczak et al. [5]. The proof of this theorem can be found in the full version of this paper. Note that the difference between $\text{E}[n_d^{(1)}]$ and $\text{E}[n_d^{(2)}]$ vanishes for large values of c_0.

Heuristic Comparison of GSPI Feature Vectors: Here we compare the expected GSPI feature vectors $\text{E}[\boldsymbol{v}_{\text{gsp}}^{(1)}]$ and $\text{E}[\boldsymbol{v}_{\text{gsp}}^{(2)}]$, and show evidence that they have some non-negligible difference. Here we focus on the distance $d = 2$ part of the GSPI feature vectors, i.e., subvectors $[\text{E}[n_{2,x}^{(z)}]]_{x \geq 1}$ for $z \in \{1, 2\}$. Since it is not so easy to analyze the distribution of the values $\text{E}[n_{2,1}^{(z)}], \text{E}[n_{2,2}^{(z)}], \ldots$, we introduce some "heuristic" analysis. The results of the analysis can be summarized as follows. The values of the feature vector for a one-cluster graph are approximately distributed as

$$\text{E}[n_{2,x}^{(1)}] \approx \text{E}[n_2^{(1)}] \cdot \Pr\left[\text{Bin}(np_1, p_1) = x\right].$$

where $\text{Bin}(N, p)$ is the binomial distribution. This means that the values of this subvector has one peak at $x_{\text{peak}}^{(1)} = np_1^2$. The values of the feature vector (at distance 2) of a two-cluster graph are approximately distributed according to the mixture distribution of the two distributions $\text{N}(n(p_2^2 + q_2^2)/2, \sigma_1^2 + \sigma_2^2)$ and $\text{N}(np_2q_2, \sigma_3^2 + \sigma_4^2)$, with weights $\text{E}[V_2^+]$ and $\text{E}[V_2^-]$. Where $\text{N}(\mu, \sigma^2)$ is the normal distribution, $\text{E}[V_2^+]$ and $\text{E}[V_2^-]$ are the expected number of nodes at distance 2 from s that are in the same cluster as s or in the other cluster respectively. The

---

[1] This smallness assumption is for our analysis, and we believe that the situation is more or less the same for any d.

two peaks of this mixture distribution are $x_{\text{peak}}^{(2,+)} = n(p_2^2 + q_2^2)/2$ and $x_{\text{peak}}^{(2,-)} = np_2q_2$. Using this, we may bound the difference between these peaks by

$$x_{\text{peak}}^{(2,+)} - x_{\text{peak}}^{(2,-)} \approx_{\text{rel}} 2n\alpha_0^2 p_1^2 \geq 2\alpha_0^2 x_{\text{peak}}^{(2,-)}.$$

This means that these peaks have a non-negligible relative difference. From this heuristic analysis we may conclude that the two vectors $[\text{E}[n_{2,x}^{(1)}]]_{x\geq 1}$ and $[\text{E}[n_{2,x}^{(2)}]]_{x\geq 1}$ have different distributions of their component values. In particular, while the former vector has only one peak, the latter vector has a double peak shape (for big enough α_0). Note that this difference does not vanish even when c_0 is big. This means that the GSPI feature vectors are different for one-cluster graphs and two-clusters graphs, even when c_0 is big, which is not the case for the SPI feature vectors, since their difference vanishes when c_0 is big. This provides evidence as to why our GSPI kernel performs better than the SPI kernel.

Though this is a heuristic analysis, we can show some examples that our observation is not so different from experimental results. In Fig. 1 we have plotted the mixture normal distribution that gives us our approximated vector $[\text{E}[n_{2,x}^{(2)}]]_{x\geq 1}$ and the corresponding experimental vector obtained by generating graphs according to our random model. In this figure the double peak shape can clearly be observed, which provides empirical evidence supporting our analysis.

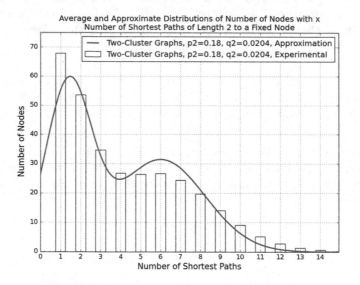

Fig. 1. Average experimental and approximate distributions of number of nodes with x number of shortest paths of length 2 from a fixed node. The experimental distribution has been averaged for each node in the graph and also averaged over 500 randomly generated graphs. Graphs used had parameters $n = 400$, $p_2 = 0.18$ and $q_2 = 0.0204$.

7   Conclusions

We have defined a new graph kernel, based on the number of shortest paths between node pairs in a graph. Calculating the GSP kernel does not take longer time than the SP kernel. The reason for this is that the number of shortest paths between node pairs is a by-product of using Dijkstra's algorithm to get the length of the shortest paths between all node pairs in a graph. We showed experimentally that the GSP kernel outperformed the SP kernel, at the task of classifying graphs as containing one or two clusters. We also gave an analysis motivating why the GSP kernel is able to correctly classify the two types of graphs when the SP kernel is not able to do so.

Acknowledgements. This work is supported in part by the ELC project (MEXT KAKENHI No. 24106008) and also in part by the Swedish Foundation for Strategic Research.

References

1. Bilgin, C., Demir, C., Nagi, C., Yener, B.: Cell-graph mining for breast tissue modeling and classification. In: 29th Annual International Conference of the IEEE Engineering in Medicine and Biology Society, EMBS 2007, pp. 5311–5314. IEEE (2007)
2. Bollobás, B.: Random Graphs. Springer, New York (1998)
3. Borgwardt, K.M., Kriegel, H.-P.: Shortest-path kernels on graphs. In: Proceedings of ICDM (2005)
4. Borgwardt, K.M., Ong, C.S., Schönauer, S., Vishwanathan, S., Smola, A.J., Kriegel, H.-P.: Protein function prediction via graph kernels. Bioinformatics **21**(suppl 1), i47–i56 (2005)
5. Fronczak, A., Fronczak, P., Hołyst, J.A.: Average path length in random networks. Phys. Rev. E **70**(5), 056110 (2004)
6. Gärtner, T., Flach, P.A., Wrobel, S.: On graph kernels: hardness results and efficient alternatives. In: Schölkopf, B., Warmuth, M.K. (eds.) COLT/Kernel 2003. LNCS (LNAI), vol. 2777, pp. 129–143. Springer, Heidelberg (2003)
7. Havlin, S., Ben-Avraham, D.: Theoretical and numerical study of fractal dimensionality in self-avoiding walks. Phys. Rev. A **26**(3), 1728 (1982)
8. Hermansson, L., Kerola, T., Johansson, F., Jethava, V., Dubhashi, D.: Entity disambiguation in anonymized graphs using graph kernels. In: Proceedings of the 22nd ACM International Conference on Conference on Information and Knowledge Management, pp. 1037–1046. ACM (2013)
9. Johansson, F., Jethava, V., Dubhashi, D., Bhattacharyya, C.: Global graph kernels using geometric embeddings. In: Proceedings of the 31st International Conference on Machine Learning (ICML-14), pp. 694–702 (2014)
10. Kolla, S.D.A., Koiliaris, K.: Spectra of random graphs with planted partitions (2013)
11. Kudo, T., Maeda, E., Matsumoto, Y.: An application of boosting to graph classification. In: Advances in Neural Information Processing Systems, pp. 729–736 (2004)
12. Liśkiewicz, M., Ogihara, M., Toda, S.: The complexity of counting self-avoiding walks in subgraphs of two-dimensional grids and hypercubes. Theor. Comput. Sci. **304**(1), 129–156 (2003)
13. Shalev-Shwartz, S., Singer, Y., Srebro, N., Cotter, A.: Pegasos: primal estimated sub-gradient solver for SVM. Math. Program. **127**(1), 3–30 (2011)
14. Shervashidze, N., Vishwanathan, S., Petri, T., Mehlhorn, K., Borgwardt, K.M.: Efficient graphlet kernels for large graph comparison. In Proceedings of AISTATS (2009)

Ensembles of Extremely Randomized Trees for Multi-target Regression

Dragi Kocev[1,2(✉)] and Michelangelo Ceci[1]

[1] Department of Computer Science, University of Bari Aldo Moro, Bari, Italy
dragi.kocev@ijs.si, michelangelo.ceci@uniba.it
[2] Department of Knowledge Technologies, Jožef Stefan Institute, Ljubljana, Slovenia

Abstract. In this work, we address the task of learning ensembles of predictive models for predicting multiple continuous variables, i.e., multi-target regression (MTR). In contrast to standard regression, where the output is a single scalar value, in MTR the output is a data structure – a tuple/vector of continuous variables. The task of MTR is recently gaining increasing interest by the research community due to its applicability in a practically relevant domains. More specifically, we consider the EXTRA-TREE ensembles – the overall top performer in the DREAM4 and DREAM5 challenges for gene network reconstruction. We extend this method for the task of multi-target regression and call the extension EXTRA-PCTs ensembles. As base predictive models, we propose to use predictive clustering trees (PCTs) – a generalization of decision trees for predicting structured outputs, including multiple continuous variables. We consider both global and local prediction of the multiple variables, the former based on a single model that predicts all of the target variables simultaneously and the latter based on a collection of models, each predicting a single target variable. We conduct an experimental evaluation of the proposed method on a collection of 10 benchmark datasets for with multiple continuous targets and compare its performance to random forests of PCTs. The results reveal that a multi-target EXTRA-PCTs ensemble performs statistically significantly better than a single multi-target or single-target PCT. Next, the performance among the different ensemble learning methods is not statistically significantly different, while multi-target EXTRA-PCTs ensembles are the best performing method. Finally, in terms of efficiency (running times and model complexity), both multi-target variants of the ensemble methods are more efficient and produce smaller models as compared to the single-target ensembles.

Keywords: Multi-target regression · Ensembles · Extremely randomized trees · Predictive clustering trees

1 Introduction

Supervised learning is one of the most widely researched and investigated areas of machine learning. The goal in supervised learning is to learn, from a set of examples with known class, a function that outputs a prediction for the class of

© Springer International Publishing Switzerland 2015
N. Japkowicz and S. Matwin (Eds.): DS 2015, LNAI 9356, pp. 86–100, 2015.
DOI: 10.1007/978-3-319-24282-8_9

a previously unseen example. However, in many real life problems of predictive modelling the output (i.e., the target) is structured, meaning that there can be dependencies between classes (e.g., classes are organized into a tree-shaped hierarchy or a directed acyclic graph) or some internal relations between the classes (e.g., sequences).

In this work, we concentrate on the task of predicting multiple continuous variables. Examples thus take the form $(\mathbf{x_i}, \mathbf{y_i})$, where $\mathbf{x_i} = (x_{i1}, \ldots, x_{ik})$ is a vector of k input variables and $\mathbf{y_i} = (y_{i1}, \ldots, y_{it})$ is a vector of t target variables. This task is known under the name of *multi-target regression* (MTR) [1] (also known as multi-output or multivariate regression).

MTR is a type of structured output prediction task which has application in many real life problems where we are interested in simultaneously predicting multiple continuous variables. Prominent examples come from ecology: predicting abundance of different species living in the same habitat [2], or predicting properties of forests [3,4]. Due to its applicability in a wide range of domains, this task is recently gaining increasing interest in the research community.

Several methods for addressing the task of MTR have been proposed [1,5]. These methods can be categorized into two groups of methods [6]: (1) local methods that predict each of the target variable separately and then combine the individual models to get the overall model and (2) global methods that predict all of the variables simultaneously (also known as 'big-bang' approaches). In the case of local models, for a domain with t target variables one needs to construct t predictive models – each predicting a single target. The prediction vector (that consists of t components) of an unseen example is then obtained by concatenating the predictions of the multiple single-target predictive models. Conversely, in the case of global models, for the same domain, one needs to construct 1 model. The prediction vector of an unseen example here is then obtained by passing the example through the model and getting its prediction.

In the past, several researchers proposed methods for solving the task of MTR directly and demonstrated their effectiveness [1,4,7,8]. The global methods have several advantages over the local methods. First, they exploit and use the dependencies that exist between the components of the structured output in the model learning phase, which can result in better predictive performance. Next, they are typically more efficient: it can easily happen that the number of components in the output is very large (e.g., hierarchies in functional genomics can have several thousands of components), in which case executing a basic method for each component is not feasible. Furthermore, they produce models that are typically smaller than the sum of the sizes of the models built for each of the components.

In [1,9], we evaluated the construction of local and global models for MTR in the context of ensemble learning. More specifically, we focus on two most widely used ensemble learning techniques: bagging [10] and random forests [11]. We show that both global and local tree ensembles perform better than the single model counterparts in terms of predictive power. Global and local tree ensembles perform equally well, with global ensembles being more efficient and producing smaller models, as well as needing fewer trees in the ensemble to achieve the maximal performance.

In this paper, we investigate a new strategy for learning MTR global models through ensemble learning. In particular, we extend the EXTRA-TREES ALGO-RITHM to the context of MTR. The EXTRA-TREES algorithm, proposed by Geurts et al. [12], is an algorithm for tree ensemble construction based on an extreme randomization of the tree construction algorithm. The algorithm at each node of the tree randomly selects k attributes and, on each of them, randomly selects a split. The k candidate splits are then evaluated and the best split is put in the node. Here, we propose an extension of the EXTRA-TREES algorithm for the task of predicting multiple continuous variables.

Geurts et al. evaluated their approach in the context of single-target regression and classification problems containing only numerical attributes. The bias/variance analysis of the error revealed that Extra-Trees decrease the variance while at the same time they increase the bias. If the level of randomization is well adjusted, then the variance almost disappears at the cost of a slight increase of the bias with respect to that of standard trees. In this study, we perform an empirical evaluation of the EXTRA-TREES algorithm extension in domains where the descriptive attributes can be continuous, categorical or mixed.

The EXTRA-TREES ALGORITHM has been successfully applied to several practically relevant domains including computer vision [13] and gene network inference [14,15]. Especially noticeable are the applications in the latter domain: a variant of the method that exploits its feature ranking mechanism (GENIE3 algorithm) has been overall top performer in the DREAM4 and DREAM5 challenges[1] for gene network inference. All of these considerations strongly motivate this study.

In this paper, we propose an extension of the EXTRA-TREES algorithm based on the predictive clustering trees (PCTs) framework [1,16]. We call this extension EXTRA-PCTs algorithm. PCTs belong to the group of global methods and can be considered as a generalization of standard decision trees towards predicting structured outputs. They offer a unifying approach for dealing with different types of structured outputs and construct the predictive models very efficiently. They are able to make predictions for several types of structured outputs: tuples of continuous/discrete variables, hierarchies of classes [17], and time series.

The remainder of this paper is organized as follows. Section 2 presents the proposed EXTRA-PCTs algorithm for MTR. Next, Sect. 3 outlines the design of the experimental evaluation. Furthermore, Sect. 4 discusses the results. Finally, Sect. 5 concludes and provides directions for further work.

2   MTR with Ensembles of Predictive Clustering Trees

The predictive clustering trees framework views a decision tree as a hierarchy of clusters. The top-node corresponds to one cluster containing all data, which

---

[1] For more information, visit http://dreamchallenges.org/.

is recursively partitioned into smaller clusters while moving down the tree. The
PCT framework is implemented in the CLUS system [18], which is available for
download at http://clus.sourceforge.net.

PCTs are induced with a standard *top-down induction of decision trees*
(TDIDT) algorithm [19]. Table 1 outlines the general algorithm for PCT induction. It takes as input a set of examples (E) and outputs a tree. The heuristic
(h) that is used for selecting the tests (t), in a regular PCT, is the reduction in
variance caused by the partitioning $(\mathcal{P})$ of the instances corresponding to the
tests (t) (see line 7 of the BestTest procedure in Table 2). Intuitively, by maximizing the variance reduction, the cluster homogeneity is maximized and the
predictive performance is improved.

Table 1. The top-down induction algorithm for PCTs.

---

procedure ExtremelyRnd PCT
Input: A dataset E, size of attribute subset k
Output: A predictive clustering tree
1: $(t^*, h^*, \mathcal{P}^*) = \text{FindTest}(E)$
2: **if** $t^* \neq none$ **then**
3:      **for each** $E_i \in \mathcal{P}^*$ **do**
4:           $tree_i = \text{PCT}(E_i)$
5:      **return** node$(t^*, \bigcup_i \{tree_i\})$
6: **else**
7:      **return** leaf(Prototype(E))

The extremely randomized variant of PCTs introduces a randomization in
the test selection (Table 2). More specifically, it requires an input parameter (k)
that controls the number of attributes considered at each node of the tree. The
test selection procedure randomly selects k attributes and from each attribute
randomly selects a split. For each of the k selected attributes, the algorithm
selects the split in two different ways, depending on the type of the attribute. If
the attribute is numeric the splitting point is selected randomly from the set of
possible splitting points. Possible splitting points are found in the set of values of
the attribute in the training set associated to the specific node. If the attribute
is categorical (i.e., nominal) then a non-empty subset of values of the attribute
in the training set associated to the specific node are randomly selected.

The k-candidate tests are then evaluated using the variance reduction heuristic and the best test is selected. In order to take multiple target variables into
account simultaneously, variance used to initialize h is defined as follows:

$$Var(E) = \sum_{j=1}^{t} Var(E, Y_j),$$

Table 2. Extremely randomized test selection for PCTs.

procedure FindTest	**procedure** selectRandomTest
Input: A dataset E	**Input:** Attribute a and partition $\mathcal{P}$
Output: the selected test (t^*), its heuristic score (h^*) and the partition $(\mathcal{P}^*)$ it induces on the dataset (E)	**Output:** A test t
1: $(t^*, h^*, \mathcal{P}^*) = (none, 0, \emptyset)$	1: $t = none$
2: $A = getAttributeList(E)$	2: $A_v = getAttributeValues(a, \mathcal{P})$
3: $A_s = selectAttributes(E, k)$	3: **if** a is numerical **then**
4: **for each** attribute $a \in A_s$ **do**	4: $a_M = getMaxValue(A_v)$
5: $t = selectRandomTest(a)$	5: $a_m = getMinValue(A_v)$
6: $\mathcal{P}$ = partition induced by t on E	6: $a_c = rndCutPoint(a_m, a_M)$
7: $h = Var(E) - \sum_{E_i \in \mathcal{P}} \frac{|E_i|}{|E|} Var(E_i)$	7: $t = a < a_c$
	8: **if** a is categorical **then**
8: **if** $(h > h^*)$ **then**	9: $A_s = rndNonEmptySet(A_v, \mathcal{P})$
9: $(t^*, h^*, \mathcal{P}^*) = (t, h, \mathcal{P})$	10: $t = a \in A_s$
10: **return** $(t^*, h^*, \mathcal{P}^*)$	11: **return** t

where $Var(E, Y_j)$ is the normalized variance (according to the $min - max$ normalization function) of the variable Y_j in the set E. The variances of the target variables are normalized so that each target variable contributes equally to the overall variance. This is due to the fact that the target variables can have completely different ranges.

Obviously, the smaller the variance reduction (h in the procedure $FindTest$ - see Table 2) the better the split. If we set the value of k to 1, this algorithm works in the same way of the Random Tree algorithm proposed in [20]. The advantage with respect to the Random Tree algorithm is that in the approach we adopt there is still a non-random selection based on some evaluation measure (i.e., variance reduction).

The extremely randomized PCTs are very unstable predictive models because of the intense randomization at each node. Consequently, such PCTs are only meaningful when used in combination with an ensemble learning framework. In this work, we construct ensembles of extremely randomized PCTs (Extra-PCTs) by following the same ensemble learning approach proposed in [12] where, however, PCTs are not used and, consequently, it is not possible to directly follow a global approach and naturally consider the multi-target regression task.

Each of the base predictive models is constructed using the complete training set and each of them uses different, randomly selected, attributes in the nodes. The number of attributes (k) that are retained is given by a function of the total number of descriptive attributes D (e.g., $k = 1$, $k = \lfloor \sqrt{D} + 1 \rfloor$, $f(D) = \lfloor log_2(D) + 1 \rfloor$, $k = D \ldots$). Depending on the application, one can select to use different values for k. In this study, we investigate the effect of the function used to initialize k on the performance of the ensemble for MTR.

In the Extra-PCTs ensemble, the prediction for a new instance is obtained by combining the predictions of all the base predictive models. For the MTR task, the prediction for each target variable is computed as the average of the predictions obtained from each tree. Note that this solution exploits possible dependencies in the output space since clusters used for prediction (and their hierarchical organization, i.e., the tree) have been built by taking into account the whole output space.

One of the strong advantages of the Extra-PCTs ensembles is their computational efficiency. In [1], we discuss the computational cost of an ordinary PCTs and ensembles of PCTs extensively. The computational cost of constructing an ordinary PCT for predicting multiple target variables can be summarized as

$$\mathcal{O}(DN \log^2 N) + \mathcal{O}(SDN \log N) + \mathcal{O}(N \log N),$$

where D is the number of descriptive attributes, N is the number of examples and S is the number of target variables.

The cost of constructing a Extra-PCTs can be derived as follows. Two procedures are executed at each node of the tree and they include: calculating the best split out of the k randomly selected, candidate splits at a cost of $\mathcal{O}(kSN)$, and applying the split to the training instances with a cost of $\mathcal{O}(N)$. Furthermore, we assume, as in [20], that the tree is balanced and bushy. This means that the depth of the tree is in the order of $\log N$, i.e., $\mathcal{O}(\log N)$. Having this in mind, the total computational cost of constructing a single tree is

$$\mathcal{O}(kS \log N) + \mathcal{O}(N \log N).$$

Comparing the two costs, we can note that Extra-PCTs have much lower computational complexity as compared to regular PCTs. The ensembles usually amplify the computational cost of the base predictive models linearly with the number of base models. Consequently, the cost of an Extra-PCTs ensemble will be much lower than the cost of a regular ensemble.

3   Experimental Design

We construct several types of trees and ensembles thereof. First, we construct PCTs that predict a separate tree for each variable from the multiple target variables. Second, we learn PCTs that predict all of the target variables simultaneously. Finally, we construct the ensemble models in the same manner by using both random forests and the Extra-PCTs algorithm.

3.1   Experimental Questions

We consider three aspects of constructing tree ensembles with the Extra-PCTs algorithm for predicting multiple target variables: convergence, predictive performance and efficiency. We first investigate the saturation/convergence of the predictive performance of global and local ensembles with respect to the number of base predictive models they consist of. Namely, we inspect the predictive performance of the ensembles at different ensemble sizes (i.e., we construct saturation curves). The goal is to check which type of EXTRA-PCTs ensembles, global or local, saturates at a smaller number of trees.

We next assess the predictive performance of global and local EXTRA-PCTs ensembles and investigate whether global and local ensembles have better predictive performance than the respective single model counterparts. Moreover, we check whether the exploitation of the multiple targets can lift the predictive performance of an EXTRA-PCTs ensemble (i.e., global versus local ensembles). Furthermore, we compare the performance of the EXTRA-PCTs ensembles with the performance of a random forest ensemble of PCTs. Random forests of PCTs are considered among the state-of-the-art predictive modelling techniques [1]. Finally, we assess the efficiency of both global and local single predictive models and ensembles thereof by comparing the running times for and the sizes of the models obtained by the different approaches.

3.2   Data Description

The datasets with multiple continuous targets used in this study are 13 in total and are mainly from the domain of ecological modelling. Table 3 outlines the properties of the datasets. The selection of the datasets contain datasets with various number of examples described with various number of attributes. For more details on the datasets, we refer the reader to the referenced literature.

Table 3. Properties of the datasets with multiple continuous targets (regression datasets); N is the number of instances, $\overline{D/C}$ the number of descriptive attributes (discrete/continuous), and T the number of target attributes.

Name of dataset	N	$|D|/|C|$	T
Collembola [21]	393	8/39	3
EDM [22]	154	0/16	2
Forestry-Slivnica-LandSat [23]	6218	0/150	2
Forestry-Slivnica-IRS [23]	2731	0/29	2
Forestry-Slivnica-SPOT [23]	2731	0/49	2
Sigmea real [24]	817	0/4	2
Soil quality [2]	1944	0/142	3
Solar-flare 1 [25]	323	10/0	3
Solar-flare 2 [25]	1066	10/0	3
Water quality [26]	1060	0/16	14

3.3   Experimental Setup

Empirical evaluation is the most widely used approach for assessing the performance of machine learning algorithms, that is based on the 10-fold cross-validation evaluation strategy. The performance of the algorithms are assessed using some evaluation measures and, in particular, since the task we consider is that of MTR, we employed three well known measures: the correlation coefficient (CC), root mean squared error ($RMSE$) and relative root mean squared error ($RRMSE$). We present here only the results in terms of $RRMSE$, but similar conclusions hold for the other two measures.

Next, we define the parameter values used in the algorithms for constructing the single trees and the ensembles of PCTs. The single trees (both for multi-target and single-target regression) are obtained using F-test pruning. This pruning procedure uses the exact Fisher test to check whether a given split/test in an internal node of the tree results in a reduction in variance that is statistically significant at a given significance level. If there is no split/test that can satisfy this, then the node is converted to a leaf. An optimal significance level was selected by using internal 3-fold cross validation, from the following values: 0.125, 0.1, 0.05, 0.01, 0.005 and 0.001.

The construction of both ensemble methods takes, as an input parameter, the size of the ensemble, i.e., number of base predictive models to be constructed. We constructed ensembles with 10, 25, 50, 75, 100, 150 and 250 base predictive models. Following the findings from the study conducted by Bauer and Kohavi [27], the trees in the ensembles were not pruned.

Both the EXTRA-PCTs ensemble and the random forests algorithm take as input the size of the feature subset that is randomly selected at each node. For the EXTRA-PCTs ensemble, we follow the recommendations from Geurts et al. [12], and set the value of k to the number of descriptive attributes, i.e., $k = D$. For the random forests of PCTs, we apply the logarithmic function of the number of descriptive attributes $\lfloor \log_2 |D| \rfloor + 1$, which is recommended by Breiman [11].

In order to assess the statistical significance of the differences in performance of the studied algorithms, we adopt the recommendations by Demšar [28] for the statistical evaluation of the results. In particular, we use the Friedman test for statistical significance. Afterwards, to check where the statistically significant differences appear (between which algorithms), we use two post-hoc tests. First, we use Bonferroni-Dunn test to compare the best performing method with the remaining methods. Second, we use Nemenyi post-hoc test when we compare all of the methods among each other. We present the results from the statistical analysis with *average ranks diagrams* [28]. The diagrams plot the average ranks of the algorithms and connect the ones whose average ranks are smaller than a given value, called critical distance. The critical distance depends on the level of the statistical significance, in our case 0.05. The difference in the performance of the algorithms connected with a line is not statistically significant at the given significance level.

4   Results and Discussion

We analyze the results from the experiments along three dimensions. First, we present the saturation curves of the ensemble methods (both for multi-target and single-target regression). We also compare single trees vs. ensembles of trees. Next, we compare models that predict the complete structured output vs. models that predict components of the structured output. Finally, we evaluate the algorithms by their efficiency in terms of running time and model size.

In Fig. 1, we present the saturation curves for the ensemble methods for multi-target regression. Although these curves are averaged across all target variables for a given dataset, they still provide useful insight into the performance of the algorithms. First, the curves show that for part of the datasets the ensembles reach their optimal performance when just as few as 25 base predictive models are constructed.

Second, we note that on majority of the datasets the proposed EXTRA-PCTs ensembles outperform the random forests of PCTs across all ensemble sizes. The most notable improvements are for the following datasets: *EDM, Forestry-Slivnica-LandSat, Forestry-Slivnica-SPOT* and *Soil quality*. The worst performance of the EXTRA-PCTs ensembles as compared with the random forests is for the dataset *Sigmea real*. For this, dataset the EXTRA-PCTs ensembles perform worse even than a single PCT. This may be due to the fact that this dataset has only 4 descriptive variables and the extreme randomization used in the EXTRA-PCTs ensembles hurts the predictive performance of the ensemble and misses on a crucial information. More specifically, the extreme randomization in this case decreases the variance only slightly while it increases the bias significantly (similarly as observed in [12]). Furthermore, on the datasets containing only categorical descriptive variables (*Solar-flare1* and *Solar-flare2*) both the EXTRA-PCTs ensembles and random forests perform poorly and their performance is worse than the performance of a single tree.Finally, in the case of mixed numeric and categorical variables (*Collembola* dataset) the multi-target random forests are the best performing method. The application of the proposed EXTRA-PCTs ensembles on datasets with categorical variables prompts further investigation.

Next, we perform statistical tests to detect up to which point the improvement is no longer statistically significant. To this end, we used Friedman test with Bonferroni-Dunn post-hoc test. We center the Bonferroni-Dunn test around the best performing ensemble size and check until which size the performance does not degrade statistically significantly. The results are presented in Fig. 2. From the diagrams, we can note that the multi-target EXTRA-PCTs ensembles achieve optimal performance with 75 base predictive models added to the ensemble. The remaining methods, multi-target and single-target random forests and single-target EXTRA-PCTs ensembles, require 100 base predictive models to achieve their optimal performance. This means that the global EXTRA-PCTs ensembles achieve their optimal performance with fewer trees added as compared with the local EXTRA-PCTs ensembles. Considering this, we perform the statistical analysis on ensembles with both 75 and 100 base predictive models.

Figure 3 gives the average rank diagrams of the different ensemble methods and the single-tree models. The results for ensembles with both 75 and 100 base predictive models show that the differences in predictive performance among

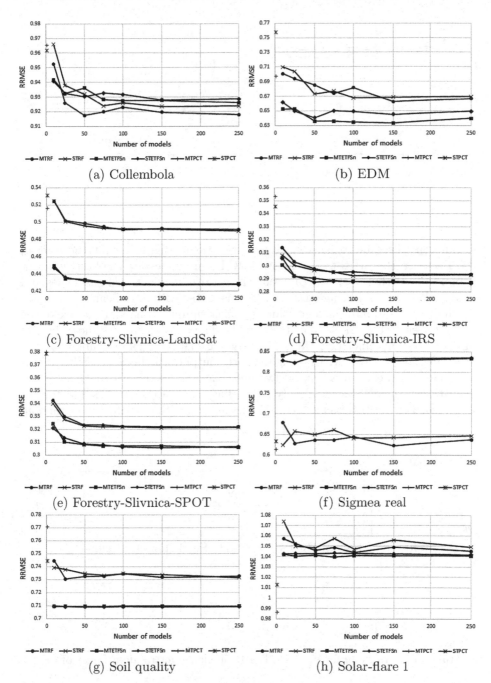

Fig. 1. Saturation curves for the two ensemble methods for MTR. Note that the scale of the y-axis is adapted for each curve. The algorithm names are abbreviated as follows: Predictive clustering trees - *PCT*, Extra-PCTs - *ET*, random forests - *RF*, multi-target prediction - *MT* and single-target prediction - *ST*.

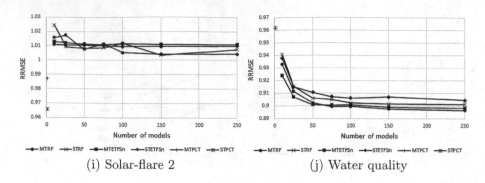

(i) Solar-flare 2   (j) Water quality

Fig. 1. (*continued*)

the different ensemble methods are not statistically significant at the level of 0.05. However, the multi-target EXTRA-PCTs ensembles are the best performing method. Furthermore, the difference in performance between ensembles and single multi-target and single-target PCTs is statistically significant.

Finally, we compare the algorithms by their running time and the size of the models for ensembles of 50 trees (see Fig. 4). The statistical tests show that, in terms of the time efficiency, the multi-target EXTRA-PCTs ensembles are the fastest method. Moreover, they significantly outperform both ensemble methods predicting the targets separately. The diagram also shows that the global (multi-target) ensembles are clearly more efficient than the local (single-target) ensembles. The multi-target EXTRA-PCTs are faster than multi-target random

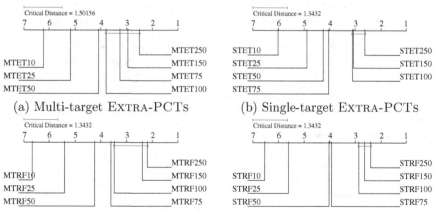

(a) Multi-target EXTRA-PCTs   (b) Single-target EXTRA-PCTs

(c) Multi-target random forest of PCTs (d) Single-target random forest of PCTs

Fig. 2. Average rank diagrams for the ensembles constructed with varying number of base predictive models. The critical distance is set for a significance level at 0.05. The differences in performance of the algorithms connected with a line are not statistically significant.

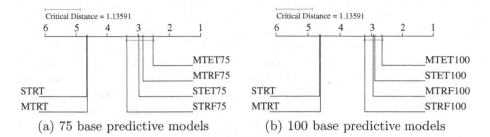

(a) 75 base predictive models   (b) 100 base predictive models

Fig. 3. Average rank diagrams for the various ensembles consisting of (a) 75 and (b) 100 base predictive models. The critical distance is set for a significance level at 0.05. The differences in performance of the algorithms connected with a line are not statistically significant.

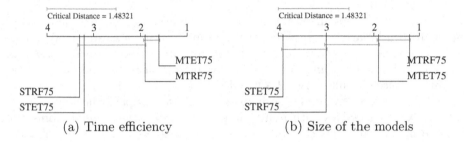

(a) Time efficiency   (b) Size of the models

Fig. 4. Efficiency (running time and model size) of the ensembles for MTR. The size of the ensembles is 75 trees, however the same conclusions hold across all ensemble sizes. The critical distance is set for a significance level at 0.05. The differences in performance of the algorithms connected with a line are not statistically significant.

forests ~ 1.77 times. The computational advantage is even more pronounced in the datasets with more examples. In terms of model size, the multi-target random forests are the best performing method. Both global ensembles are clearly better than their local counterparts. The results for the efficiency of the methods given in Fig. 4 show that the computational efficiency of the multi-target EXTRA-PCTs ensembles comes at the price of constructing slightly larger models. Namely, due to the increased randomness as compared to the random forests method fewer test are being evaluated (i.e., smaller computational cost) but, in the same time, this means that the constructed (PCT) models will grow larger.

5   Conclusions

In this work, we address the task of learning ensembles of predictive models for predicting multiple continuous variables, i.e., multi-target regression. In contrast to standard regression, where the output is a single scalar value, in MTR the output is a data structure – a tuple/vector of continuous variables. We consider both global and local prediction of the multiple variables, the former based on a single model that predicts all of the target variables simultaneously and the latter based on a collection of models, each predicting a single target variable.

Ensembles have proved to be highly effective methods for improving the predictive performance of their constituent models, especially for classification and regression tree models. In particular, we consider the EXTRA-TREE ensembles as predictive models. EXTRA-TREE ensembles are a well established method for predictive modelling that has been successfully applied to computer vision and, especially, gene network inference. This approach has been the overall top performer in the DREAM4 and DREAM5 challenges for gene network reconstruction. Following this, we extend this method for the task of multi-target regression and call the EXTRA-PCTS ensembles. As base predictive models, we propose to use predictive clustering trees (PCTs). These can be considered as a generalization of decision trees for predicting structured outputs, including multiple continuous variables.

We conduct an experimental evaluation of the proposed method on a collection of 10 benchmark datasets for with multiple continuous targets. We make several comparisons. First, we investigate the influence of the number of base predictive models in an ensemble to its predictive performance. Second, we compare the performance of multi-target EXTRA-PCTS ensembles with the performance of single-target EXTRA-PCTS ensembles. Next, we compare the multi-target EXTRA-PCTS ensembles with multi-target and single-target random forests of PCTs. Random forests are considered among the state-of-the-art modelling techniques. Furthermore, we compare the efficiency of the different approaches.

The results reveal the following. First, the performance of the multi-target EXTRA-PCTS ensembles starts to saturate as soon as even only 25 base predictive models are added to the ensemble. Moreover, after adding 75 base predictive models, the performance of a multi-target EXTRA-PCTS ensemble does not change statistically significantly. Second, a multi-target EXTRA-PCTS ensemble performs statistically significantly better than a single multi-target or single-target PCT. Next, the performance among the different ensemble learning methods is not statistically significantly different, while multi-target EXTRA-PCTS ensembles are the best performing method. Finally, in terms of efficiency (running times and model complexity), both multi-target variants of the ensemble methods are more efficient and produce smaller models as compared to the single-target ensembles.

We plan to extend the work along four major dimensions. First, we will extend the proposed algorithm to cover other tasks of structured output prediction, such as multi-target classification, multi-label classification and hierarchical multi-label classification. Second, we will adapt the feature ranking mechanism of the EXTRA-TREES algorithm for different types of structured outputs. Next, we will perform a more extensive study on the sensitivity of the algorithm of its parameter k and the influence of categorical variables in the dataset to the ensembles' performance. Finally, we will perform a more extensive experimental evaluation by using a larger number of benchmarking datasets.

Acknowledgments. We acknowledge the financial support of the European Commission through the grants ICT-2013-612944 MAESTRA and ICT-2013-604102 HBP.

References

1. Kocev, D., Vens, C., Struyf, J., Džeroski, S.: Tree ensembles for predicting structured outputs. Pattern Recogn. **46**(3), 817–833 (2013)
2. Demšar, D., Džeroski, S., Larsen, T., Struyf, J., Axelsen, J., Bruns-Pedersen, M., Krogh, P.H.: Using multi-objective classification to model communities of soil. Ecol. Model. **191**(1), 131–143 (2006)
3. Stojanova, D., Panov, P., Gjorgjioski, V., Kobler, A., Džeroski, S.: Estimating vegetation height and canopy cover from remotely sensed data with machine learning. Ecol. Inform. **5**(4), 256–266 (2010)
4. Kocev, D., Džeroski, S., White, M., Newell, G., Griffioen, P.: Using single- and multi-target regression trees and ensembles to model a compound index of vegetation condition. Ecol. Model. **220**(8), 1159–1168 (2009)
5. Tsoumakas, G., Spyromitros-Xioufis, E., Vrekou, A., Vlahavas, I.: Multi-target regression via random linear target combinations. In: Calders, T., Esposito, F., Hüllermeier, E., Meo, R. (eds.) ECML PKDD 2014, Part III. LNCS, vol. 8726, pp. 225–240. Springer, Heidelberg (2014)
6. Bakır, G.H., Hofmann, T., Schölkopf, B., Smola, A.J., Taskar, B., Vishwanathan, S.V.N.: Predicting Structured Data. Neural Information Processing. The MIT Press, Cambridge (2007)
7. Struyf, J., Džeroski, S.: Constraint based induction of multi-objective regression trees. In: Bonchi, F., Boulicaut, J.-F. (eds.) KDID 2005. LNCS, vol. 3933, pp. 222–233. Springer, Heidelberg (2006)
8. Appice, A., Džeroski, S.: Stepwise induction of multi-target model trees. In: Kok, J.N., Koronacki, J., Lopez de Mantaras, R., Matwin, S., Mladenič, D., Skowron, A. (eds.) ECML 2007. LNCS (LNAI), vol. 4701, pp. 502–509. Springer, Heidelberg (2007)
9. Kocev, D., Vens, C., Struyf, J., Džeroski, S.: Ensembles of multi-objective decision trees. In: Kok, J.N., Koronacki, J., Lopez de Mantaras, R., Matwin, S., Mladenič, D., Skowron, A. (eds.) ECML 2007. LNCS (LNAI), vol. 4701, pp. 624–631. Springer, Heidelberg (2007)
10. Breiman, L.: Bagging predictors. Mach. Learn. **24**(2), 123–140 (1996)
11. Breiman, L.: Random forests. Mach. Learn. **45**(1), 5–32 (2001)
12. Geurts, P., Ernst, D., Wehenkel, L.: Extremely randomized trees. Mach. Learn. **36**(1), 3–42 (2006)
13. Maree, R., Geurts, P., Piater, J., Wehenkel, L.: Random subwindows for robust image classification. In: IEEE Computer Society Conference on Computer Vision and Pattern Recognition (CVPR), vol. 1, pp. 34–40 (2005)
14. Ruyssinck, J., Huynh-Thu, V.A., Geurts, P., Dhaene, T., Demeester, P., Saeys, Y.: NIMEFI: gene regulatory network inference using multiple ensemble feature importance algorithms. PLoS ONE **9**(3), 1–13 (2014)
15. Huynh-Thu, V.A., Irrthum, A., Wehenkel, L., Geurts, P.: Inferring regulatory networks from expression data using tree-based methods. PLoS ONE **5**(9), 1–10 (2010)
16. Kocev, D.: Ensembles for Predicting Structured Outputs. Ph.D. thesis, Jožef Stefan International Postgraduate School, Ljubljana, Slovenia (2011)

17. Stojanova, D., Ceci, M., Malerba, D., Deroski, S.: Using PPI network autocorrelation in hierarchical multi-label classification trees for gene function prediction. BMC Bioinform. **14**, 285 (2013)

18. Blockeel, H., Struyf, J.: Efficient algorithms for decision tree cross-validation. J. Mach. Learn. Res. **3**, 621–650 (2002)

19. Breiman, L., Friedman, J., Olshen, R., Stone, C.J.: Classification and Regression Trees. Chapman & Hall/CRC, Boca Raton (1984)

20. Witten, I.H., Frank, E.: Data Mining: Practical Machine Learning Tools and Techniques. Morgan Kaufmann, San Francisco (2005)

21. Kampichler, C., Džeroski, S., Wieland, R.: Application of machine learning techniques to the analysis of soil ecological data bases: relationships between habitat features and Collembolan community characteristics. Soil Biol. Biochem. **32**(2), 197–209 (2000)

22. Karalič, A.: First order regression. Ph.D. thesis, Faculty of Computer Science, University of Ljubljana, Ljubljana, Slovenia (1995)

23. Stojanova, D.: Estimating forest properties from remotely sensed data by using machine learning. Master's thesis, Jožef Stefan International Postgraduate School, Ljubljana, Slovenia (2009)

24. Demšar, D., Debeljak, M., Džeroski, S., Lavigne, C.: Modelling pollen dispersal of genetically modified oilseed rape within the field. In: The Annual Meeting of the Ecological Society of America, p. 152 (2005)

25. Asuncion, A., Newman, D.: UCI - Machine Learning Repository (2007). http://www.ics.uci.edu/mlearn/MLRepository.html

26. Džeroski, S., Demšar, D., Grbović, J.: Predicting chemical parameters of river water quality from bioindicator data. Appl. Intell. **13**(1), 7–17 (2000)

27. Bauer, E., Kohavi, R.: An empirical comparison of voting classification algorithms: bagging, boosting, and variants. Mach. Learn. **36**(1), 105–139 (1999)

28. Demšar, J.: Statistical comparisons of classifiers over multiple data sets. J. Mach. Learn. Res. **7**, 1–30 (2006)

Clustering-Based Optimised Probabilistic Active Learning (COPAL)

Georg Krempl[✉], Tuan Cuong Ha, and Myra Spiliopoulou

Knowledge Management and Discovery, Otto-von-Guericke University,
Universitätsplatz 2, 39106 Magdeburg, Germany
georg.krempl@iti.cs.uni-magdeburg.de
http://kmd.cs.ovgu.de/res/pal

Abstract. Facing ever increasing volumes of data but limited human annotation capacities, active learning approaches that allocate these capacities to the labelling of the most valuable instances gain in importance. A particular challenge is the active learning of arbitrary, user-specified adaptive classifiers in evolving datastreams. We address this challenge by proposing a novel clustering-based optimised probabilistic active learning (COPAL) approach for evolving datastreams. It combines established clustering techniques, inspired by semi-supervised learning, which are used to capture the structure of the unlabelled data, with the recently introduced probabilistic active learning approach, which is used for the selection among clusters. The labels actively selected by COPAL are then available for training an arbitrary adaptive stream classifier. The performance of our algorithm is evaluated on several synthetic and real-world datasets. The results show that it achieves a better accuracy for the same budget than other recently proposed active learning approaches for such evolving datastreams.

Keywords: Probabilistic active learning · Selective sampling · Evolving datastreams · Nonstationary environments · Concept drift · Adaptive classification · Clustering

1 Introduction

In the face of ever increasing volumes of data [6] that contrast limited human annotation and supervision capacities, approaches for the efficient allocation of these capacities are of increasing interest. Active machine learning approaches address this by providing strategies for determining and selecting the most valuable information. In classification tasks, this corresponds to selecting the instance from a set of candidates, whose label is expected to improve a classifier's performance the most [23]. Active learning is considered a particularly challenging problem [15] in evolving datastreams, where instances arrive continuously over time and distributions may change and require adaptation.

© Springer International Publishing Switzerland 2015
N. Japkowicz and S. Matwin (Eds.): DS 2015, LNAI 9356, pp. 101–115, 2015.
DOI: 10.1007/978-3-319-24282-8_10

We address this challenge by proposing a novel active learning approach for evolving datastreams. Inspired from semi-supervised learning, our clustering-based optimised probabilistic active learning (COPAL) approach uses established clustering techniques to capture the structure of the unlabelled data. We combine this with the recently introduced optimised probabilistic active learning [13] approach that we use for selecting the cluster, and with a diversity-maximising criterion for selecting the instance within that cluster. This actively selected label is then used for the training of a user-specified adaptive stream classifier.

We contribute two such clustering-based probabilistic active learning approaches. The first is an incremental clustering variant that maintains and adapts its clustering model over time. The second is an amnesic clustering variant that iteratively learns new clustering models from scratch on each chunk and discards it after updating the classifier. We evaluate both variants by comparing them against each other and several competitors on six different datastreams, among them four real-world datasets. The results of the experimental evaluation indicate an overall superior performance of our incremental clustering-based probabilistic active learning approach.

We first review the related work in Sect. 2, before presenting our clustering-based optimised probabilistic active learning approach in its incremental and amnesic clustering variant in Sect. 3. These and other recently proposed AL-approaches are evaluated in Sect. 4, followed by a conclusion in Sect. 5.

2   Background and Related Work

Active machine learning [23] aims to optimise the selection of labels when they are costly to obtain. The scenario addressed in this paper is stream-based selective sampling [22], where instances arrive continuously and an active classifier has to decide for each instance upon its arrival once-and-forever whether to acquire its label. Compared to the rich literature on selective sampling in streams in general, active learning in *nonstationary, evolving* datastreams has received far less attention, although it is considered as a challenging, relevant task [7,15].

One line of research [16,17,21,24,25] has investigated ensemble-based active learning approaches for evolving datastreams. The approach in [24] processes instances in chunks, such that in each chunk a certain initial percentage of instances are labelled. These initial labels in the chunk are used to learn a new base classifier, which is added to the ensemble. The disagreement within the updated ensemble is then used to select iteratively a given number of instances within the remaining unlabelled ones in the chunk. This is extended in [25] by a criterion that determines when to stop the active learning process on a chunk, and by an adaptive weighting of the base classifiers in the ensemble. The ActMiner-algorithm proposed in [17] processes data also in chunks, but clusters the data into spherical micro-clusters, which represent base classifiers of an ensemble. New instances that are not covered by any micro-cluster (so-called F-outliers) or instances with disagreeing micro-clusters are labelled and saved in a buffer for later inspection. If the instances in the buffer form a new cluster, this cluster is added to the ensemble. The approach suggested in [21] extends this in two directions. First, by processing the stream instance-wise, and second, by using decision trees as base classifiers. Like [17], it also summarises the distribution of each base classifier's training data by a spherical cluster centred at its mean. This clustering is then used for the weighting of base classifiers and for the identification of suspicious instances outside all clusters. The labels of these suspicious instances are then requested to train a new base classifier. In contrast to these works, a different combination of query-by-committee and clustering for instance-wise active learning is proposed in [16]: upon the arrival of a new instance, a new ensemble of Gaussian mixture models is created by sampling from a normal inverse Wishart distribution, such that each Gaussian component corresponds to one class. The GMMs in the ensemble converge as the number of acquired labels increases, reducing the areas of disagreement between the GMMs and balancing exploration and exploitation.

More recently, other authors [1,10,11,20] have investigated the idea of combining clustering and stream-based active learning further. They extend the older clustering-based active learning approach in [19], which addressed a pool-based setting but already used the clustering information to select the most representative instances for labelling, thereby reducing the required number of labelled instances in each cluster. In contrast, the newer StreamAR approach in [1] is actually a semi-supervised stream classification approach that uses a micro-clustering ensemble to assign labels and uses active learning solely to resolve ties due to votes from opposing classes in the ensemble. Thus, while reducing the requested number of labels, it provides no means for controlling its budget.

In [20], a clustering-based approach is proposed for evolving datastreams, the so-called Concurrent Semi-supervised Learning of Data Streams (CSL-Stream). It maintains a clustering and assumes the posterior distribution within a cluster to be homogeneous, i.e. statistically independent of the feature position given the cluster membership. Its active learning step differentiates between clusters with and without any labelled instances. In the latter case, the algorithm seeks to obtain the label of the centremost instance. In the former case, the algorithm checks for a skewed label distribution: if all labels are on one side of the cluster, an additional label at the opposite side of the cluster is requested. Otherwise, if the class of the labels differs between the sides of the cluster, the cluster is split. If the distribution of labels is homogeneous, the cluster is kept as it is. Concept drift is addressed by using a fading model such that instances age over time. Unfortunately, the author's informed us that an implementation for this algorithm is no longer available.

While the clustering-based approaches above integrate clustering and classification, the aim of Clustering Based Active Learning for Evolving Data Streams (ACLStream) proposed in [10] is to be usable with any stream classifier technology. On each arriving chunk of instances a new clustering is performed and the most informative instances from each cluster are selected for labelling. For this selection, the approach distinguishes between a macro and a micro step. The macro step is used to rank clusters according to their homogeneity in terms of their predicted class distribution. This distribution is estimated by the model learnt from all the labelled instances from previous chunks of instances. The later micro step determines the most useful instance within a given cluster. Thus, it ranks instances by combining geometrical information inside their cluster and the maximum a posteriori classification probability. After selected instances are labelled, the clustering information is discarded.

The most recent active learning approach for evolving datastreams is DBAL-Stream [11]. This instance-wise approach combines density-weighting with uncertainty sampling. The density-weighting is used in a preselection step, such that solely instances within dense areas are considered as labelling candidates. Among those preselected candidates from dense regions, a margin-based uncertainty sampling approach is used to select one-by-one instances for labelling. This is done by comparing an instance's margin against a threshold, which is adjusted depending on the consumed and available budget and combined with random noise to improve exploration. This approach was reported to perform best in the evaluation by [11], making it an interesting candidate for our experimental evaluation.

The active learning algorithms for evolving datastreams discussed above are all based either on the disagreement in a query-by-committee approach, or the uncertainty in an uncertainty sampling approach, with known shortcomings [14,23]. Recently, the probabilistic active learning (PAL) approach has been proposed to overcome these shortcomings in the pool-based setting [14]. PAL summarises the labelled information in an instance's neighbourhood and evaluates the impact of acquiring a label therein in terms of the expected performance change. Expectation is not only done over the possible realisations of a candidate's label as in error reduction, but also over the true posterior in the candidate's neighbourhood. In [12], combining this approach with budgeting for datastreams is investigated. In [13], a fast closed-form solution is proposed that combines the qualities of uncertainty sampling and error reduction, namely being fast and optimising directly a performance measure. Thus, it seems worth exploring this fast approach in combination with clustering in a stream-based setting.

3   Clustering-Based Probabilistic Active Learning

Our approach combines ideas from clustering-based semi-supervised learning and probabilistic active learning. More precisely, we use the clustering model to define the neighbourhoods for the label statistics in a probabilistic active learning approach [14]. Our Clustering-based Optimised Probabilistic Active Learning (COPAL) algorithm consists of four steps, which are pre-clustering, macro and micro selection, and updating. To complete the big picture, we briefly summarise them before providing their details in the Subsects. 3.1 to 3.3. Finally, in Subsect. 3.4, we present two variants of COPAL with their pseudocode.

The *pre-clustering* step starts with a pool of unlabelled instances. In this step, all instances are divided into some initial clustering. While more elaborate clustering algorithms can be used, we opted for conventional K-means because it builds spherical clusters and because our focus is on assessing the neighbours of a data point and not on achieving a good partitioning of the data space.

The task of the *macro* step is to determine the most important cluster to select instances from for labelling. Therefore, we need an approach to measure the value of additional labels for a cluster. For that purpose, we adapt the OPAL-gain formula from probabilistic active learning [13] to our clustering model. The *micro* step then selects an instance from the previously chosen cluster for labelling, such that the *diversity* among the labelled instances within a cluster is maximised. After a new instance is labelled, the class distribution in the selected cluster may have changed. Thus, an *updating* step is used to adjust the clustering model. In this last step, we examine the homogeneity of the posterior distribution within the selected cluster. We split the cluster in case it has become inhomogeneous.

3.1 Macro Step: Determining the Most Valuable Cluster

The OPAL-gain formula in [13] is designed to compute the expected average misclassification loss reduction from obtaining m additional labels within a candidate's neighbourhood. It relies on label statistics ls, which summarise the number of already obtained labels n within its neighbourhood and the share of positives therein $\hat{p}$. For COPAL, the cluster of an instance defines its neighbourhood, thus n equals the number of labels acquired in that cluster, and $\hat{p}$ equals the share of positives therein. Because all instances in the cluster share the same neighbourhood, their probabilistic gains are equal. Following [13], the resulting expected average misclassification loss reduction in the cluster (G_{OPAL}) is calculated as:

$$G_{OPAL}(n,\hat{p},\tau,m) = \frac{(n+1)}{m} \cdot \binom{n}{n.\hat{p}} \cdot \left(I_{ML}(n,\hat{p},\tau,0,0) \right) - \sum_{k=0}^{m} I_{ML}(n,\hat{p},\tau,m,k)$$

(1)

Here, $\tau \in [0,1]$ is given by the application and corresponds to the relative cost of each false positive classification, normalised such that the costs of a false positive and a false negative sum to one. For example, when the objective is maximising the classifier's accuracy, $\tau = 0.5$ and G_{OPAL} is proportional to the gain in accuracy. Likewise, $m > 0$ is the application-given remaining budget for the currently processed chunk. Thus, τ and m are the same for each cluster. Equation 1 uses the function I_{ML}, introduced in [13], to compute a value that is proportional to the expected misclassification loss within a cluster, given that k additional positives among the m additional labels are sampled:

$$I_{ML}(n,\hat{p},\tau,m,k) = \binom{m}{k} \cdot \begin{cases} (1-\tau) & \cdot \frac{\Gamma(1-k+m+n-n\hat{p})\Gamma(2+k+n\hat{p})}{\Gamma(3+m+n)} & \frac{n\hat{p}+k}{m+n} < \tau \\ (\tau-\tau^2) & \cdot \frac{\Gamma(1-k+m+n-n\hat{p})\Gamma(1+k+n\hat{p})}{\Gamma(2+m+n)} & \frac{n\hat{p}+k}{m+n} = \tau \\ \tau & \cdot \frac{\Gamma(2-k+m+n-n\hat{p})\Gamma(1+k+n\hat{p})}{\Gamma(3+m+n)} & \frac{n\hat{p}+k}{m+n} > \tau \end{cases}$$

(2)

In a clustering model, the expected misclassification loss reduction for a cluster depends not only on the probabilistic gain, but also on the size of the cluster. Since a larger cluster affects more future classifications than a smaller one, the larger is favoured if their probabilistic gains are (nearly) equal. Therefore, for estimating the importance of a cluster $Cluster_i$ with $N_{Cluster_i}$ (labelled and unlabelled) instances therein, we propose to compute a cluster-size-weighted probabilistic gain $G_{Cluster_i}$ by the following formula:

$$G_{Cluster_i} = G_{OPAL_i} \cdot \frac{N_{Cluster_i}}{\sum_{j=1}^{N} N_{Cluster_j}} \tag{3}$$

The algorithm in Algorithm 1 describes this macro step in detail. For each cluster c, we calculate the values of n and $\hat{p}$ for this cluster, before computing its weighted gain by using the formulas 1, 2 and 3 (line 3–5). Finally, we select the cluster with the largest weighted gain and return it as the output (line 7–8).

Algorithm 1. Select the Best Cluster for Budget m and Cost-Ratio τ

1: **procedure** SELECTCLUSTER(C, m, τ)   ▷ C: Pool of clusters
2:   **for** $c \in C$ **do**
3:     $(n, \hat{p}) \leftarrow labelstatistic(c)$
4:     $G_{OPAL} \leftarrow getOPALGain(n, \hat{p}, m, \tau)$ ▷ Use Eq. 1
5:     $G_c \leftarrow getWeightedGain(G_{OPAL}, c)$ ▷ Use Eq. 3
6:   **end for**
7:   $c^* \leftarrow \arg\max_{c \in C}(G_c)$
8:   **return** c^*
9: **end procedure**

3.2   Micro Step: Selecting an Instance Within the Cluster

The micro step selects an instance within the cluster c^* that was previously chosen in the macro step. We aim to maximise diversity among the label that are requested within that cluster. Thus, by using Eq. 4, we select the instance for labelling, whose nearest labelled neighbour is the furthest away. Here, $c_{\mathcal{U}}^*$ is the subset of the current chunk's unlabelled instances within c^*, $c_{\mathcal{L}}^*$ is the subset of labelled ones, and $|\cdot|_2$ is the l^2-Norm:

$$x^* \leftarrow \arg\max_{x_i \in c_{\mathcal{U}}^*} \left(\min_{x_l \in c_{\mathcal{L}}^*} |x_i - x_l|_2 \right). \tag{4}$$

3.3   Updating Step: Adjusting the Clustering Model

Upon having obtained the label for the instance selected in the micro step, the cluster it belongs to is updated. In this step, two alternating hypotheses are considered: The first hypothesis H_1 is that the cluster is homogeneous with respect to its posterior distribution and, as a consequence, should not be split further. The second, alternative hypothesis H_2 is that the cluster is inhomogeneous and the instances therein originate from two spatially separable subpopulations with different posteriors. In the second case, splitting the cluster should improve homogeneity. Therefore, we calculate the current error rate E_1 (under H_1) and compare it to the error rate after splitting[1](E_2 under H_2). However, due to the limited number of remaining labels in each subcluster, the simple approach to use directly the training error is prone to overfitting. Instead, we perform a leave-one-out cross-validation on the labelled instances and calculate E_2 as the average error rate over each fold. In the case of ties due to an equal number of positives and negatives, we use an error rate of 50 %. If the error rate decreases by splitting (i.e. $E_1 > E_2$), we retrain the classifier on all labels in the cluster and partition the instances based on their assigned labels into two new clusters.

3.4 Variants of COPAL

We propose two variants of COPAL for combining the modules above. The first uses a sliding window and an incremental clustering, the second an amnesic clustering that forgets the clustering model after processing its chunk. The latter is inspired by the discussions of the authors of [10], who observed good performance with their amnesic clustering approach. However, COPAL worked better with an incremental rather than an amnesic clustering in our experiments.

Incremental Clustering Variant (COPAL-I). The pseudocode of the proposed incremental variant *COPAL-I* is provided in Algorithm 2. It uses the four steps above in combination with an incremental clustering model to actively select instances for labelling, which are then passed to an arbitrary incremental classifier. Using a sliding window approach, a fixed number of the most recent instances is kept in a *Cache*. These instances are used to maintain the cluster model. Consequently, the pre-clustering step is only applied for the first chunk (line 6), which also initialises the *Cache* (line 7). For subsequent chunks, the new instances are matched to the closest cluster of the current model (lines 9–11), and appended to the *Cache*, eventually replacing the oldest ones therein (lines 13–17). Afterwards, the macro, micro and update steps are applied iteratively to select the most valuable instances for labelling (lines 19–25). The update in each iteration comprises updating the dedicated classifier by the new label (line 23), and updating the clustering model (line 24).

Amnesic Clustering Variant (COPAL-A). This variant of COPAL uses an amnesic clustering model, as outlined in Algorithm 3. As above, after a chunk has been processed, the labels selected by *COPAL-A* therein are used to train an incremental classifier. The clustering model, however, is forgotten. Therefore, the pre-clustering step is repeated on each chunk (line 4). Afterwards, the macro and micro steps are applied to get the most valuable instance for labelling (lines 6–7). Its label is used to update the incremental classifier (line 9), and the process is repeated until the budget for this chunk is exhausted. Then, if necessary, the clustering is updated by splitting the cluster of the new instance (line 10).

---

[1] For speed, we used logistic regression for determining the preliminary splits.

Algorithm 2. Incremental Clustering Variant COPAL-I

Require: S: Stream of Instances
Require: b: Budget (per Chunk)
Require: w: Window Size (of Cache)
1: $cl \leftarrow initClassifier$
2: $Cache \leftarrow null$ ▷ Initialise cache of recent instances
3: **while** $hasMoreInstances(S)$ **do**
4:      $S_t \leftarrow nextChunk(S)$
5:      **if** $Cache == null$ **then**
6:          $C \leftarrow preClustering(S_t)$ ▷ C: Pool of clusters with centroids $\bar{c}$
7:          $Cache \leftarrow S_t$
8:      **else**   ▷ Cluster pool and cache maintainance
9:          **for** $x_i \in S_t$ **do**
10:             $c^* \leftarrow \arg\min_{\bar{c} \in C} |x_i - \bar{c}|_2$ ▷ l^2-Norm(instance x_i,centroid $\bar{c}$ of C)
11:             addInstance(c^*, x_i)   ▷ Add x_i to cluster c^*
12:          **end for**
13:          $Cache.append(S_t)$ ▷ Add new instances to cache
14:          **while** $Cache.size() > w$ **do**
15:             $x \leftarrow Cache.removeOldest()$ ▷ Remove oldest instance from cache
16:             removeInstance(C, x)   ▷ Remove oldest instance from clustering
17:          **end while**
18:      **end if**
19:      **for** $k \in \{1, 2, \cdots, b\}$ **do**
20:          $c^* \leftarrow selectCluster(C, b + 1 - k, \tau)$ ▷ Marco step, Alg. 1
21:          $x_i \leftarrow selectInstance(c^*)$ ▷ Micro step, Eq. 4
22:          $y_i \leftarrow askLabel(x_i)$
23:          $trainClassifier(cl, x_i, y_i)$ ▷ Classifier update
24:          $updateCluster(c^*, x_i, y_i)$ ▷ Cluster update
25:      **end for**
26: **end while**

Algorithm 3. Amnesic Clustering Variant COPAL-A

Require: S: Stream of instances
Require: b: Budget (per chunk)
Require: τ: false positive misclassification cost
1: $cl \leftarrow initClassifier$
2: **while** $hasMoreInstances(S)$ **do**
3:      $S_t \leftarrow nextChunk(S)$
4:      $C \leftarrow preClustering(S_t)$ ▷ C: Pool of clusters with centroids $\bar{c}$
5:      **for** $k \in \{1, 2, \cdots, b\}$ **do**
6:          $c^* \leftarrow selectCluster(C, b + 1 - k, \tau)$ ▷ Marco step, Alg. 1
7:          $x_i \leftarrow selectInstance(c^*)$ ▷ Micro step, Eq. 4
8:          $y_i \leftarrow askLabel(x_i)$
9:          $trainClassifier(cl, x_i, y_i)$ ▷ Classifier update
10:         $updateCluster(c^*, x_i, y_i)$ ▷ Cluster update
11:      **end for**
12: **end while**

---

4 Experimental Evaluation

In the following Subsect. 4.1, we describe the setting for our experimental evaluation, including the datasets and the compared active learning approaches. This is followed by a presentation and discussion of the results in Subsect. 4.2.

4.1 Experimental Setup

The objective in active learning is the selection of the most beneficial labels for the training of the classifier, such that for a given budget the classification performance is maximised. How well a strategy handles this trade-off between classification performance and consumed budget is usually evaluated in learning curves, which plot the performance in dependence of the budget. For stream-based scenarios, this requires to aggregate the performance over time, as done for example in [10,11]. However, for evolving datastreams the variance in the performance over time is also an important aspect, as it indicates whether and how quickly an algorithm adapts to drift. For passive stream classifiers, the standard approach is prequential evaluation [5], which uses newly arrived instances first for testing the current classifier, before using them for updating the classifier.

We consider both aspects in our experimental evaluation: following the prequential evaluation paradigm, we evaluate the classifier first on newly arriving instances, before we consider them as candidates for the active learning and classifier updating step. For studying the active learning strategies' effect on the adaptation of the classifier to drift, we provide curves that show the accuracy of the algorithms over time. For evaluating how well the strategies perform in the trade-off between accuracy and budget size, we provide learning curves that plot the aggregated accuracy over time for different budget shares. This is the most informative common evaluation method, as there is no consensus on an approach for statistically testing such active learning results in evolving datastreams yet.

Using this setup, we compare the incremental variant *COPAL-I* and the amnesic variant *COPAL-A* of our approach against several other active learning strategies: first, we use complete labelling (denoted as *Complete*), which requests all labels and serves as a proxy for the upper bound of the achievable performance, thereby indicating the complexity of the datastream. Second, we use random selection (denoted as *Random*) as a baseline, where instances are chosen randomly with equal selection probabilities. Third, we compare our approach to *ACLStream*, the most recently proposed [10] *clustering-based* active learning strategy for evolving datastreams. Finally, we compare against *DBAL-Stream*, to our knowledge the *most recently proposed* active learning strategy for *evolving* datastreams. This strategy was reported in [11] to outperform several other active learning strategies for evolving datastreams, including the ones proposed in [26]. Other active learning approaches for evolving datastreams discussed in Sect. 2 integrate a specific classifier into their algorithm. Since this conflicts with a differentiated evaluation between the impact of the AL-component alone and the used classifier technology, they were not included into the evaluation. Furthermore, to ensure a fair evaluation, all algorithms are run within the MOA framework in Java, using the original implementations and recommended parameter settings of their authors. For the non-deterministic strategies ACLStream and Random, we average the performance over 5 runs. For better comparison, we use for COPAL the same k-means pre-clustering technique with $k = 5$ as in ACLStream, and the same type of classifier (adaptive Naive Bayes with drift detection, see [4]) that was proposed for DBALStream in the evaluation of all approaches. The chunk- and sliding window size is set to 100 instances for all approaches. We measure accuracy gain in COPAL by setting $\tau = 0.5$.

The experimental evaluation is done on six datastreams, including four real-world ones. The first synthetic datastream is based on the *Moving Hyperplane* generator proposed in [9]. The concept therein is based on a hyperplane, which rotates over time to generate drift. The implementation of the HyperplaneGenerator class in MOA was used with default settings to generate the data. The second synthetic datastream, random radial basis function (*Random RBF*), is based on the randomRBFGeneratorDrift class in MOA [3]. It uses a mixture of Gaussians with a fixed number of components, such that each component generates instances from a single class. Drift is induced by moving the centroids of the components in the featurespace. Except for the number of components, which was set to 20, the default parameter settings were used. The first real-world datastream is the *Airline* dataset by the US Bureau of Transportation Statistics, Research and Innovative Technology Administration (RITA), with the task being to predict whether a flight will be delayed based on the information of its scheduled departure. The second one is the *Bank Marketing* dataset by [18], where the task is to predict whether the client will subscribe to a term deposit subsequently to a direct marketing campaign. The third one is the *Electricity* dataset by [8], with the task to predict an increase or decline of the electricity prices in New South Wales (Australia). The fourth datastream is the *EEG Eye State* dataset from [2], with the task to repeatedly predict over the experiment's duration of 117 s whether a proband's eyes are opened or closed.

4.2   Results and Discussion

We first discuss the results of the evaluation of the active learning strategies' performances under different budgets. These are shown in the learning curves in Fig. 1, which plot the accuracy (aggregated over time) for different budget shares. Overall, *COPAL-I* performs best for the most datastreams and budget sizes. It is always better than *ACLStream*, better than *Random* except for a budget of 0.05 % on Bank Marketing, and better than *DBALStream* except for a single budget share on the Airline, the Bank Marketing, the EEG Eye State and the Electricity datastreams. Compared to its amnesic counterpart *COPAL-A*, it performs better on Airline and Moving Hyperplane, and comparably on the remaining datastreams, except for its worse performance on the EEG Eye State. Compared to their competitors, *COPAL-A* performs also well, being better than *Random* on all datastream-budget combinations except for one particular budget share on Electricity and Bank Marketing, and performing always better than *ACLStream* except for the budget share of 0.05 on the

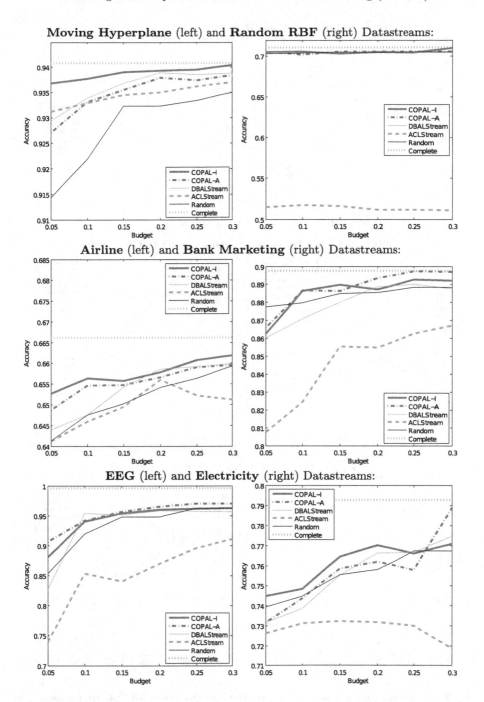

Fig. 1. Learning curves in budget share against accuracy for different datastreams. Complete (black dotted line) corresponds to an upper bound of the performance with all instances being labelled. Early convergence to high values is favourable.

Moving Hyperplane (left) and **Random RBF** (right) Datastreams:

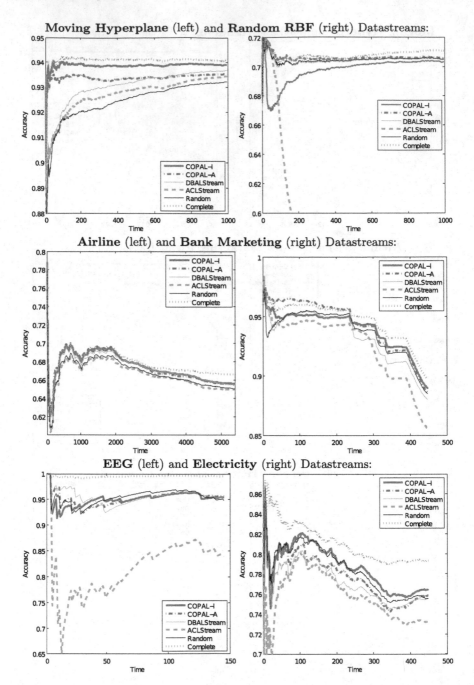

Airline (left) and **Bank Marketing** (right) Datastreams:

EEG (left) and **Electricity** (right) Datastreams:

Fig. 2. Performance (in accuracy) over time (in steps of 100 instances). On all datastreams prequential evaluation and a budget share of 15 % were used. Complete (black dotted line) corresponds to an upper bound of the performance with all instances being labelled. Higher values are favourable.

Moving Hyperplane datastream. However, on the latter datastream, *COPAL-A* is worse than *DBALStream*, while being on the other datastreams still better in the majority of tested budget shares. Concerning the superiority of *DBALStream* over *ACLStream*, which was indicated in [11], our results confirm that overall the former is the better strategy of the two. Due to the label sets becoming more and more similar with increasing budgets, one would expect the differences between the strategies to diminish with increasing budget shares. This is indeed the case in most of our results, except for *ACLStream* on Electricity.

The performance and adaptivity over time are reported in Fig. 2, which shows the active classifier's accuracy for the budget share of 0.15. The trend of the black-doted curve of the *Complete*-baseline indicates changes in the classification task's complexity over time. Except for initially low performance (compared to *Complete*) on the Random RBF and Electricity datastreams, the curves of *COPAL-I* and *COPAL-A* follow this trend, indicating a quick adaptation. However, on the Random RBF datastream all approaches initially perform poorly, and the *ACLStream* approach completely fails to improve over time (for better visibility of the other strategies' performance, its curve was cut below an accuracy of 0.6, but its downward trend continued). Thus, except for *ACLStream* all active learning approaches were able to recover from drift.

In summary, our experimental evaluation indicates a mostly superior performance of the incremental variant *COPAL-I* compared to all other tested approaches including its amnesic counterpart *COPAL-A*, while the latter shows comparable performance to the most recently proposed *DBALStream* approach. In our experiments, the clustering-based active learning strategy *ACLStream* proposed in [10] performed in most test-cases not better than random sampling.

5   Conclusion

In this paper, we have proposed a clustering-based optimised probabilistic active learning approach (COPAL) for selective sampling in evolving datastreams. Inspired from semi-supervised learning, it combines established clustering techniques, which it uses to capture the structure within the unlabelled data, with the recently proposed probabilistic active learning [14] approach, which serves for selecting the best among the clusters. Our approach is designed for selecting labels actively in nonstationary environments, and is usable to actively train any adaptive stream classifier. We studied two variants of this approach: *COPAL-I* uses incremental clustering and windowing to maintain and adapt a single clustering model over time. *COPAL-A* is an amnesic clustering variant that iteratively learns a new clustering model on each chunk and discards it after classifier training. The experimental evaluation against competitors that include two recently proposed approaches for evolving datastreams shows an overall superior performance of the proposed COPAL approach. The incremental variant performs overall the best, while the amnesic variant performs at least on par with competitors and in three out of six datasets best for large budget sizes. While for better comparison with competitors the same combination of clustering and classifier technique was used in this paper, the performance with other combinations is subject of ongoing research.

Furthermore, COPAL uses the obtained clustering model solely in the active sampling process. Thus, the information from the structure of the unlabelled data is not considered explicitly during classifier training. Future work will focus on extending COPAL by semi-supervised techniques in the classification step, for example by self-labelling of the unlabelled instances or by using the clustering directly in the classification process.

Acknowledgments. We thank our colleagues, in particular Daniel Kottke, from University of Magdeburg, Christian Beyer from IBM Germany, and Vincent Lemaire from Orange Labs France, as well as Dino Ienco, Albert Bifet and Bernhard Pfahringer and the anonymous reviewers.

References

1. Abdallah, Z., Gaber, M., Srinivasan, B., Krishnaswamy, S.: Streamar: incremental and active learning with evolving sensory data for activity recognition. In: Proceedings of the 24th IEEE International Conference on Tools with Artificial Intelligence (2012)
2. Asuncion, A., Newman, D.J.: UCI machine learning repository (2015)
3. Bifet, A., Holmes, G., Kirkby, R., Pfahringer, B.: MOA: massive online analysis. J. Mach. Learn. Res. **11**, 1601–1604 (2010)
4. Gama, J., Medas, P., Castillo, G., Rodrigues, P.: Learning with drift detection. In: Bazzan, A.L.C., Labidi, S. (eds.) SBIA 2004. LNCS (LNAI), vol. 3171, pp. 286–295. Springer, Heidelberg (2004)
5. Gama, J., Sebastião, R., Rodrigues, P.P.: On evaluating stream learning algorithms. Mach. Learn. **90**, 317–346 (2013)
6. Gantz, J., Reinsel, D.: The digital universe in 2020: Big data, bigger digital shadows, and biggest growth in the far east, December 2012
7. Gopalkrishnan, V., Steier, D., Lewis, H., Guszcza, J.: Big data, big business: Bridging the gap. In: Proceedings of the 1st International Workshop on Big Data, Streams and Heterogeneous Source Mining: Algorithms, Systems, Programming Models and Applications, BigMine 2012, pp. 7–11. ACM, New York (2012)
8. Harries, M.: Splice-2 comparative evaluation: Electricity pricing. University of New South Wales, Australia, Technical report (1999)
9. Hulten, G., Spencer, L., Domingos, P.: Mining time-changing data streams. In: KDD 2001: Proceedings of the seventh ACM SIGKDD International Conference on Knowledge discovery and data mining, pp. 97–106. ACM, New York (2001)
10. Ienco, D., Bifet, A., Žliobaitė, I., Pfahringer, B.: Clustering based active learning for evolving data streams. In: Fürnkranz, J., Hüllermeier, E., Higuchi, T. (eds.) DS 2013. LNCS (LNAI), vol. 8140, pp. 79–93. Springer, Heidelberg (2013)
11. Ienco, D., Pfahringer, B., Zliobaitė, I.: High density-focused uncertainty sampling for active learning over evolving stream data. In: Proceedings of the 3rd International Workshop on Big Data, Streams and Heterogeneous Source Mining: Algorithms, Systems, Programming Models and Applications, pp. 133–148 (2014)
12. Kottke, D., Krempl, G., Spiliopoulou, M.: Probabilistic active learning in data streams. In: De Bie, T., Fromont, E. (eds.) Advances in Intelligent Data Analysis XIV - 14th International Symposium (IDA 2015). LNCS. Springer (2015)

13. Krempl, G., Kottke, D., Lemaire, V.: Optimised probabilistic active learning (OPAL) for fast, non-myopic, cost-sensitive active classification. Mach. Learn. Spec. Issue ECML PKDD **2015**, 1–28 (2015)
14. Krempl, G., Kottke, D., Spiliopoulou, M.: Probabilistic active learning: towards combining versatility, optimality and efficiency. In: Džeroski, S., Panov, P., Kocev, D., Todorovski, L. (eds.) DS 2014. LNCS, vol. 8777, pp. 168–179. Springer, Heidelberg (2014)
15. Krempl, G., Zliobaitė, I., Brzeziński, D., Hüllermeier, E., Last, M., Lemaire, V., Noack, T., Shaker, A., Sievi, S., Spiliopoulou, M., Stefanowski, J.: Open challenges for data stream mining research. SIGKDD Explor. **16**(1), 1–10 (2014). special Issue on Big Data
16. Loy, C.C., Hospedales, T.M., Xiang, T., Gong, S.: Stream-based joint exploration-exploitation active learning. In: 2012 IEEE Conference on Computer Vision and Pattern Recognition (CVPR), pp. 1560–1567 (2012)
17. Masud, M.M., Gao, J., Khan, L., Han, J., Thuraisingham, B.: Classification and novel class detection in data streams with active mining. In: Zaki, M.J., Yu, J.X., Ravindran, B., Pudi, V. (eds.) PAKDD 2010. LNCS, vol. 6119, pp. 311–324. Springer, Heidelberg (2010)
18. Moro, S., Laureano, R., Cortez, P.: Using data mining for bank direct marketing: an application of the crisp-dm methodology. In: Novais, P. (ed.) Proceedings of the European Simulation and Modelling Conference (ESM'2011), pp. 117–121. EUROSIS, Guimarães (2011)
19. Nguyen, H.T., Smeulders, A.: Active learning using pre-clustering. In: Proceedings of the 21st International Conference on Machine Learning, ICML 2004, Banff, Alberta, Canada, pp. 79–86. ACM Press (2004)
20. Nguyen, H.-L., Ng, W.-K., Woon, Y.-K.: Concurrent semi-supervised learning with active learning of data streams. In: Hameurlain, A., Küng, J., Wagner, R., Cuzzocrea, A., Dayal, U. (eds.) TLDKS VIII. LNCS, vol. 7790, pp. 113–136. Springer, Heidelberg (2013)
21. Ryu, J.W., Kantardzic, M.M., Kim, M.-W., Ra Khil, A.: An efficient method of building an ensemble of classifiers in streaming data. In: Srinivasa, S., Bhatnagar, V. (eds.) BDA 2012. LNCS, vol. 7678, pp. 122–133. Springer, Heidelberg (2012)
22. Settles, B.: Active learning literature survey. Computer Sciences Technical Report 1648, University of Wisconsin-Madison, Madison, Wisconsin, USA (2009)
23. Settles, B.: Active Learning. Synthesis Lectures on Artificial Intelligence and Machine Learning, vol. 18. Morgan and Claypool Publishers, San Rafael (2012)
24. Zhu, X., Zhang, P., Lin, X., Shi, Y.: Active learning from data streams. In: Proceedings of the 2007 Seventh IEEE International Conference on Data Mining, ICDM 2007, pp. 757–762. IEEE Computer Society, Washington, DC (2007)
25. Zhu, X., Zhang, P., Lin, X., Shi, Y.: Active learning from stream data using optimal weight classifier ensemble. IEEE Trans. Syst. Man. Cybern. Part B Cybern. **40**(6), 1607–1621 (2010)
26. Zliobaitė, I., Bifet, A., Pfahringer, B., Holmes, G.: Active learning with drifting streaming data. IEEE Trans. Neural Netw. Learn. Syst. **25**(1), 27–39 (2013)

Predictive Analysis on Tracking Emails for Targeted Marketing

Xiao Luo[1](✉), Revanth Nadanasabapathy[1], A. Nur Zincir-Heywood[1],
Keith Gallant[2], and Janith Peduruge[2]

[1] Faculty of Computer Science, Dalhousie University, Halifax, Canada
{luo,zincir}@cs.dal.ca, rv974562@dal.ca
[2] EmailOpened, Halifax, Canada
{keith,janith}@emailopened.com

Abstract. In this work, we present our experiences using a learning
model on predicting the "opens" and "unopens" of targeted marketing
emails. The model is based on the features extracted from the emails and
email recipients profiles. To achieve this, we have employed and evaluated
two different classifiers and two different data sets using different feature
sets. Our results demonstrate that it is possible to predict the rate for a
targeted marketing email to be opened or not with approximately 78 %
F1-measure.

Keywords: Email · Targeted marketing · Feature extraction ·
Prediction

1 Introduction

Over the last two decades, online marketing has grown to a $70 billion industry
worldwide annually [1]. Online marketing aims to promote products or brands
via one or more forms of electronic media. email is one of the important electronic
marketing media. In this media, the rate of opened emails is a critical factor to
evaluate the effectiveness of targeted marketing via email. If a system can predict
the opened rate for a marketing email before it is sent out to the recipients,
that could be very beneficial to the sender to improve the effectiveness of the
marketing.

In 2002, May et al. [2] investigated the success factors for email marketing.
They recommended that features derived from the email and the recipients'
historical profile should be used for the task of prediction. However, they did
not investigate what and how the features should be extracted for predicting
which e-mail would be opened.

Additionally, the Bag of Words approach has been widely used and has shown
its effectiveness for many information management tasks [18,19], network intru-
sion detection tasks [8] and bioinformatics tasks [9]. We used this approach in
our work too. We treat the open and unopen instances of emails as documents.
For each open or unopen, a bag of features is constructed based on the emails

© Springer International Publishing Switzerland 2015
N. Japkowicz and S. Matwin (Eds.): DS 2015, LNAI 9356, pp. 116–130, 2015.
DOI: 10.1007/978-3-319-24282-8_11

and the recipients' historical profiles. On the other hand, we also realize that by including all the features from emails, such as including all the email recipients' domains, the number of features could grow dramatically with the growth of the number of email recipients. Hence, the real time prediction of the email opened rate becomes computationally costly. Thus, different feature selection methods are investigated to select the most relevant features for predicting the email opened rate.

Predicting the rate of the "opens" and "unopens" of an email is similar to classifying an email recipient's behavior into two categories: open or unopen, once the email recipient sees the email in the Inbox. Hence, in this research, we deployed two classification models for the task of prediction. Specifically, they are the Decision Tree (C4.5) [26] and the Support Vector Machines (SVMs) classifiers [21]. To fully evaluate our prediction models, we train the models on different combinations of scenarios. These include: (i) all data and all features; (ii) different organizations' data, with (as well as without) feature selection; and (iii) including (as well as not including) the recipients' email domains. The results show that the C4.5 classifier outperforms the SVMs classifier and achieves approximately 81 % F1-measure prediction rate on "opens". With feature selection, the prediction rate decreases a little. However, employing the proposed prediction system in practice (real time) becomes feasible since there are less number of features. With feature selection, the proposed prediction model can process 29206 emails within 34 seconds on a regular personal laptop. The F1-measure prediction rate on "opens" can reach up to 78 % (10-fold cross validation). The results also show that it is better to train the prediction models for each client given that the interest or characteristics of the email recipients are different.

The rest of this paper is organized as follows. Section 2 summarizes the related work and methodologies in the area of email marketing; Sect. 3 presents the features used, how they are extracted and derived from the raw log data. Section 4 shows the two classification models that are employed for the prediction purposes in this research. Experimental setup and results are provided in Sect. 5. Finally, conclusions are drawn and the future work is discussed in Sect. 6.

2   Related Work

The application of email as a marketing media, as well as how powerful this media can help building customer relationship has been explored in Jim Sterne's book in 2000 [3]. Sternes also demonstrated the strategies and the techniques to help improve email response rates and forge lasting customer relationships. However, his work is purely from the marketing point of view, there is no discussion on information technology related issues such as data collection or automatic prediction of the response rate given a targeted marketing email. In 2004, Phelps et al. studied the different aspects that increase the probability of emails to be forwarded [17]. In 2006, Bonfrer et al. [16] developed a methodology that could be used to predict the performance of email marketing campaigns in real time.

In that research, only features derived from the clicks as well as the timestamp of the email sent and the timestamp of the email most likely would be opened were used to build the prediction model. Pavlov et al. [15] analyzed how to build an infrastructure towards a sustainable email marketing. Some researchers discussed that the subject line of an email has a strong influence on the consumers decision to open or not to open the email [4]. Unlike our research, their work employs features only from the email itself without considering the email recipients' profile, which we believe plays a very important role in the email "opens".

On the other hand, employing user profiles for prediction has been researched in other areas. For example, Pazzani et al. [6] studied the identification of interesting websites based on learning and revising user profiles. Espinosa et al. employed user profiles to predict students' GPAs [5]. User profiles have also been used to identify users' intentions for many different activities such as online shopping etc. on the Internet [10]. Moreover, they have been widely employed in personalized information retrieval, personalized web ranking and so on [7]. With the increase of the popularity of social media, the prediction of information diffusion in social networks has also employed user profiles [11]. However, to the best of our knowledge, no material has been published on the usage of recipient profiles in the area of email marketing.

3   Feature Extraction and Data Representation

In this research, we have employed real targeted marketing data in order to study the usage of emails for this purpose. Our data sets include the emails that were sent to the recipients and the logs that recorded the recipients' actions such as opening an email or clicking on an email and so on. Based on the log data, we extract features from emails and build the profiles of the recipients. Figure 1 shows the process of feature extraction.

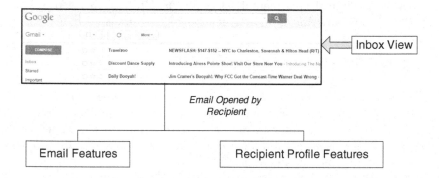

Fig. 1. Feature extraction and email recipient profile construction

The email features employed in this work include the subject of an email as well as the day and time when an email is sent. Additionally, the recipient profile features employed in this work include the location of the recipient, the computing environment of the recipient, the response time and the email domain of the recipient. However, it should be noted here that some of the recipient profile features can only be extracted when there is an action on the email.

3.1 Extracting the Email Features

In this work, we extracted three features from each email. These are: the words of the subject line of the email, the day the email is sent and the time the email is sent. The time and the day are two features that are very straight forward. We use the time stamp of the sent email, and then extracted these features. For the day feature, we use the day of the week which runs from Sunday to Saturday. As for the time feature, we divided the time of a day into four groups: Morning (06:00 am–12:00 pm), Night (00:00 am–06:00 am), Evening (18:00 pm–24:00 am), Afternoon (12:00 pm–18:00 pm). Hence, given a time stamp as: 2014-07-08 16:39:16, the sent day feature is: Tuesday, and the sent time feature is: Afternoon.

Prediction based on the individual words within the subject line of an email could be very challenging, because the number of words in a subject line is not that many to start with. In general, we can say that prediction by using a bag of words approach relies on how many times a word has been used in the email. If a word has been repeatedly used in the subject, it is easier to predict than the case, where a word has never been used before in the subject line. In the latter case, it is difficult to associate that word with the open rate of an email. Other than these two extreme cases, the words, which are used rarely in the past, and the ones similar to the previously used, have varying degrees of difficulty in using them for the prediction. Hence, based on the generic knowledge on the popular words that are used for marketing purposes, we categorize the words in the subject lines into the following six categories. In this project, we classified all the words in the subject lines into these categories (Table 1).

Table 1. Categories of the words used in the subject line of an email

Category	Description	Examples
Client	Related to the business name of the sender	McDonalds
Business	Related to the business product of the sender	Hamburg, Fries, Pop, ...
Time	A name of a time or day	Holidays, July, Weekend, ...
Location	A name of a location	New York, Amherst, ...
Highlight	An adjective to highlight information	Happy, Brilliant, ...
CAP	Word in the upper case form	COUPON, THURSDAY, ...

3.2 Extracting the Recipient Profile Features

Other than the characteristics of an email itself, the properties of email recipients are also very important in determining if the email will be opened by the recipient or not. Researches show that the geographical location is important for many Internet applications, such as online advertisement [20]. It helps to understand the customer distribution and enables location-based advertising services. Thus, in this work, we explored whether the same will apply in the email marketing field too. Many researches [12,13] have been done on identifying the geolocation from IP addresses. We employed the popular GeoIP open source [14] tool to identify the country, the state and the city of the recipient based on his/her IP address, and include this as one of the features of the recipient.

We hypothesize that whether a recipient opens an email or not will also rely on the computing environment that the recipient has when the email hits his/her Inbox. Imagine that when an email recipient is using a mobile device during the day, and there is a maximum data limit on the mobile device, then the probability of not opening an email on the mobile device might be higher than opening it. Instead, he/she might choose to open it while he/she is at home using a PC with a higher network bandwidth. Thus, we hypothesize that the computing environment of the recipient at a certain time and location plays an important role in predicting the rate of opened emails, especially for targeted marketing purposes. On the other hand, based on an individual's preferences, some recipients might like to open the marketing email using some specific browsers. In this work, we use the user-agent string header embedded in the HTTP to extract the recipient's OS type, browser type and device type as part of the profile features. Table 2 presents the features used for recipient profiling as well as some of their sample values.

Table 2. Features for email recipient profiling

Type	Data source	Examples
Country	IP address	Canada, US, ...
State	IP address	Ontario, California, ...
OS type	User-Agent string	iOS, Android, ...
Device type	User-Agent string	iPad, PC, ...
Browser type	User-Agent string	IE, Chrome, ...
Domain	The domain part of recipient's email address	hotmail.com, rogers.com, ...

3.3 Data Representation

The Bag-of-Word model or the vector space model approach has been widely used in the information retrieval, data mining and even bioinformatics fields [8,9,18,19]. It provides the foundation for the data representation in these fields. In this work, we use this approach to represent the opened and the unopened emails. Vectors are built for all the instances of "opens" and "unopens". The length of the vector is the total number of features seen in all the instances. It is easy to get all the features of each instance of opened emails. However, for each instance of the unopened emails, only the set of email features are present. We can not capture the actual geolocation and computing environment information since there is no action from the recipient. In this case, we use the most highly used values of each of the features from the recipient's profile to populate unopen instances. For example, if a recipient has an unopened instance, we will employ mostly used location from the "opens" in his/her past history for the location of this unopen instance. We are aware that populating the missing values using this approach may introduce some errors. However, this is one of the simplest ways to overcome the missing values problem. In the future, we will also investigate other approaches to study this problem. An example of such a data representation is shown in Table 3.

Table 3. Example of data representation

EmailSentEvening	CAPTitle	CIESIN.COLUMBIA.EDU	Canada	. . . . .
0.14	0.45	0	2.67	. . . . .

Instead of using 1 and 0 to present the occurrence of the feature, we used the traditional $tfidf$ function to provide a weight for each feature. The $tfidf$ weight calculation is given in Eq. 1:

$$W_{ij} = tf_{ij}idf_i = tf_{ij}log_2(\frac{N}{idf_j}).$$ (1)

tf_{ij}: Feature tf_{ij} frequency in open or unopen instance j
idf_i: Instance frequency of feature i.

4   Prediction

In the early 1990s, classification techniques such as Decision Tree learners, Naive Bayes classifiers and Neural Networks were applied to many data mining tasks. Since late 1990s, new machine learning methods, the so called large margin classifiers such as Support Vector Machines (SVMs) have been proposed. One of the differences between decision trees and other classification methods is that the trained model of the decision tree can be interpreted as a readable form, while others are more like black boxes, the trained model can not be easily interpreted. In this paper, we employ both the Decision Tree and the SVMs classifiers for our task.

4.1   Decision Tree

ID3 and C4.5 are the most popular ones among the decision tree inductive learning algorithms. C4.5 is a representative and a software extension of ID3 algorithm [26]. Unlike ID3, C4.5 handles both continuous as well as discrete attribute values. It also handles missing attributes and does pruning after the tree has been created. Due to pruning in C4.5, as the branches that do not contribute much on the classification are removed, tree size is considerably reduced and memory usage is improved in comparison to ID3. We use C4.5 decision tree learning algorithm to classify the features based on their numeric $tfidf$ scores [24]. In our experiments, the C4.5 is applied to the list of features described in Sect. 3. The C4.5 algorithm can be summarised as the following:

1. Take all features in the training set. Calculate its entropy information.
2. Choose maximum entropy feature as tree node.
3. Split the tree node into two branches for examples with $tfidf$ scores that are $= 0$ and > 0.
4. For each branch, find the child node from the sublists of features.
5. Repeat the steps until the examples of the node are empty and directly predicts single class.

4.2 Support Vector Machines

Another classification algorithm we use is the Support Vector Machines (SVMs) [21]. The SVMs classifier aims to separate the input data linearly using hyperplanes. For a good generalization for the classification and a less complex hyperplane function, the maximum margin between the hyperplane and the support vectors is required.

When the samples are not linearly separable, support vector machines is used to non-linearly transform the training features from a two dimensional space 'x' to a higher dimensional feature space 'φ (x)' using a factor $\phi : x \rightarrow \varphi$ (x) and a function called 'Kernel', an inner product of two examples in the feature space. The kernel function used in our experiments is given in Eq. 2:

$$K(x_i, x_j) = (1 + x_i^T x_j)^p. \tag{2}$$

p : Degree of polynomial function K.

This ability to learn from large feature spaces and the dimensionality independence make the support vector machines a universal learner for text classification. Another characteristic of SVMs is the use of soft margins to protect from overfitting caused by the misclassification of noisy data in such large feature spaces [22].

5 Experimental Setup and Results

The data sets employed in this work are the log files received from our industrial partner - EmailOpened. These log files include data from two different organizations that are from two different business sectors. One is from the restaurant business, we call it as Restaurant-A in the rest of the paper. The other one is from the publishing business, we call it as Journal-B in the rest of the paper. The data sets include 34 emails sent by Restaurant-A to 2221 recipients, and 20 emails sent by Journal-B to 4703 recipients. For all these emails sent, other than the bounced back emails, Restaurant-A has 10845 opened instances ("opens") and 12636 unopened instances ("unopens") whereas Journal-B has 18987 "opens" and 10219 "unopens". After the data cleaning and the feature extraction, there are totally 891 features over all data sets. Figure 2 shows the device distribution of the "opens" within all the data. Approximately 65 % of "opens" happen on PCs compared to the other device types. The "opens" on the iPhone devices seem to be 19 % more than the rest of the mobile phones based on the data used for the experiments. One reason could be that more email recipients in these email campaigns were iPhone users.

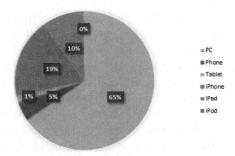

Fig. 2. Percentage of open instances on different device types

Figures 3 and 4 present the location distributions of the "opens" for Restaurant-A and Journal-B. The figures show that the data distributions for different business clients are different. For Journal-B, 94 % of the "opens" happen in United States, while 2 % of the "opens" happen in the Canada. However, for Restaurant-A, 85 % of the "opens" are from Canada and 13 % are from the United States. The different distributions represent the difference of the businesses and the targeted audiences, i.e. email recipients.

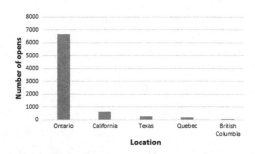

Fig. 3. Top location distributions of the "opens" : Restaurant-A

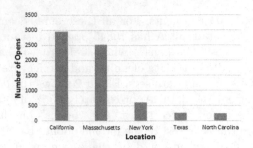

Fig. 4. Top location distributions of the "opens" : Journal-B

In this work, Waikato Environment for Knowledge Analysis (WEKA) [25] system, which is an open source software that consists of a collection of machine learning algorithms for data mining tasks, is used for training and testing our classifiers for the purpose of prediction. In WEKA, the J48 classifier, which is the C4.5 Decision Tree classifier, and the SMO type under functions category, which is the SVMs classifier, are used. Specifically, for C4.5, we used three fold pruned C4.5 classifier with confidence factor of 0.25. As for SVMs, we used polynomial kernel function with degree 3, exponent and random seed set to 1.0 on normalized training data. In order to fully analyze the prediction performance on the data sets without any biases, ten-fold cross-validation is used to evaluate the prediction models. After the feature extraction and data representation steps, we train the two prediction models on all the data of Restaurant-A and Journal-B, respectively. The decision tree trained from the data set is fairly large, because it makes use of a high number of features from the training data. Hence, it is not included in the paper. The results based upon the three traditional evaluation measurements, namely: precision, recall and F1-Measure are shown in Tables 4 and 5.

Table 4. Prediction results of the decision tree

	All Data			Restaurant-A			Journal-B		
	Recall	Precision	F1	Recall	Precision	F1	Recall	Precision	F1
Opens	0.797	0.801	0.799	0.773	0.681	0.724	0.806	0.877	0.840
UnOpens	0.739	0.734	0.737	0.752	0.828	0.788	0.727	0.609	0.663
Weighted average	0.772	0.772	0.772	0.762	0.760	0.759	0.779	0.783	0.778
Number of features	891			354			654		

The results show that with all the data, the proposed prediction analysis models achieved a prediction F1-measure rate of approximately 80 % on the "opens" when the decision tree classifier is employed and a prediction F1-measure rate of approximately 74 % on the "opens" when the SVMs classifier is employed.

Table 5. Prediction results of the support vector machines

	All Data			Restaurant-A			Journal-B		
	Recall	Precision	F1	Recall	Precision	F1	Recall	Precision	F1
Opens	0.717	0.760	0.738	0.782	0.538	0.637	0.703	0.914	0.795
UnOpens	0.660	0.608	0.633	0.687	0.871	0.768	0.639	0.283	0.393
Weighted Average	0.692	0.694	0.692	0.731	0.717	0.708	0.681	0.693	0.654
Number of features	891			354			654		

The results show that the Decision Tree performs better on all data cases. One of the reasons for the lower performance of the SVMs classifier could be, we deployed a polynomial kernel for the SVMs. However, the instances of "unopens" and "opens" might not be separable by deploying the polynomial kernel of degree 3. On the other end, both prediction analysis models performs slightly better on the "opens" for Journal-B data. The highest prediction F1-measure for the "opens" on the Journal-B data is 84 %.

Tables 4 and 5 demonstrate that Restaurant-A and Journal-B have different number of features. We investigated all the features, and identified that a high percentage of the features are domains and locations derived from the email recipients. We then take a further step to investigate how much the email domains of the recipients contribute to the prediction performance. To this end, we filtered out all the domain features and repeated experiments on the rest of the data from Restaurant-A and Journal-B data sets, respectively. Tables 6 and 7 show the results without including domains as features for the prediction.

Table 6. Prediction results of the decision tree classifier without using the domains

	All Data			Restaurant-A			Journal-B		
	Recall	Precision	F1	Recall	Precision	F1	Recall	Precision	F1
Opens	0.757	0.773	0.765	0.735	0.578	0.647	0.768	0.893	0.826
UnOpens	0.696	0.676	0.694	0.821	0.752	0.765	0.715	0.500	0.589
Weighted Average	0.730	0.731	0.731	0.713	0.709	0.704	0.750	0.755	0.743
Number of features	267			136			234		

Based on the weighted average F1-measure, without including the domains as features, the performance of the classifiers on Restaurant-A and Journal-B drops 2 % to 4 %. This implies that the domains features do contribute to the prediction performance at some level.

Table 7. Prediction results of the SVMs classifier without using the domains

	All Data			Restaurant-A			Journal-B		
	Recall	Precision	F1	Recall	Precision	F1	Recall	Precision	F1
Opens	0.677	0.766	0.719	0.741	0.486	0.587	0.681	0.973	0.801
UnOpens	0.631	0.523	0.572	0.660	0.854	0.744	0.755	0.152	0.253
Weighted average	0.657	0.661	0.655	0.697	0.684	0.672	0.707	0.686	0.609
Number of features	267			136			234		

5.1   Online Prediction and Feature Selection

In the email marketing area, it is very important to effectively send the marketing emails and gain a high opening rate. This will help building healthy (without invading the privacy) relationships between the marketing businesses and the email recipients. So, if the prediction can happen in real time, after the sender constructs the email but before the email is sent out, it will be very useful for marketers to forecast the effectiveness of the email. This can also help the email sender to revise the email before it is sent. On the other hand, for the email recipients, they would receive less unwanted marketing emails. Figure 5 shows the proposed online prediction process.

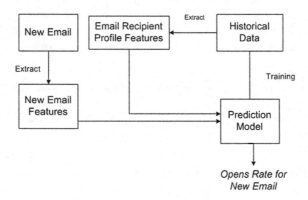

Fig. 5. Proposed online email opens prediction process

The prediction model is trained based on the historical data, emails sent before. The features are extracted from the "opens" and "unopens" instances of the emails that were sent previously. Email recipient features for the "unopens" are constructed based on the "opens". After the email sender finishes writing the subject line, sets up the planned sending time, and selects the group of email recipients, the online prediction process can start to extract the features based upon these information and predict the email opens rate for the email to be sent. If the predicted opens rate of the email to be sent is low, then the email sender can modify the email subject line or the planned sending time to start the prediction process again till the email sender is satisfied with the predicted opens rate. Figure 6 shows the user interface of the tool developed by our industrial partner for their business clients to form their marketing emails.

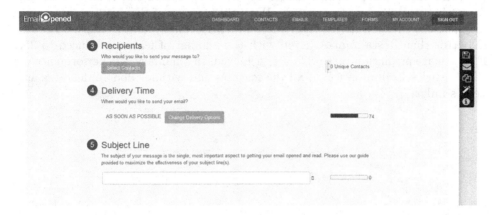

Fig. 6. User interface of the marketing emails sending tool of EmailOpened

From the previous results, we concluded that the profile features of the email recipients of different businesses, e.g. Restaurant-A or Journal-B, might be totally different. Hence, training the prediction model for each business client seems to be more effective. In Fig. 5, the prediction model is trained by using the historical data of a business client.

We realize that the training process can be done offline periodically. However, the prediction process, which is the testing process, needs to be efficient. In this work, we explored the efficiency and the prediction performance based on different numbers of features input to the prediction models. To achieve this, Chi Square feature selection method is deployed to remove redundant and irrelevant features. The Chi Squared attribute evaluation is used in statistics to test the independence of two events [23]. In this case, it is used to test whether the occurrence of a specific feature and the occurrence of a specific class are independent. Thus, Eq. 3 is used to calculate the chi-square value ($\chi^2(f, C_j)$) for each feature (f). High scores indicate that the null hypothesis of independence should be rejected and thus that the occurrence of the feature and the class are dependent. Hence, we select the features with higher scores, and use those selected features for class prediction.

$$\chi^2(f, C_j) = \frac{(P(f, C_j)P(\overline{f}, \overline{C_j}) - P(\overline{f}, C_j)P(f, \overline{C_j}))^2}{P(f)P(\overline{f})P(C_j)P(\overline{C_j})}. \tag{3}$$

f: a feature
C_j: the category j in the data set.

We ranked the features according to the score, and then selected eight different feature sets with an increasing number of features in each set for Restaurant-
A and Journal-B, respectively. In the experiments conducted, we identified that
it took less time to train the Decision Tree classifier and also to test the instances
with the trained Decision Tree model. Figure 7 shows the F1-measure rate on the
"opens" of the different feature sets. With the increase of the number of features
in a set, the F1-measure rate increases. The performance on the Journal-B data
is better than Restaurant-A. Even with the number of features reduced to 25,
F1-measure prediction rate of 78 % is achieved. In summary, the performance of
the feature selection on top of all the features and without domain features are
very similar.

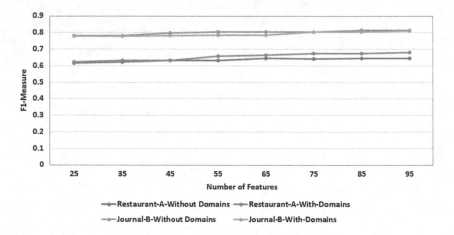

Fig. 7. Prediction results of decision tree with feature selection

Experimental results also show that as the number of features increase, the
computational cost for testing increases too. With 23481 instance, the testing
time for 25 features by using C4.5 classifier is about 25 s. While the training
time for 95 features is about 160 s. Hence, when building the online prediction
model, the number of features selected needs to be balanced with the prediction
performance.

6   Conclusions and the Future Work

Through this work, we employed a learning model for predicting the "opens"
and "unopens" of targeted marketing emails. The model is based on the features extracted from the emails and email recipients profiles. Two classification
methods are compared for predicting whether an email written will be opened

by a potential recipient or not. They are the C4.5 Decision Tree classifier and the SVMs classifier. The results show that the Decision Tree classifier performs better in all the scenarios. However, the parameter sensitivity of these two algorithms for this task is left for future work.

We have also investigated the feasibility of online prediction by using this model. Chi square feature selection technique has been used to select the most relevant features and to improve the efficiency of the prediction process. The results show that with the feature selection, the prediction performance drops a little. However, the proposed model can be deployed for online prediction in a timely fashion. The number of selected features can be adjusted based upon the prediction accuracy requirements and the computational power of the system.

Based on the literature review, this is the first research work that builds an email recipient profile. In this work, the email recipient profile contains only the location and the computational environment features. In the future, more sophisticated profiles can be build to include email features when the recipient have "opens" and "clicks" on the email. Similar to the user's profile for personalized search, the profile can be ever adaptive and changing to reflect the email recipient preferences. On the other hand, more email features can be extracted to include the message objective defined by the sender. We also believe that similar models can be explored to predict the "clicks" and "unclicks" on the URLs after an email is opened.

Acknowledgments. This research is supported by EmailOpened and the Natural Science and Engineering Research Council of Canada Engage Grant, and is conducted as part of the Dalhousie NIMS Lab at: https://projects.cs.dal.ca/projectx/.

References

1. Farahat, A., Shanahan, J.: Econometric analysis and digital marketing: how to measure the effectiveness of an ad. In: ACM WSDM (2013)
2. May, P., Ehrlich, H.C., Steinke, T.: Email marketing: success factors (2002). http://eprints.kingston.ac.uk/2108/1/paper.html
3. Sterne, J., Priore, A.: Email marketing: using email to reach your target audience and build customer relationships (2000). ISBN-10: 0471383090
4. Balakrishnan, R., Parekh, R.: Learning to predict subject-line opens for large-scale email marketing. In: IEEE International Conference on Big Data (2014)
5. Espinosa, A.F., Regts, M., Tashiro, J., Martin, M.V.: Prediction model based on user profile and partial course progress for a digital media learning environment. In: The Fourth International Conference on Advances in Databases, Knowledge, and Data Applications, pp. 120–123 (2012). ISBN: 978-1-61208-185-4
6. Pazzani, M., Billsus, D.: Learning and revising user profiles: the identification of interesting web sites. Mach. Learn. **27**, 313–331 (1997)
7. O'Brien, P., Luo, X., Abou-Assaleh, T., Li, S. : Personalization of content ranking in the context of local search. In: Proceedings of 2009 IEEE/WIC/ACM International Conferences on Web Intelligence and Intelligent Agent Technology, pp. 532–539 (2009)

8. Kang, D-K., Fuller, D., Honavar, V. : Learning classifiers for misuse and anomaly detection using a bag of system calls representation. In: Proceedings of the 2005 IEEE Workshop on Information Assurance and Security United States Military Academy, pp. 118–125. IEEE Press (2005)
9. Perina, A., Lovato, P., Jojic, N.: Bags of words models of epitope sets: HIV viral load regression with counting grids. In: Proceedings of International Pacific Symposium on Biocomputing, pp. 288–299 (2014)
10. Amazon.com, Inc. http://www.amazon.com
11. Lagnier, C., Denoyer, L., Gaussier, E., Gallinari, P.: Predicting information diffusion in social networks using content and user's profiles. In: Serdyukov, P., Braslavski, P., Kuznetsov, S.O., Kamps, J., Rüger, S., Agichtein, E., Segalovich, I., Yilmaz, E. (eds.) ECIR 2013. LNCS, vol. 7814, pp. 74–85. Springer, Heidelberg (2013)
12. Wang, Y., Burgener, D., Flores, M. : Towards street-level client-independent IP geolocation. In: Proceedings of the 8th USENIX conference on Networked Systems Design and Implementation, pp. 365–379 (2011)
13. Eriksson, B., Barford, P., Maggs, B., Nowak, R. : Posit: a lightweight approach for IP geolocation. In: Newsletter ACM SIGMETRICS Performance Evaluation Review Archive, vol. 40, pp. 2–11 (2012)
14. MaxMind GeoIP2 Java API. https://github.com/maxmind/geoip-api-java
15. Pavlov, O.V., Melville, N., Plice, R.K.: Toward a sustainable email marketing infrastructure. J. Bus. Res. **61**, 1191–1199 (2008)
16. Bonfrer, A., Drze, X.: Real-time evaluation of email campaign performance. J. Mark. Sci. **28**(2), 251–263 (2008)
17. Phelps, J.E., Lewis, R., Mobilos, L., Perry, D., Raman, N.: Viral marketing or electronic word-of-mouth advertising: examining consumer responses and motivations to pass along email. J. Advertising Res. **44**(4), 333–348 (2004)
18. Yang, Y.: An evaluation of statistical approaches to text categorization. Inf. Retrieval **1**, 69–90 (1999)
19. Lewis, D.D., Ringuette, M.: A comparison of two learning algorithms for text categorization. In: Proceedings of the 3rd Annual Symposium on Document Analysis and Information Retrieval, pp. 81–93 (1994)
20. Kaddeche, K., Wang, J. : Method and system for targeting internet advertisements and messages by geographic location. US Patent, US 20030036949 A1 (2013)
21. Vapnik, V.N.: The Nature of Statistical Learning Theory. Springer, NewYork (1995)
22. Joachims, T.: Text categorization with support vector machines: learning with many relevant features. In: Nédellec, C., Rouveirol, C. (eds.) ECML 1998. LNCS, vol. 1398, pp. 137–142. Springer, Heidelberg (1998)
23. Yang, Y., Pedersen, J.O.A. : Comparative study on feature selection in text categorization. In: Proceedings of the Fourtheenth International Conference on Machine Learning, pp. 412–420 (1997)
24. Duan, F., Zhao, Z., Zeng, X. : Application of decision tree based on C4.5 in analysis of coal logistics customer. In: Third International Symposium on Intelligent Information Technology Application, pp. 380–383 (2009)
25. Dash, R.: Selection of the best classifier from different datasets using WEKA. HERT **2**(3), 1–7 (2013)
26. Quinlan, J.R.: C4.5: Programs for Machine Learning. Morgan Kaufmann Publishers, USA (1993)

Semi-supervised Learning for Stream Recommender Systems

Pawel Matuszyk$^{(\boxtimes)}$ and Myra Spiliopoulou

Otto-von-Guericke-University Magdeburg, Universitätsplatz 2,
39106 Magdeburg, Germany
{pawel.matuszyk,myra}@iti.cs.uni-magdeburg.de

Abstract. Recommender systems suffer from an extreme data sparsity that results from a large number of items and only a limited capability of users to perceive them. Only a small fraction of items can be rated by a single user. Consequently, there is plenty of unlabelled information that can be leveraged by semi-supervised methods. We propose the first semi-supervised framework for stream recommender systems that can leverage this information incrementally on a stream of ratings. We design several novel components, such as a sensitivity-based reliability measure, and extend a state-of-the-art matrix factorization algorithm by the capability to extend the dimensions of a matrix incrementally as new users and items occur in a stream. We show that our framework improves the quality of recommendations at nearly all time points in a stream.

Keywords: Recommender systems · Semi-supervised learning · Matrix factorization · Collaborative filtering · Stream mining

1 Introduction

Data sparsity is a known problem in recommenders. It is amplified by the introduction of new items and the appearance of new users, on which and whom little is known. In [11], Zhang et al. proposed to deal with this problem with semi-supervised learning. In this study, we demonstrate the potential of semi-supervised learning (SSL) as cure to data sparsity in **stream** recommenders. The streaming context poses several challenges on semi-supervised algorithms, which do not show in the static context: on which data of the stream should the learning be done, on which data should the learner be tested before being applied on the **ongoing** stream, how should an algorithms treat new users and items? To deal with these challenges, we propose a semi-supervised stream recommender that deals with data sparsity by deriving predictions from part of the stream (unlabelled information) and using them for learning. To deal with new users and items we extend a state-of-the-art matrix factorization algorithm BRISMF [10] by the ability to deal with growing dimensions of the matrix on the ongoing stream. Our framework encompasses novel reliability measures, selectors for unlabelled data and further components specified in Sect. 3. To our knowledge, this is the first such framework for **stream** recommender systems.

© Springer International Publishing Switzerland 2015
N. Japkowicz and S. Matwin (Eds.): DS 2015, LNAI 9356, pp. 131–145, 2015.
DOI: 10.1007/978-3-319-24282-8_12

Sparsity and cold start problems are often tackled by using context or external sources of information (e.g. demographics of users, characteristics of items, etc.). These approaches, however, narrow down the palette of applicable algorithms to the few ones able to use them and it excludes many practitioners, who do not have the required data. Our framework does not rely on any additional source of information, but only on the user-item-rating matrix, which makes it general and applicable to any collaborative filtering algorithm.

In an empirical study on real-world datasets we show that our SSL framework improves the quality of recommendations at nearly all time points in the stream.

Contributions. To summarize, our contributions are as follows:

- we propose the **first SSL framework for stream recommenders** including novel reliability measures, selectors for unlabelled instances, etc.
- we extend the BRISMF algorithm by the ability to deal with growing dimensions of the matrix
- we show that SSL for stream recommenders improves the quality of recommendations.

Organization. This paper is structured as follows. In Sect. 2 we discuss related work. Section 3 gives an overview over our framework and explains the interplay of its components. The following section describes an instantiation of the components of the general framework. Evaluation protocol is described in Sect. 5. Our results are explained in Sect. 6. Finally, in Sect. 7, we conclude our work and discuss open issues.

2   Related Work

Recommender systems have been researched thoroughly in the recent years. State-of-the-art in the group of collaborative filtering approaches are nowadays matrix factorization methods. Their predictive performance has been shown in several publications [5,6,10]. We focus on their incremental version, since those methods are applicable to streams of ratings. In this work we extend the BRISMF algorithm (biased regularized incremental simultaneous matrix factorization) proposed by Takács et al. [10]. BRISMF exists in two versions. One of them was developed for batch processing. Takács et al., however, also developed an incremental version of it (cf. Algorithm 2 in [10]). In this version the latent item vectors are fixated and updated as new ratings occur in the stream. Latent user vectors are updated, however, no new users are added to the matrix. In our work we lift those limitations of the BRISMF algorithm.

While semi-supervised classification has been investigated thoroughly in the field of data mining, semi-supervised regression, a discipline that matrix factorization belongs to, is a less researched problem [12]. Recommender system domain is even more specific due to its idiosyncrasies, such as dealing with large matrices that are typically up to 99 % empty, clod start problem and many more. Due to those challenges only little work was done on semi-supervised learning in recommender systems. The work by Zhang et al. [11] belongs to the few ones. They proposed a co-training method for stationary, batch-based recommender systems. Their approach, however, is not incremental and not appropriate for streams of ratings and, therefore, would require a frequent retraining of the models. The framework proposed by us lifts those limitations by incrementally incorporating new rating information into the models and adapting to changes.

One of the biggest challenges in recommender systems is an extreme sparsity of data. Many techniques have been developed in order to tackle this problem. One of the most straightforward techniques is filling of the missing values in the matrix with default values (e.g. averages). This method, however, is very time and memory consuming and it lacks personalization. Another approach involves active learning techniques, where an algorithm chooses what label (rating) to request from a user in order to maximize a predefined gain for the model [4]. Active Learning techniques base on the assumption that a user knows the requested label and is willing to share it. This is often not the case in real applications. Semi-supervised learning provides here an important advantage of not having to relay on users' input.

3   Semi-supervised Framework for Stream Recommenders

In this section we present our main contribution - a semi-supervised framework for stream recomemnders with a description of the components and their function. Our new components are marked in red in the figures below. This section gives definitions and an overview of how the components are interrelated. An instantiation and implementation of the components is provided in Sect. 4.

3.1   Incremental Recommendation Algorithm

The core of our framework is a recommendation system algorithm. Figure 1 depicts two modes of a stream-based recommendation algorithm. The entire rectangle in the figure represents a dataset consisting of ratings. The dataset is split between a batch mode (blue part) and a stream mode (yellow part). The stream mode is the main mode of an algorithm, where information about new ratings is incorporated incrementally into the model, so that it can be used immediately in the next prediction. Semi-supervised learning also takes place in this phase (green bars).

Before the stream mode can start, the algorithm performs an initial training in the batch mode. The batch mode data is, therefore, split again into training and test set. On the training dataset latent factors are initialized and trained. The corresponding prediction error is then calculated on the test dataset (second blue rectangle) and the latent factors are readjusted iteratively. Once the initial training is finished, the algorithm switches into the streaming mode, where learning and prediction take place simultaneously. Our extended version of the BRISMF algorithm, etxBRISMF, is described in Sect. 4.1.

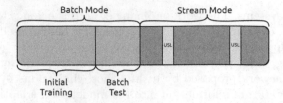

Fig. 1. Division of a dataset (entire rectangle) into batch (blue part) and stream mode (yellow part). The stream mode is the main part of an algorithm with incremental learning and predictions. Batch mode is used for initial training (Color figure online).

3.2   Stream Co-Training Approach

For semi-supervised learning we use the co-training approach. We run in parallel multiple stream-based recommendation algorithms that are specialized on different aspects of a dataset and can teach each other. Due to this specialization an ensemble in the co-training approach can outperform a single model that uses all information.

Initial Training. The specialization of the algorithms takes place already in the initial training. In Fig. 2 we present a close-up of the batch mode from Fig. 1. Here, the initial training set is divided between N co-trainers from the set $C = \{Co - Tr_1, ..., Co - Tr_N\}$, where $N \geq 2$.

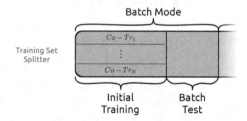

Fig. 2. Different co-trainers are trained on different parts of the initial training set. The component responsible for splitting the training set is **training set splitter** (Color figure online).

The component that decides, how the initial training set is divided between the co-trainers is called **training set splitter** (marked in red in Fig. 2; cf. Sect. 4.2 for instances of this component). Formally, a training set splitter function that relates all co-trainers to subsets of all ratings in the initial training set $R_{initialTrain}$:

$$f(C, R_{initialTrain}) : \forall n\{(Co - Tr_n \in C) \rightarrow R_{initialTrain}^{Co-Tr_n}\} \tag{1}$$

with $n = 1, ..., N$ and $R_{initialTrain}^{Co-Tr_n} \subseteq R_{initialTrain}$. This function is not a partitioning function, since overlapping between different $R_{initialTrain}^{Co-Tr_n}$ is allowed and often beneficial. Implementations of this component are provided in Sect. 4.2.

Streaming Mode - Supervised and Unsupervised Learning. After the initial training is finished all co-trainers switch into the streaming mode. In this mode a stream of ratings r_t is processed incrementally. First, a prediction is made and evaluated, then the models are updated using the new information according to the prequential evaluation (cf. Sect. 5).

Figure 3 is a close-up of the stream mode from Fig. 1. It represents a stream of ratings $r_1, r_2, ...$ The yellow part of the figure depicts the supervised learning, whereas the green part symbolizes the unsupervised learning (cf. next section). For each rating r_x in the stream all co-trainers calculate a prediction:

$$\forall n : Co - Tr_n(r_x) = \hat{r}_{xCo-Tr_n} \tag{2}$$

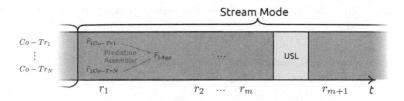

Fig. 3. Close-up of the stream mode from Fig. 1. The yellow part represents the supervised learning and the green one unsupervised learning. Predictions made by co-trainers are aggregated by a **prediction assembler** (Color figure online).

In order to aggregate all predictions made by co-trainers into one prediction $\hat{r}_{xAgg}$ we use a component called **prediction assembler**. The most simple implementation is arithmetical average (further implementations in Sect. 4.3). The function of prediction assembler is as follows:

$$predictionAssembler(r_x, C) = \hat{r}_{xAgg} \tag{3}$$

In Fig. 3 this process is visualized only for the rating r_1 due to space constraints, however in real application, it is repeated for all ratings in the stream with known ground truth (supervised learning). For instances with no ground truth the procedure is different.

Unsupervised Learning. USL takes place periodically in the stream. After every m-th rating (m can be set to 1) our framework executes the following procedure. First, a component called **unlabelled instance selector** selects z unlabelled instances (cf. Fig. 4). Unlabelled instances in recommender systems are user-item-pairs that have no ratings. We indicate those instances with r_z^u ("u" for unsupervised). The unlabelled instance selector is important, because the number of unsupervised instances is much larger then the number of supervised ones. Processing all unsupervised instances is not possible, therefore, with this component we propose several strategies of instance selection (cf. Sect. 4.4).

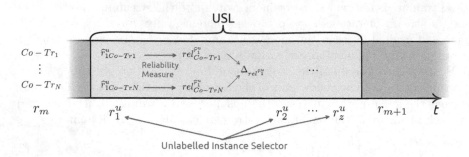

Fig. 4. The procedure of unsupervised learning. User-item-pair without ratings are selected using an **unlabelled instance selector**. Predictions and their **reliabilities** are estimated. The most reliable predictions are used as labels for the least reliable co-trainers.

Once the unlabelled instances $r_1^u, ..., r_z^u$ are selected, co-trainers are used again to make predictions:

$$\forall n, i : Co - Tr_n(r_i^u) = \widehat{r}_{iCo-Tr_n}^u \tag{4}$$

where $i = 1, ..., z$ and $n = 1, ..., N$. After this step we use a **reliability measure** to assess in an unsupervised way, how reliable is a prediction made by each co-trainer. Formally, a reliability measure is the following function:

$$reliability : (Co - Tr_n, \widehat{r}_{iCo-Tr_n}^u) \rightarrow [0; 1] \tag{5}$$

This function takes a co-trainer and its prediction as arguments and maps them into a value range between 0 and 1, where 1 means the maximal and 0 the minimal reliability. Subsequently, we calculate pairwise differences of all reliabilities of the predictions for r_i^u:

$$\Delta = |rel(\widehat{r}_{iCo-Tr_a}^u) - rel(\widehat{r}_{iCo-Tr_b}^u)| \tag{6}$$

for all $a, b = 1, ..., N$ and $a \neq b$. All values of Δ are stored temporarily in a list, which is then sorted. From this list we extract the top-q highest differences of reliability i.e. cases, where one co-trainer was very reliable and the second one very unreliable. In such cases the reliable co-trainer provides a label to r_i^u and the unreliable co-trainer learns incrementally using the provided label.

4   Instantiation of Framework Components

While, in previous section, we provided definitions of our components, here we present several **instances** of each component and their implementations.

4.1   extBRISMF - Dimensionality Extending BRISMF

The core of our framework is a matrix factorization algorithm. We extended the BRISMF algorithm by Takács et al. [10] by the ability to deal with changing dimensions of the matrix over time. We named this new variant of the algorithm extBRISMF for dimensionality **ext**ending **BRISMF**. The original BRISMF keeps the dimensions of the matrix fixed and does not update latent item factors. In our algorithm we lift those limitations. This ability is important in SSL, because the algorithms often encounter items and users not seen before.

For decomposition of the rating matrix R into two latent matrices $R \approx PQ$ we use stochastic gradient descent (SGD). P is a matrix of latent user factors with elements p_{uk}, where u is a user and k is a latent dimension. Similarly, Q is a matrix of latent item factors with elements q_{ik}, where i is an item. That results in the following update formulas for SGD [10]:

$$p_{u,k} \leftarrow p_{u,k} + \eta \cdot (predictionError \cdot q_{i,k} - \lambda \cdot p_{u,k})$$
$$q_{i,k} \leftarrow q_{i,k} + \eta \cdot (predictionError \cdot p_{u,k} - \lambda \cdot q_{i,k})$$

(7)

where η is a learn rate and λ a regularization parameter that prevents overfitting. A rating prediction can be obtained by multiplying the corresponding item and user vector from latent matrices $\widehat{r}_{ui} \approx p_u \cdot q_i$.

In Algorithm 1 we present our extBRISMF. Apart from expanding dimensions of latent matrices, we also introduced a different type of initialization for new user/item vectors. We initialize them with an average vector of the corresponding matrix plus a small random component instead of just a random vector.

4.2   Training Set Splitter

We propose three types of training set splitter (cf. Fig. 1). All of them have one parameter p that controls the degree of overlapping between the co-trainers.

User Size Splitter. This splitter discriminates between users of different sizes. Size of a user is defined as the number of rating he/she has provided. Users are divided into segments based on their sizes and assigned to co-trainers. In case of only two co-trainers, for instance, one of them will be trained on so called "power users" and the other one on small users. This method is based on a histogram of user sizes. It creates N segments (N = number of co-trainers) using equal density binning (each segment has the same number of users).

Random Splitter. Ratings are divided between co-trainers randomly. This method serves as a baseline for comparisons.

Dimensions Preserving Random Splitter. This splitter also assigns ratings randomly to co-trainers, however, in contrast to the previous method, it guarantees that all co-trainers have a matrix with same dimensions. This means that all co-trainers have at least one rating from all users and items from the initial training set. This might be beneficial for methods not able to extend the dimensions of their matrices over time.

Algorithm 1. extBRISMF - trainIncrementally($r_{u,i}$)

Input: $r_{u,i}, P, Q, \eta, k, \lambda$
1: $\vec{p_u} \leftarrow$ getLatentUserVector(P, u)
2: $\vec{q_i} \leftarrow$ getLatentItemVector(Q, i)
3: **if** $\vec{p_u} = null$ **then**
4:    $\vec{p_u} \leftarrow$ getAverageVector(P) + randomVector
5:    $P \leftarrow P.append(p_u)$
6: **end if**
7: **if** $\vec{q_i} = null$ **then**
8:    $\vec{q_i} \leftarrow$ getAverageVector(Q) + randomVector
9:    $Q \leftarrow Q.append(q_i)$
10: **end if**
11: $\widehat{r}_{u,i} = \vec{p_u} \cdot \vec{q_i}$ //predict a rating for $r_{u,i}$
12: evaluatePrequentially($\widehat{r}_{u,i}, r_{u,i}$) //update evaluation measures
13: $epoch = 0$
14: **while** $epoch < optimalNumberOfEpochs$ **do**
15:    epoch++; //for all retained ratings
16:    $\vec{p_u} \leftarrow$ getLatentUserVector(P, u)
17:    $\vec{q_i} \leftarrow$ getLatentItemVector(Q, i)
18:    $predictionError = r_{u,i} - \vec{p_u} \cdot \vec{q_i}$
19:    **for all** latent dimensions k **do**
20:       if $k \neq 1$: $p_{u,k} \leftarrow p_{u,k} + \eta \cdot (predictionError \cdot q_{i,k} - \lambda \cdot p_{u,k})$
21:       if $k \neq 2$: $q_{i,k} \leftarrow q_{i,k} + \eta \cdot (predictionError \cdot p_{u,k} - \lambda \cdot q_{i,k})$
22:    **end for**
23: **end while**

4.3 Prediction Assembler

Prediction assembler aggregates rating predictions from all co-trainers into a single value. We propose three ways of calculating this aggregation.

Recall-based Prediction Assembler assembles predictions of N co-trainers using a weighted average with weights depending on their past recall values.

$$\widehat{r}_{u,iAgg} = \frac{\sum_{j=0}^{N} recall(Co - Tr_j) \cdot \widehat{r}_{u,iCo-Tr_j}}{\sum_{j=0}^{N} recall(Co - Tr_j)} \tag{8}$$

In the above formula recall is measured globally for an entire co-trainer. Alternatively, recall can be measured also on user or item level. In this case $recall(Co - Tr_j)$ can be substituted with $recall(Co - Tr_j, u)$ or $recall(Co - Tr_j, i)$.

RMSE-based Prediction Assembler. Similarly to the previous method, this prediction assembler uses a weighted average, however, here the RMSE measures (root mean square error) serve as weights. Also here, measuring RMSE on user and item levels are possible.

$$\widehat{r}_{u,iAgg} = \frac{\sum_{j=0}^{N} RMSE(Co - Tr_j) \cdot \widehat{r}_{u,iCo-Tr_j}}{\sum_{j=0}^{N} RMSE(Co - Tr_j)} \tag{9}$$

Reliability-weighted Prediction Assembler. This prediction assembler uses a reliability measure to give more weight to more reliable co-trainers.

$$\widehat{r}_{u,iAgg} = \frac{\sum_{j=0}^{N} rel_{Co-Tr_j}^{\widehat{r}_{u,i}} \cdot \widehat{r}_{u,iCo-Tr_j}}{\sum_{j=0}^{N} rel_{Co-Tr_j}^{\widehat{r}_{u,i}}}. \tag{10}$$

4.4 Selector of Unlabelled Instances

This component is used in the unsupervised learning to select unlabelled instances as candidates for training. Due to a large number of unlabelled instances a methods for selecting them is needed. We propose two such methods that as parameter take the number of instances to be selected.

Latent Disagreement Selector. For all known users each co-trainer stores a latent vector that is specific for this co-trainer. We denote this vector as $p_u^{Co-Tr_n}$. In this method we search for users, where the disagreement of the latent user vectors among the co-trainers is the highest. We define the disagreement among two co-trainers upon a user u as follows:

$$disagreement(Co - Tr_a, Co - Tr_b, u) = |p_u^{Co-Tr_a} - p_u^{Co-Tr_b}| \tag{11}$$

This measure can be computed for all known users and all co-trainer pairs. Users with highest disagreement are then selected as candidates together with a random selection of items. The motivation behind this method is that the instances with highest disagreement can contribute the most to the learners. This method can be analogously applied onto latent item vectors.

Random Selector. Random combinations of known users and items are generated. This method is used as a baseline for comparisons.

4.5 Reliability Measure

Reliability measures are used in our framework to assess the reliability of a rating prediction in an unsupervised way. Based on prediction reliability decisions on which co-trainer teaches which one are made.

Sensitivity-based Reliability Measure. This is a novel measure of reliability for recommender systems that is based on local sensitivity of a matrix factorization model. As a user model in matrix factorization we understand a latent user vector p_u. This vector changes over time as new rating information that occurs in the stream is incorporated incrementally into the model. The changes of this vector can be captured using the following formula:

$$\Delta_{p_u} = \sum_{i=0}^{k} (p_{u,i}^{t+1} - p_{u,i}^{t})^2 \tag{12}$$

where $p_{u,i}^{t+1}$ and $p_{u,i}^{t}$ are user vectors at different time points. If Δ_{p_u} is high, then it means that the user model is not stable and it changes considerably over time. Therefore, predictions made by this model can be trusted less. Similarly to the user sensitivity we can also measure a global sensitivity of the entire model as a different variant of this measure. Since Δ_{p_u} has a value range $[0, \infty)$ a normalization is needed (cf. last paragraph of this section).

Popularity-based Reliability Measure. Zhang et al. proposed in [11] a reliability measure based on popularity. This measure uses the idea that the quality of recommendations increases as the recommender system accumulates more ratings. They used the absolute popularity of users and items normalized by a fixed term. We implemented this reliability measure in our framework for comparison, however, with a different normalization method. Normalisation on streams is different and more challenging (cf. last paragraph of this section).

Random Reliability Measure. A random number from the range $[0, 1]$ is generated and used as reliability. This measure is used as a baseline.

Normalization of Reliability Measures. As defined in Sect. 3.2, a reliability measure is a function with value range of $[0; 1]$. With many aforementioned reliability measures this is not the case, therefore, a normalization is necessary. Normalization on a stream, however, is not trivial. Division by a maximal value is not sufficient, since this value can be exceeded in a stream and a retrospective re-normalization is not possible. In our framework we use the following sigmoid function for normalization:

$$f(reliability) = \frac{1}{1 + e^{\alpha \cdot (reliability - \mu)}} \tag{13}$$

where α controls the slope of the function and μ is the mean of the distribution. The parameters can be set either manually, or automatically and adaptively in a self-tuning approach. While the adaptive calculation of μ in a stream is trivial, the calculation of α requires more effort. For that purpose we store 1000 most recent arguments of this function and determine their fifth percentile. We define that the value of the sigmoid function for this percentile should be equal to 0.9. From that, the optimal value of α can be derived. Note that α also controls if the function is monotonically increasing or decreasing.

5   Evaluation Setting

Incremental matrix factorization algorithm works on a stream of ratings, however, it requires a short initialization phase that also needs to be taken into account, while evaluating our framework. Therefore, a simple split into a training and test datasets is not sufficient. Also peculiarities of a stream based evaluation should be considered. In [8] we developed an evaluation framework suitable for this type of algorithms. We adopt it also here to measure the performance of our SSL algorithms and explain it shortly in this section.

5.1   Evaluation Protocol

Our evaluation protocol consists of two modes that are presented in Fig. 5. The figure represents an entire dataset and splitting of it into the two aforementioned modes, as well as into training and test datasets within the modes. The first mode is a batch mode used for initial training and tuning (blue colour in Fig. 5). The batch mode itself consists of two further parts. Part (1) of a dataset is used as initial training set. Part (2) is used for testing of the initial training and for adjusting parameters of the model. This part is crucial especially for matrix factorization algorithms using stochastic gradient descent. Once the initial training is finished, the algorithm switches into the stream mode (yellow part). From this moment on, evaluation and training are performed simultaneously as proposed by Gama et al. in prequential evaluation [2].

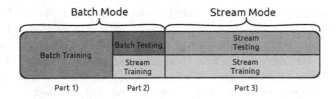

Fig. 5. Splitting of the dataset between the batch and streaming mode. Separation of training and test datasets in each of the modes [9] (Color figure online).

In Fig. 5 we can see that part (1) and (3) are user for learning and part (2) for testing in batch mode. Excluding part (2) from learning would create a temporal gap in the stream. This gap can be problematic for many incremental methods that rely on time aspects and are sensitive to ordering of data instances. In order to avoid this gap, we use part (2) for learning in the streaming mode as well, but not for testing, since it has been used for batch testing already. Testing in the stream mode starts in part (3).

In our framework the streaming mode is the main part of the preference learning, where SSL methods are used. Therefore, our results refer always to the streaming mode of an algorithm. To investigate the effect of SSL, we compare results of exrBRISMF **with the SSL setting and without it.**

5.2   Evaluation Measure - Incremental Recall

In our experiments we use an incremental recall measure proposed by Cremonesi et al. [1]. In the incremental setting precision can be derived from incremental recall (cf. [1]) and, therefore, we do not present it. In contrast to purely rating-based measures, such as RMSE or MAE, the incremental recall measure also considers ranking of a predicted item. Another problem of RMSE and MAE is that they weight all predictions uniformly, regardless of relevance of items, i.e. predictions on all irrelevant items (typically $> 99\%$) influence the error measure equally strong as predictions on relevant items. In incremental recall the ranking of relevant items only counts.

The procedure of measuring $incrementallRecall@N$ is as follows. At each new rating in a stream the relevance of the corresponding item is determined using a rating threshold (e.g. $r_{ui} > 4$ is considered relevant). For each relevant item additional 1000 random irrelevant items, are selected. For all of those items rating predictions are made and sorted. Finally, the rank p of the relevant item among the irrelevant ones is determined. If the rank p is lower than N, a hit is counted. The value of $incrementalRecall@N$ is set to $\frac{\#hits}{|Testset|}$.

6   Experiments

To show the effect of our SSL framework we compared the results of the extBRISMF algorithm alone (abbreviated hereafter as **NoSSL**) and extBRISMF with our SSL framework (abbrv. as **SSL**). Both, the algorithm and the framework require setting parameters, such as learn rate η in gradient descent, etc. Therefore, in order to find approximately best parameter setting we performed a grid search in the parameter space. The grid search was performed on a cluster running the (Neuro) Debian operating system [3]. In total we conducted more than 350 experiments. For brevity we present here only the best result achieved by the SSL and NoSSL method on each dataset.

Datasets. In our experiments we use four real-world datasets from the recommender systems domain. We stress out that our framework is applicable to all datasets in form of a user-item-rating matrix unlike similar SSL frameworks that rely on external sources of information. The datasets we used encompass Movielens 100 k, Movielens 1M[1] datasets, as well as random samples of 1000 users of the Netfilx[2] and Epinions (extended) [7] datasets. We used sampling on the large datasets due to a large numbers of experiments in the grid search. The percentage of labelled data out of all possible user-item-pairs in those datasets amounts to values between 0.03 % and 6.3 %. This shows how much of the unlabelled information is available in the process of preference learning.

Results. Figure 6 shows the incremental recall@10 over time on the vertical axis (higher values are better) and the time dimension on the horizontal axis. The red curves represent the SSL method, whereas the blue ones stand for the NoSSL method. The dashed lines in colours of both curves represents the median value of the incremental recall. They correspond to the medians in the boxplots in the right part of the figures that visualize the distribution of incremental recall in a simplified way. In all parts of Fig. 6 we can see that the **SSL method dominates the NoSSL** one at nearly all time points. More precise results with the corresponding settings are presented in Table 1. The columns 2–5 of the table represent the components (e.g. a reliability estimator based on user popularity in the first row). Rows with no components, but with "NoSSL" stand for the

---

[1] www.movielens.org.

[2] www.netflix.com.

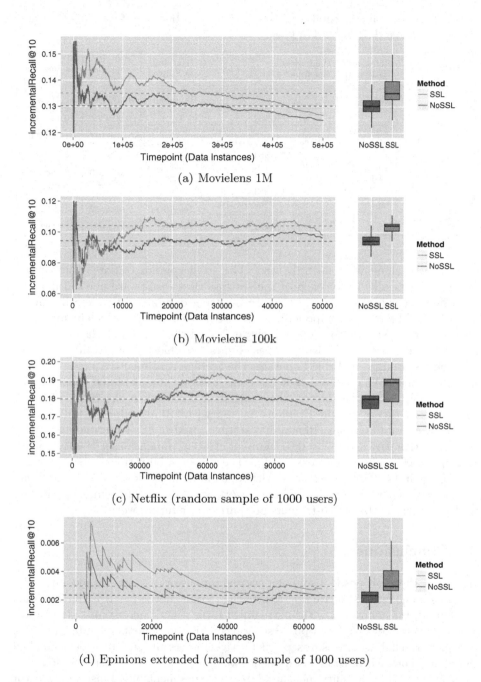

(a) Movielens 1M

(b) Movielens 100k

(c) Netflix (random sample of 1000 users)

(d) Epinions extended (random sample of 1000 users)

Fig. 6. *Incremental Recall* on four real datasets (higher values are better). Application of SSL techniques yields an improvement on all datasets at nearly all time points (Color figure online).

Table 1. Average incremental recall@10 and computation time for a single instance. Our SSL framework is marked in blue.

Dataset	Reliability Estimator	Prediction Assembler	Unlabelled Instance Selector	Training Set Splitter	Avg. In- cremental Recall@10	Avg. Time for Instance (ms)
ML100k	User Popularity	User Recall	Latent User Disagreement	Random	**0.101235**	2.417228
	NoSSL				0.095099	0.138642
ML1M	Sensitivity Global	User Recall	Latent User Disagreement	Dimensions Preserving	**0.136564**	10.88977
	NoSSL				0.130721	0.250437
Netflix 1000 users	Sensitivity Global	RMSE Global	Latent Item Disagreement	Dimensions Preserving	**0.184150**	6.888166
	NoSSL				0.177380	0.400382
Epinions 1000 users	Sensitivity Global	RMSE Global	Latent User Disagreement	User Size	**0.003312**	131.7290
	NoSSL				0.002289	0.979214

extBRISMF alone with no SSL used. The sixth column contains the average incremental recall@10 for each of the setting. Best results are marked in bold. Also here we can recognize that the SSL setting dominated the NoSSL one on all datasets. From the components there are no clear winners, except for latent disagreement instance selectors, which performed the best on all datasets. From reliability measures the sensitivity-based ones were mostly successful.

The last column in Table 1 contains the average runtime for a single instance in milliseconds. We observed that the computation time increased considerably, when using SSL. Nevertheless, the runtime still stayed in the range of a few milliseconds, except for the Epinions dataset with 131 ms, which is still feasible in real-world applications. Remaining settings used in the experiments are the regularization parameter $\lambda = 0.01$, number of latent dimensions $k = 40$ and learn rate $\eta = 0.003$. The framework used USL every $m = 50$ ratings, where $z = 100$ unlabelled instances were selected. Although our framework was developed for an arbitrary number of co-trainers, in this work we used two of them. Experiments with a larger number of co-trainers are part of our future work.

7   Conclusions

In this work we proposed a semi-supervised framework for stream recommender systems based on the co-training approach. To our knowledge, it is the first such framework that can deal with a stream of ratings and incremental algorithms. Within the framework we proposed several generic components including training set splitter, reliability measures, prediction assemblers and selectors for unlabelled instances. For each of those components we developed several instantiations e.g. sensitivity-based reliability measure, latent disagreement-based instance selector and many more. Furthermore, we extended the BRISMF algorithm [10] by the ability to extend the dimensions of the matrix incrementally.

In experiments on four real datasets we showed that **our SSL framework outperforms the non-SSL method at nearly all time points**. This is, because our framework is able to leverage the unlabelled information, which in recommender systems is abundant. Using this information allows us alleviate the problem of sparsity even without using any context information. The improvement is, however, at the cost of longer computation time. Nevertheless, the computation time of a single instance still remains in a range of a few milliseconds (normally around 6 ms, except for Epinions dataset - ca. 131 ms). Therefore, this framework applicable in real-world recommenders.

Our immediate next steps are to investigate, how to make this framework faster by e.g. sharing parts of the matrix among co-trainers and by using efficient data structures. Furthermore, we plan to experiment with more than two co-trainers and to implement further instances of the aforementioned components.

Acknowledgements. The authors would like to thank Daniel Kottke and Dr. Georg Krempl for suggestions regarding self-tuning normalization for streams and also the Institute of Psychology II at the University of Magdeburg for making their computational cluster available for our experiments.

References

1. Cremonesi, P., Koren, Y., Turrin, R.: Performance of recommender algorithms on top-N recommendation tasks. In: RecSys 2010. ACM (2010)
2. Gama, J., Sebastião, R., Rodrigues, P.P.: Issues in evaluation of stream learning algorithms. In: KDD. ACM (2009)
3. Halchenko, Y.O., Hanke, M.: Open is not enough. Let's take the next step: an integrated, community-driven computing platform for neuroscience. Front. Neuroinform. **6**, 1–4 (2012)
4. Karimi, R., Freudenthaler, C., Nanopoulos, A., Schmidt-Thieme, L.: Towards optimal active learning for matrix factorization in recommender systems. In: ICTAI, pp. 1069–1076. IEEE (2011)
5. Koren, Y.: Collaborative filtering with temporal dynamics. In: KDD 2009. ACM (2009)
6. Koren, Y., Bell, R., Volinsky, C.: Matrix factorization techniques for recommender systems. Computer **42**(8), 30–37 (2009)
7. Massa, P., Avesani, P.: Trust-aware bootstrapping of recommender systems. In: ECAI Workshop on Recommender Systems, pp. 29–33. Citeseer (2006)
8. Matuszyk, P., Spiliopoulou, M.: Selective forgetting for incremental matrix factorization in recommender systems. In: Džeroski, S., Panov, P., Kocev, D., Todorovski, L. (eds.) DS 2014. LNCS, vol. 8777, pp. 204–215. Springer, Heidelberg (2014)
9. Matuszyk, P., Vinagre, J., Spiliopoulou, M., Jorge, A.M., Gama, J.: Forgetting methods for incremental matrix factorization in recommender systems. In: Proceedings of the SAC 2015 Conference. ACM (2015)
10. Takács, G., Pilászy, I., Németh, B., Tikk, D.: Scalable collaborative filtering approaches for large recommender systems. J. Mach. Learn. Res. **10**, 623–656 (2009)
11. Zhang, M., Tang, J., Zhang, X., Xue, X.: Addressing cold start in recommender systems: a semi-supervised co-training algorithm. In: SIGIR. ACM (2014)
12. Zhou, Z.-H., Li, M.: Semisupervised regression with cotraining-style algorithms. IEEE Trans. Knowl. Data Eng. **19**(11), 1479–1493 (2007)

Detecting Transmembrane Proteins
Using Decision Trees

Mohammad Hossein Nikravan, Ashwani Kumar, and Sandra Zilles[✉]

Department of Computer Science, University of Regina,
Regina, SK S4S 0A2, Canada
{nikravam,zilles}@cs.uregina.ca, kumar26a@uregina.ca

Abstract. Transmembrane (TM) proteins are proteins that span a cell membrane; their segments crossing the membrane are called TM domains. TM domain and TM protein detection are important problems in computational biology, but typical machine learning approaches yield classifiers that are difficult to interpret and hence yield no biological insight. We study both TM domain and TM protein detection with easy to interpret decision trees. For TM domain detection, the use of decision trees is already reported in the literature, but we provide a critical study of the existing approach, resulting in improved feature sets as well as observations on how to avoid biased training and test sets. In particular, we discover a motif known to be common to TM domains that was not discovered in previous research using machine learning. For TM protein detection, we propose a 2-layer learning method. This method can be generalized to deal with a large class of string classification problems. The method achieves sensitivity and specificity values of up to 92 % on the settings we experimented with, while providing intuitive classifiers that are easy to interpret for the domain expert.

1 Introduction

A transmembrane (TM) protein is a protein that is located partly inside and partly outside a cell. Such proteins usually cross the cell membrane several times, and each protein segment that spans the cell membrane is called a TM domain, see Fig. 1. Among the diverse functions of TM proteins, cellular communication with the external environment and transportation of ions and molecules are the most important. TM proteins assist in host-pathogen interactions and play key roles in the host's immune response and drug resistance [3,10]. A major fraction of clinically approved drugs target TM proteins, which indicates their importance in drug design and discovery [4]. It is hence critical to devise efficient and accurate computational methods for TM protein detection.

TM proteins are well-studied, and a number of classification approaches with which to distinguish them from other types of proteins have been suggested in the literature. In particular, we focus on methods that identify TM proteins based

This work was supported by the Natural Sciences and Engineering Research Council of Canada (NSERC).

© Springer International Publishing Switzerland 2015
N. Japkowicz and S. Matwin (Eds.): DS 2015, LNAI 9356, pp. 146–160, 2015.
DOI: 10.1007/978-3-319-24282-8_13

on their primary structure alone, i.e., based only on their underlying amino acid sequences. Most existing machine learning approaches, e.g., using neural networks [7,8] or hidden Markov models [12], typically use highly specific expert knowledge in selecting training data or designing the model structure. While the resulting classifiers perform well (see [2] for a survey), they are difficult to interpret and do not provide the biologist with amino acid patterns that are common in TM proteins or in non-TM proteins. As opposed to that, statistical analyses yield patterns that are common in TM domains and pat-

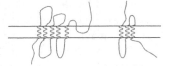

Fig. 1. A cell membrane (delimited by the two straight lines) hosting two TM proteins: one with five TM domains and one with three TM domains.

terns that are uncommon in TM domains [11], but they do not immediately yield classifiers that can be used to identify TM proteins among a set of proteins.

The most common example of a type of classifier that can easily be interpreted by the domain specialist is the decision tree. In this paper, we train and test decision trees for the purpose of classifying proteins into TM proteins and non-TM proteins. To the best of our knowledge, the only related studies using decision trees were conducted by Arikawa et al. [1] and He et al. [5]. Both focused only on the problem of deciding whether or not a given amino acid segment is a TM domain. While solving this problem can help to solve the problem of identifying TM proteins, it does not immediately yield a solution to the latter. Furthermore, He et al. [5] tested the rules generated by their system on only 165 hand-selected TM proteins, which cannot be considered a representative sample.

We propose a machine learning approach, called 2LDT (2-Layer Decision Tree), for classifying proteins into TM and non-TM proteins with a classifier that is easy to interpret. Our machine learning approach is based solely on decision trees and works in two layers. In the first layer, we train a decision tree for identifying TM domains among short protein segments. We then apply the learned decision tree to a series of subsegments of actual TM and non-TM proteins and record for each protein how many segments were classified as TM domains by our trees learned in the first layer. This information is then used in the second layer of training to build a decision tree for identifying TM proteins. 2LDT follows a framework that works in principle with training any kind of classifier—it does not require the use of decision trees.

In our experiments, the training in the first layer of 2LDT is quite similar to the method suggested by Arikawa et al. [1], but we propose to use different features and to evaluate the first layer decision trees more carefully than they did. In a systematic analysis of the approach by Arikawa et al., we test which parameters of their learning environment and their testing environment might lead to improvements when modified.

Overall, our contributions and findings can be summarized as follows:

– We present 2LDT, the first learning method for identifying TM proteins that results in easy-to-interpret classifiers. It achieves sensitivity and specificity values of up to 92 % in our experiments, while the classifiers trained with this method provide more insights into patterns common and uncommon in TM proteins than previously published machine learning methods do. The two-layer learning approach of 2LDT requires no domain knowledge whatsoever and may potentially be applicable to string classification problems of any kind. When applying the method to amino acid sequences for TM protein identification, the only domain knowledge we use is that the hydropathy of amino acids plays a role in this context. For other applications, one can simply exchange the feature sets based on hydropathy by other features, i.e., one would change the presentation of the data, not the learning method.

- The TM domain identification method deployed in the first layer of our approach outperforms the method by Arikawa et al., after which it was modeled. In this context, we demonstrate that (i) the frequency of certain types of amino acids (especially when using hydropathy indices [6]) in a sequence is more useful for classification than testing their mere existence or testing the existence of substrings of arbitrary length, (ii) to identify TM domains is easier when the given sequence has a fixed length of 30 (as in Arikawa et al.'s experiments) than when its length is chosen uniformly at random from the length distribution of TM domains, (iii) when training a decision tree, it does not make much difference whether the negative training sequences are sampled from non-TM proteins or from the non-TM domain segments of TM proteins, and (iv) sequences sampled from non-TM proteins are difficult to distinguish from those sampled from the non-TM domain segments of TM proteins, at least when using decision trees over a variety of intuitive feature sets.

- The decision trees trained by Arikawa et al. showed that amino acids from the set $\{D, E, H, K, N, Q, R\}$ occur often *outside* TM domains, i.e., they form a *negative motif* for TM domain detection. Our experiments suggest that these negative motifs alone may be too weak when lifting TM domain detection to TM protein detection. The trees we train provide an additional *positive motif* to check for, namely the frequent occurrence of amino acids from the set $\{A, C, F, I, L, M, V\}$. While it is known that such amino acids (which are non-polar) are frequent in TM domains [9], existing machine learning approaches have not explicitly suggested this pattern (mostly because they do not yield intuitive classifiers).

2   Problem Formulation

We are concerned with the problem of detecting TM proteins in a set of proteins of various kinds, and we are looking for an easy-to-interpet classifier that achieves good values both for sensitivity (true positive rate) and for specificity (true negative rate).

The lead question we try to answer in our study is the following:

Question 1. Can one obtain reasonable sensitivity and specificity values when learning decision trees that detect TM proteins among a set of proteins of various kinds, while using as input only amino acid sequences?

The 2LDT approach described in Sect. 3 proposes a solution to this classification problem. It is based on a first step of detecting TM domains among a set of amino acid sequences. TM domain detection was the focus of work by He et al. [5], who used a combination of decision trees and SVM to obtain a set of rules—to obtain a classifier one would further have to pick support and confidence thresholds. They test only on sequences taken from 165 TM proteins, not on any sequences taken from non-TM proteins, and they provide no information on the training data. Because of these issues, we do not compare the results of our first step to the accuracy values reported by He et al.

Arikawa et al. [1] also focused solely on detecting TM domains rather than TM proteins, and they only mention in passing that they used their method to design a classifier for proteins, achieving accuracy values between 85 % and 90 %. They describe neither this classifier nor the data used for its evaluation, so it remains unclear what these accuracy values actually mean. For example, if the test data is imbalanced (as is the case in nature, where the majority of proteins are non-TM), even a classifier labelling every instance negative can achieve very high accuracy. Hence, when comparing to Arikawa et al.'s work, we restrict ourselves to the problem of detecting TM domains. We critically analyze their approach along the following aspects.

Feature Sets. All the features that Arikawa et al. allowed for training their trees were binary and represented whether or not the input string contained certain substrings in a specified order. A substring of a string s is a string of consecutive symbols in s. For example, AAE is a substring of $FGFAAE$, but not of $FGAAFE$. Due to runtime problems though, in their actual training phase, they restricted these features further and considered only those that represent whether or not the input string contained a single substring. These features were computed once over the raw amino acid sequences (using an alphabet of 20 letters), and in a separate experiment over indexed sequences, in which every amino acid was replaced by one of three possible symbols representing its hydropathy [6]. The trees Arikawa et al. displayed among their results used as attributes only substrings of length 1 in case of raw sequences and substrings of length 2 or 3 in case of indexed sequences. We ask the following question:

> *Question 2.* Does the best sensitivity and specificity for detecting TM domains necessarily require testing for substrings of arbitrary length in the inner nodes of the decision trees?

Length of Negative Example Sequences used for Training and Testing. Arikawa et al. trained and tested their trees with TM domains as positive examples and other sequences consisting of around 30 amino acids as negative examples. They do not report how well their classifiers perform when trained or tested with negative examples of a different length. We ask:

> *Question 3.* How does the length of negative training and test sequences affect the sensitivity and specificity of TM domain detection?

Source of Negative Example Sequences used for Training and Testing. The negative examples Arikawa et al. used for training were substrings of the non-TM parts of TM proteins. Their test cases included substrings both from the non-TM parts of TM proteins and from non-TM proteins. We ask:

> *Question 4.* How does the type of protein used as a source for the negative examples affect the sensitivity and specificity of TM domain detection?
>
> *Question 5.* Using standard methods, can one learn a decision tree that distinguishes between substrings of the non-TM parts of TM proteins and substrings of non-TM proteins?

3   2-Layer String Classification Method

A core contribution of this paper is 2LDT, a 2-layer learning method for protein classification. We use this method in an attempt to answer Question 1, but the method is much more generally applicable.

Compare the following two string classification problems. For each problem, one wants to classify strings into classes A and B, where strings of class A typically contain substrings of class S and strings of class B do not. In Problem 1, suppose S is the class of strings typically containing substrings from a fixed set S'. In Problem 2, S is the class of strings typically *not* containing substrings from S'. To solve Problem 1, a successful classifier may simply label every string containing a substring from S' with A and all others with B. In Problem 2, S is characterized by the *absence* rather than the presence of S'. The presence of an S' pattern in a substring of a string s means that s is most likely not of class S, but that doesn't imply s itself is most likely of class B—s may have substrings that do not contain any S' pattern and are thus considered of class S; so s might still be of class A. The TM protein detection problem is an example of Problem 2: A is the class of TM proteins, B the class of all others, and S is the class of TM domains. It has been pointed out in the literature, see, e.g. [1], that TM domains are characterized not so much by the presence but rather by the absence of certain amino acid patterns. So, a classifier that detects TM domains (based on the absence of certain patterns) does not immediately yield a classifier that detects TM proteins, since both protein classes have parts that are not TM domains and thus typically have the critical patterns present.

We propose the following 2-layer decision tree (2LDT) learning framework for Problem 2:

input: a set D of strings, with class labels in $\{A, B\}$, where each data feature gives information on the containment of a specific substring

output: a decision tree that assigns any string to class A or B

1. use part of D to train a decision tree $\mathcal{T}_1$ that decides whether a string is in class S
2. fix a set M of pairs (*windowsize*, *stepsize*) with *windowsize* ≥ 1 and $1 \leq$ *stepsize* $\leq$ *windowsize*
3. for each string $d \in D$ not used in Step 1 for each pair (*windowsize*, *stepsize*)

- apply T_1 to all substrings of d of length *windowsize*, starting with the one beginning at the first letter of d and always proceeding with the substring beginning *stepsize* many positions further to the right
- record the results of T_1 as features in a new data item d'
4. use the newly generated data items to train a decision tree T_2 that decides whether a string is in class A or in class B; output T_2

2LDT trains a decision tree T_1 detecting class S at layer 1. At layer 2, it uses T_1 to change the features of the remaining pool of training data. It does so by sliding a window of a fixed length across the string, moving by a fixed stepsize, and recording information on when T_1 labels the string in the current window with S. Note that the same 2-layer approach can be used with any kind of classifier instead of decision trees.

In our instantiation, at the second layer (learning T_2) we also used the length of a protein as a feature. We will give more details on the features and the values for *windowsize* and *stepsize* in Sect. 5.

4   Experiments: Identifying Transmembrane Domains

This section is concerned with Questions 2 through 5 posed in Sect. 2. All these questions concern TM domain detection, i.e., the first layer of 2LDT.

4.1   Experimental Setup

Datasets. Our training data and test data are all extracted from the UniProt/ Swiss-Prot database (Release 2014-09, http://www.uniprot.org/). This database contains 546,439 proteins (represented as amino acid sequences in which TM domains, if existent, are marked) of which we extracted the first 400,000. If an amino acid sequence occurred more than once, only the first occurrence was kept, thus reducing the set by 64,888 sequences. Further, a total of 2,041 sequences containing any of the letters B, J, O, U, X, Z[1] were removed.

Every protein containing at least one TM domain was considered a TM protein, all others were considered non-TM proteins.

Positive data at layer 1 were sampled from the full TM domains (substrings of TM proteins) marked as such. We first extracted all TM domains and sampled 200,000 of them at random, to be used as positive instances for layer 1.

For negative instances at layer 1, we collected three types, each in two length categories. For each length category, (i) type Neg-TM contains 200,000 sequences of consecutive amino acid symbols taken from TM proteins, yet outside the TM domains, (ii) type Neg-NonTM contains 200,000 sequences of consecutive amino acid symbols taken from non-TM proteins, and (iii) type Neg-Combined contains 200,000 randomly chosen instances from the union of Neg-TM and Neg-NonTM. In the first length category ('30'), we collected only sequences of length

---

[1] U and O are two rare amino acids found in some species. B, J, X, and Z are used in case of inconclusive identification of residues in the protein sequences.

30 (comparable to what Arikawa et al. did, who took sequences "of length around 30" ([1], p. 367), while a negative dataset of the second length category ('D') has sequences following the same distribution of lengths as the positive data at layer 1, see Fig. 2, restricted to length at most 40 (only 18 out of 200,000 TM domains, i.e., 0.009 %, have length greater than 40).

We generated these negative instances in the following way: First, we extracted sequences of length 50. For Neg-NonTM we took positions 1–50, 51–100, 101–150, and so on, from a protein's amino acid sequence, for as long as full non-overlapping sequences of length 50 could be selected; for Neg-TM we did the same in each single one of the substrings neighbouring a TM domain. Second, we generated sequences of length 40 by randomly picking a start point among the first 11 amino acids in any sequence of length 50

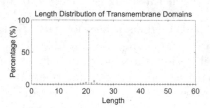

Fig. 2. Distribution of lengths of TM domains in the positive data at layer 1. The peak is at length 21.

previously generated. These sequences of length 40 were then used to generate instances in either length category as follows. For length category 30, we randomly chose the start point of the sequence to be generated. For length category D, we randomly picked a positive example, recorded its length L and took the first L symbols in a randomly chosen negative sequence of length 40 (in case $L > 40$, we would pick a new positive example).

Feature Sets. Arikawa et al. [1] demonstrated that replacing the symbol for an amino acid by a symbol for its hydropathy class (i.e., using the symbol $+$ for G, P, S, T, W, Y, the symbol $*$ for A, C, F, I, L, M, V, and the symbol $-$ for D, E, H, K, N, Q, R, cf. [6]) is helpful for TM domain detection. We used the following feature sets in our experiments:

1. AA(binary): 20 binary features, for each amino acid x a feature representing whether or not x is contained in the sequence
2. Hyd(fraction): 3 numerical features, for each hydropathy class x a feature representing the number of occurrences of x in the sequence, divided by the length of the sequence
3. 2Hyd(binary): 9 binary features, for each pair x of hydropathy classes a feature representing whether or not x is contained in the sequence
4. 3Hyd(binary): 27 binary features, for each triple x of hydropathy classes a feature representing the number of occurrences of x in the sequence, divided by the length of the sequence.

Note that a binary version of the 'Hyd' feature set would not be helpful for distinguishing TM domains from other protein segments, since a very large fraction of protein segments that are as long as TM domains contain amino acids from each of the three hydropathy classes.

$QNKYRENNK(GPMMDFLATAVFAFMWLVSSSAWA)KGLSD$
$VKMATDPENIIKEMPMCRQTGNT$

Fig. 3. A substring of a protein sequence, once represented with amino acids and once represented with hydropathy indices. The part of the sequence inside the parentheses is a TM domain, while the other parts are outside of TM domains. Taken from [1], Figs. 3 and 4.

Learning Method. For the first layer of our learning method, we used the decision tree learning method in Matlab, with the following parameters: split criterion = 'deviance', prune criterion = 'error', prior = 'uniform', algorithm for categorical predictor = 'exact'. We tried pruning to levels 0, 1, and 2. In some subexperiments, pruning was switched off.

Evaluation. We evaluate the resulting decision trees with respect to their sensitivity (TPR) and their specificity (TNR). The set of 200,000 positive examples, as well as each of the six sets of 200,000 negative examples (3 types, 2 length categories) were randomly shuffled. From each shuffled set, the first 10,000 examples were chosen for training, the remaining 190,000 were set aside for testing. We ran 5-fold cross-validation on some experiments, showing very little variance for either of these two values across the 5 folds (the differences were about 0.1 %). Hence, the results reported below are all from experiments run without cross-validation.

For each feature set, we trained 18 trees, namely one for each combination of type and length category of negative training data (6 combinations) with 3 different levels of pruning. Each tree was tested twice on negative data of type Neg-Combined, once for length category '30' and once for category 'D'.

4.2   Results

Pruning to levels 1 and 2 generally yielded fairly similar performance, and both typically slightly outperformed trees pruned at level 0. Table 1 hence reports only results for level 2.

Question 2. Does the best sensitivity and specificity for detecting TM domains necessarily require testing for substrings of arbitrary length in the inner nodes of the decision trees? Despite the fact that Arikawa et al. [1] allowed substrings of arbitary length as tests in the inner nodes, the best trees they obtained contained only short strings (length up to 3 when using hydropathy indices, length 1 when using amino acids). They hand-crafted a classifier that tests for a minimum of 5 (not necessarily consecutive) occurrences of hydrophilic amino acids (hydropathy index '-'), which showed excellent performance on their test data. This suggests that long substrings are not of interest for TM domain detection,

Table 1. Results (in percent, rounded) on TM domain detection. Results on test data using the length category 'D' ('30', resp.) for the negative examples are reported in the top (bottom, resp.) half of the table. Each row represents a specific type/length category of negative data used in training. The first four columns refer to the feature sets we tested. The last two columns refer to the best trees trained by Arikawa et al. [1].

Test: (D)		AA(b)	Hyd(f)	2Hyd(b)	3Hyd(b)	[1], Fig. 5a	[1], Fig. 5c
Neg-TM (D)	TPR	86.0	96.6	88.2	88.0		
	TNR	84.1	95.3	91.6	90.0		
Neg-TM (30)	TPR	94.7	97.3	94.2	88.9	79.0	90.0
	TNR	78.1	93.0	82.1	88.6	83.0	85.1
Neg-NonTM (D)	TPR	84.4	97.6	88.2	88.8		
	TNR	89.0	94.4	91.6	88.8		
Neg-NonTM (30)	TPR	94.6	97.3	94.2	88.8		
	TNR	77.9	93.0	82.1	88.8		
Neg-Comb (D)	TPR	84.4	94.6	88.3	88.0		
	TNR	89.0	96.8	91.2	90.0		
Neg-Comb (30)	TPR	94.6	97.3	93.3	88.8		
	TNR	78.0	93.0	82.6	88.8		
Test: (30)		AA(b)	Hyd(f)	2Hyd(b)	3Hyd(b)	[1], Fig. 5a	[1], Fig. 5c
Neg-TM (D)	TPR	86.0	96.6	88.2	88.0		
	TNR	93.1	96.4	98.3	94.6		
Neg-TM (30)	TPR	94.7	97.3	94.2	88.9	79.0	90.0
	TNR	90.6	98.2	92.6	96.4	93.1	93.2
Neg-NonTM (D)	TPR	84.4	97.6	88.2	88.8		
	TNR	95.8	95.8	98.3	95.8		
Neg-NonTM (30)	TPR	94.6	97.3	94.2	88.8		
	TNR	90.4	98.4	92.6	95.8		
Neg-Comb (D)	TPR	84.4	94.6	88.3	88.0		
	TNR	95.8	98.7	98.2	94.6		
Neg-Comb (30)	TPR	94.6	97.3	93.3	88.8		
	TNR	90.7	98.2	92.6	95.8		

and that the frequency of symbols rather than just a binary feature testing the mere existence of certain symbols provides useful knowledge. Our experiments support this conclusion. Using hydropathy indices, tests for pairs mostly yield better classifiers than tests for triples (in several cases, triples yield better specificity, but overall pairs outperform triples). Further, single hydropathy symbols are by far the best features we tested, as long as their fraction rather than their mere existence is tested. They yield sensitivity values from 94.6 % to 97.6 %, and specificity values from 93.0 % to 98.7 %. In particular, the best trees learned by

Arikawa et al.'s method, applied to our test data, cannot compete with our tree based on the fraction of hydropathy symbols. Hence, we answer Question 2 negatively: we suggest to restrict feature sets to very short substrings or even just single symbols, but to use frequencies of symbols rather than binary features. We do not claim optimality of such features, but rather that the strongest feature sets do not necessarily include tests for "long" substrings.

To obtain further evidence for our conclusion, we tested Arikawa et al.'s handcrafted classifier, which only checks for at least 5 occurrences of hydrophilic amino acids (sequences *not* passing the check are classified as TM domains, those that pass the check are considered not to be TM domains). This classifier maintains its excellent performance across the test data sets: for our test data, its (TPR,TNR) values are (98.4 %,86.4 %) on data with negative examples of length category 'D' and (98.4 %,97.7 %) on data with negative examples of length category '30'. Again this suggests that the frequency of hydropathy symbols makes for very strong features in TM domain detection.

Question 3. How does the length of negative training and test sequences affect the sensitivity and specificity of TM domain detection? An immediate observation from Table 1 is that *all* classifiers tested here perform better on length category '30' than on length category 'D', independent of the length distribution they were trained with. This suggests that it is easier to distinguish TM domains from other sequences when the latter are of length 30 than when the latter are shorter (the vast majority of sequences in the category 'D' have length 21). The classifier performance reported by Arikawa et al., who tested on negative sequences of length around 30 only, might not be stable across other lengths of protein segments, even on the proteins they used for testing. Concerning the effect of the length of negative training sequences, no clear trend is observed in Table 1. That is, while the length of negative test data shows clear patterns in how it affects the performance of the classifiers, no similar patterns have been observed concerning the length of negative training data.

Question 4. How does the type of protein used as a source for the negative examples affect the sensitivity and specificity of TM domain detection? Interestingly, the answer seems to be 'not at all', for the features we used. The classifier performance reported in Table 1 is quite stable with respect to changes in the kind of protein from which the negative training sequences were taken. Consequently, the choice of training sequences used by Arikawa et al. [1] has probably not biased their system, in contrast to their choice of test data.

Question 5. Using standard methods, can one learn a decision tree that distinguishes between substrings of the non-TM parts of TM proteins and substrings of non-TM proteins? The answer to that question seems to be 'no', at least when restricted to the feature sets we tested, as can be seen in Table 2. The classifiers obtained perform barely better than random guessing. We also tried some other feature sets not reported here, but never obtained any substantially better results. This seems to support our answer to Question 4—for the purposes of TM domain detection, there truly is no difference between negative data taken from TM proteins and negative data taken from other proteins.

Table 2. Results (in percent, rounded) on detecting non-TM domains taken from TM proteins among non-TM domains taken out of any kind of protein. (All training and test sequences were of length 30.)

	AA(b)	Hyd(f)	2Hyd(b)	3Hyd(b)
TPR	53.1	62.5	82.5	50.5
TNR	59.5	51.4	22.9	60.5

A Positive Motif. All trees presented by Arikawa et al. [1] that are based on hydropathy indices use only negative motifs (strings assumed to occur outside of TM domains) in their inner nodes. Our best trees use both a negative and a positive motif. The negative motif is similar to the one Arikawa et al. observed: TM domains typically do not contain a large number (in our case fraction) of '−' symbols. Many of our experiments yield trees, which, in case a sequence contains a small fraction of '−' symbols, additionally check for the fraction of '*' symbols. Only if that fraction is large enough, the trees label a sequence as TM domain. See Fig. 4 (left) for illustration.

5   Experiments: Identifying Transmembrane Proteins

Using 2LDT requires (i) a feature set used for training at layer 1, (ii) a set of (*windowsize, stepsize*) pairs, and (iii) a feature set used for training at layer 2.

Concerning (i), we experimented with the following decision trees at layer 1:

- T_{full}, the unpruned tree trained with negative data from Neg-Combined(D), using Hyd(f) features,
- T_{level2}, the tree pruned at level 2, trained with negative data from Neg-Combined(30), using Hyd(f) features,
- T_{5-}, the hand-crafted classifier labelling as TM domains exactly those sequences that contain fewer than 5 '−' hydropathy symbols.

Note that even 100 % accuracy at layer 1 would not guarantee that 2LDT performs well at layer 2, due to a crucial difference in the test data we used for TM domain detection and the segments on which the trees in layer 1 of 2LDT are deployed. For the former, positive examples are always full TM domains that do not overlap with non-TM domains and negative examples are always segments that do not overlap with TM domains. As opposed to that, the latter may cover part of a TM domain and part of an adjoining non-TM domain.[2]

---

[2] This suggests that it might be helpful to train and test the trees used at the first layer using segments that cover part of a TM domain and part of a non-TM segment. However, in this case it is not straightforward how to assign labels to the training and test data, i.e., to decide how much overlap a segment needs to have with a TM domain in order to be considered a positive example.

Concerning (ii), we only experimented with sets of size 1, i.e., each time we trained a tree at layer 2 of 2LDT, we used information only from one parameter setting of *windowsize* and *stepsize*. We varied both *windowsize* and *stepsize* across experiments. We tried *windowsize* = 30 because our first layer trees performed well on non-TM domain segments of length 30, see Sect. 4. However, note that T_{full}, T_{level2}, and T_{5-} all check for a negative motif in the root (symbols that are common in protein parts that are *not* TM domains). Such negative motifs often occur close to the TM domain boundary; see the segment to the left of the TM domain in Fig. 3 for illustration. Therefore the risk of false negatives at layer 1 may be very high if the window size is too large. Hence we also tested *windowsize* = 20. We report results on *stepsize* values of 1, 2, 3, and 5.

Concerning (iii), we used as features in all our experiments

- the number of hits at layer 1, i.e., the number of times the tree at layer 1 classifies a substring of the given string as a TM domain,
- the number of 'blocks' of predicted TM domains in the given string,
- the length of the protein,

To explain the notion of a 'block' of predicted TM domains, consider the sequence of predicitons the tree trained at layer 1 makes when using 2LDT. Assume the tree at layer 1 predicts 'non-TM' for the first two windows, then 'TM' at the next window, then 'non-TM' at the next window, then eight times 'TM' and afterwards only 'non-TM'. Then the protein has two blocks of predicted TM domains, namely one of length 1 and one of length 8. So, a block is any largest consecutive sequence of TM domain prediciitions made at layer 1.

A straightforward idea would be to classify a protein as TM if ever a TM domain is predicted on any of its substrings of a certain length, independent of the number of blocks or of the length of the protein. However, a small percentage of false positives predicted at layer 1 would then result in a large number of false positives at layer 2. This is one reason for our broader choice of features.

5.1   Experimental Setup

The first set of training and test examples is from UniProt/Swiss-Prot (Release 2014-09). To minimize overlap with the data used at layer 1, we took the proteins in positions 400,001 through 450,000. As in layer 1, we removed duplicates and sequences containing any of the letters B, J, O, U, X, Z. Every protein containing at least one TM domain was considered a TM protein, all others were considered non-TM proteins. We randomly picked 100 positive and 100 negative examples for training. We used two test data sets. To allow for a comparison with PRED-CLASS [7], a neural network method, our first test set was a subset of the test data used in the original PRED-CLASS study (we had to remove 83 from a total of 387 original test examples, since their IDs were either not available or obsolete). This test set contains 140 TM proteins and 164 non-TM proteins. The second test set consists of 400 positive examples and 400 negative examples randomly selected from the 50,000 proteins described above, disjoint from the training data set (referred to as UniProt test set below; the corresponding protein IDs are listed at http://www2.cs.uregina.ca/~zilles/proteinIDs.txt).

As in the case of TM domain detection, we report TPR and TNR values.

5.2   Results

One important observation is that even minor changes in the performance of the layer 1 trees for TM domain detection may drastically affect their usefulness when deployed in 2LDT. This is best illustrated when comparing the results obtained using $\mathcal{T}_{level2}$ to those obtained using $\mathcal{T}_{5-}$. Both trees show an excellent performance at TM domain detection: TPR/TNR on data of length category '30' are 96.3 %/98.4 % for $\mathcal{T}_{level2}$ and 98.4 %/97.7 % for $\mathcal{T}_{5-}$. However, when testing 2LDT on the set from the PRED-CLASS study, $\mathcal{T}_{level2}$ turns out to be much more useful than $\mathcal{T}_{5-}$, see Table 3. While $\mathcal{T}_{level2}$ yields an acceptable classification performance of 2LDT, $\mathcal{T}_{5-}$ results in TNR values of around 31 % for 2LDT (for $windowsize = 30$ and $stepsize \in \{1, 2, 3\}$). Intuitively, this is due to the fact that $\mathcal{T}_{level2}$ tests for both a negative motif and a positive motif: it classifies a segment as TM domain if it contains not too many occurrences of '−' and at the same time at least a certain fraction of occurrences of '*'. As opposed to that, $\mathcal{T}_{5-}$ already classifies a segment as TM domain if it does not contain too many occurrences of '−', i.e., it tests only for a negative motif. The slightly higher sensitivity of $\mathcal{T}_{5-}$ in TM then translates into a noticeably lower TNR in 2LDT.

Note though that the TNR values for 2LDT using $\mathcal{T}_{5-}$ increase drastically (to 87 %–90 %) on the UniProt data set, while all other values stay comparable to those on the PRED-CLASS data, see Table 3. This suggests that the hand-selected data in the PRED-CLASS set may not be a representative sample.

To show the impact of the $windowsize$ parameter, we tested 2LDT with $stepsize = 2$ using $\mathcal{T}_{full}$ on the PRED-CLASS data, for $windowsize = 30$ and $windowsize = 20$. The TPR/TNR values obtained were 87.1 %/89.0 % for $windowsize = 30$ and 96.4 %/85.4 % for $windowsize = 20$. On the UniProt data, TPR and TNR both varied by about 4 % in the same experiment, where again TPR was lower for $windowsize = 30$. As explained above, the lower TPR value for

Table 3. Results (in percent, rounded) from 2LDT using two different trees at layer 1. All results were obtained with $windowsize = 30$, either on the PRED-CLASS test set ('PRED') or the UniProt test set ('UniP').

		$stepsize = 1$		$stepsize = 2$		$stepsize = 3$	
		PRED	UniP	PRED	UniP	PRED	UniP
$\mathcal{T}_{level2}$	TPR	75.7	78.0	80.7	79.3	88.6	88.3
	TNR	95.7	92.8	92.7	93.25	89.0	93.0
$\mathcal{T}_{5-}$	TPR	86.4	88.8	83.6	87.0	85.7	86.0
	TNR	32.9	87.0	31.1	90.0	30.5	90.0

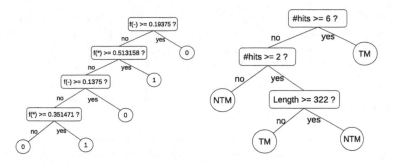

Fig. 4. The left tree was trained at layer 1 and then used to train the right tree at layer 2 with *windowsize* = 20 and *stepsize* = 5. By f(s) we refer to the fraction of occurrences of symbol s in a sequence. #hits is the number of hits at layer 1, and Length is the length of the protein. The resulting classifier labels a sequence as a TM protein if the tree on the left classifies at least 6 windows as TM domains, or if it classifies at least 2 as TM domains and the protein has fewer than 322 amino acids.

windowsize = 30 may be due to the fact that the dominant negative motif (a large number of '−' symbols outside of a TM domain) often occurs close to a TM domain boundary.

Finally, we report the best result we obtained in our 2LDT experiments on the UniProt test set. At layer 1, we trained a tree as follows. To generate positive examples, we used full TM domains of length up to 20. If a TM domain was longer than 20, we selected a random substring of length 20 from that sequence. Similarly, we created negative examples starting from the set Neg-Combined(30). We then randomly picked 10,000 positive and 10,000 negative examples for training. The resulting tree was pruned to level 3 and used at layer 1. (This tree achieved TPR/TNR values of 94.3 %/97.2 % for TM domain detection on the remaining 380,000 test samples. Again it checks for the dominant negative motif, fraction of '−', and the dominant positive motif, fraction of '∗'; this time each of these is checked twice with different thresholds, see Fig. 4, left.) At layer 2, we used *windowsize* = 20 and *stepsize* = 5, to obtain TPR/TNR values of 93.5 %/90.8 %. Firstly, this classifier performs almost as well on randomly chosen data as PRED-CLASS does on carefully selected data (with values around 96 %). Secondly, it beats PRED-CLASS in that it provides intuitive rules for classification, see Fig. 4, right. Note that this particular tree does not use the 'blocks' feature, but several of the other trees we trained do.

6   Conclusions

We presented 2LDT, a learning method for classifying proteins into TM and non-TM proteins that is potentially applicable to a large variety of string classification problems. For detecting TM proteins, it yields easy-to-interpret decision trees, while performing nearly as good as the best existing classifiers in terms of TPR/TNR. Concerning TM domain classification, we obtained a number of deep insights that have to date not been discussed in the literature. In particular, as opposed to existing machine learning approaches, our method found a positive motif of TM domains that is known to be important for TM protein detection, namely the frequent occurrence of amino acids from the set $\{A, C, F, I, L, M, V\}$.

References

1. Arikawa, S., Miyano, S., Shinohara, A., Kuhara, S., Mukouchi, Y., Shinohara, T.: A machine discovery from amino acid sequences by decision trees over regular patterns. New Gener. Comput. **11**, 361–375 (1993)
2. Chen, C.P., Kernytsky, A., Rost, B.: Transmembrane helix predictions revisited. Protein Sci. **12**, 2774–2791 (2002)
3. Flores-Mireles, A.L., Walker, J.N., Caparon, M., Hultgren, S.J.: Urinary tract infections: epidemiology, mechanisms of infection and treatment options. Nat. Rev. Microbiol. **13**, 269–284 (2015)
4. Fruh, V., Zhou, Y., Chen, D., Loch, C., Ab, E., Grinkova, Y.N., Verheij, H., Sligar, S.G., Bushweller, J.H., Siegal, G.: Application of fragment-based drug discovery to membrane proteins: identification of ligands of the integral membrane enzyme DsbB. Chem. Biol. **17**, 881–891 (2010)
5. He, J., Hu, H., Harrison, R., Tai, P.C., Pan, Y.: Transmembrane segments prediction and understanding using support vector machine and decision tree. Expert Syst. Appl. **30**, 64–72 (2006)
6. Kyte, J., Doolittle, R.F.: A simple method for displaying the hydropathic character of a protein. J. Mol. Biol. **157**, 105–132 (1982)
7. Pasquier, C., Promponas, V.J., Hamodrakas, S.J.: PRED-CLASS: cascading neural networks for generalized protein classification and genome-wide applications. Proteins **44**, 361–369 (2001)
8. Pasquier, C., Promponas, V.J., Palaios, G.A., Hamodrakas, J.S., Hamodrakas, S.J.: A novel method for predicting transmembrane segments in proteins based on a statistical analysis of the SwissProt database: the PRED-TMR algorithm. Protein Eng. **12**, 381–385 (1999)
9. Ramasarma, T., Joshi, N.V., Sekar, K., Uthayakumar, M., Sherlin, D.: Transmembrane domains. Wiley, In Encyclopedia of Life Sciences (2012)
10. Ribet, D., Cossart, P.: How bacterial pathogens colonize their hosts and invade deeper tissues. Microbes Infect. Inst. Pasteur **17**, 173–183 (2015)
11. Senes, A., Gerstein, M., Engelman, D.M.: Statistical analysis of amino acid patterns in transmembrane helices: the GxxxG motif occurs frequently and in association with beta-branched residues at neighboring positions. J. Mol. Biol. **296**, 921–936 (2000)
12. Tusnady, G.E., Simon, I.: Principles governing amino acid composition of integral membrane proteins: application to topology prediction. J. Mol. Biol. **283**, 489–506 (1998)

Change Point Detection for Information Diffusion Tree

Kouzou Ohara[1]([⊠]), Kazumi Saito[2], Masahiro Kimura[3], and Hiroshi Motoda[4,5]

[1] Department of Integrated Information Technology,
Aoyama Gakuin University, Kanagawa, Japan
`ohara@it.aoyama.ac.jp`
[2] School of Administration and Informatics, University of Shizuoka, Shizuoka, Japan
`k-saito@u-shizuoka-ken.ac.jp`
[3] Department of Electronics and Informatics, Ryukoku University, Shiga, Japan
`kimura@rins.ryukoku.ac.jp`
[4] Institute of Scientific and Industrial Research, Osaka University, Osaka, Japan
`motoda@ar.sanken.osaka-u.ac.jp`
[5] School of Computing and Information Systems,
University of Tasmania, Hobart, Australia
`hmotoda@utas.edu.au`

Abstract. We propose a method of detecting the points at which the speed of information diffusion changed from an observed diffusion sequence data over a social network, explicitly taking the network structure into account. Thus, change in diffusion is both spatial and temporal. This is different from most of the existing change detection approaches in which all the diffusion information is projected on a single time line and the search is made in this time axis. We formulate this as a search problem of change points and their respective change rates under the framework of maximum log-likelihood embedded in MDL. Time complexity of the search is almost proportional to the number of observed data points and the method is very efficient. We tested this using both a real Twitter date (ground truth not known) and the synthetic data (ground truth known), and demonstrated that the proposed method can detect the change points efficiently and the results are very different from the existing sequence-based (time axis) approach (Kleinberg's method).

Keywords: Social networks · Information diffusion · Change point detection

1 Introduction

Recent technological innovation and popularization of high performance mobile/ smart phones has drastically changed our communication style and the use of various social media such as Twitter[1] and Facebook[2] has been substantially

---

[1] https://twitter.com/.
[2] https://www.facebook.com/.

© Springer International Publishing Switzerland 2015
N. Japkowicz and S. Matwin (Eds.): DS 2015, LNAI 9356, pp. 161–169, 2015.
DOI: 10.1007/978-3-319-24282-8_14

affecting our daily lives. It is fresh to our memory that Twitter played a very important role as the information infrastructure during the recent natural disaster, both domestic and abroad, including the 2011 To-hoku earthquake and tsunami in Japan.

In reality, the way information diffuses depends on both the content and the interest of the people. Being able to detect changes in the way information propagates allows us to analyze peoples behavior, e.g. finding a community of people with a similar interest, and deepens our understanding of the world around us. This brings in an important and interesting problem, which is to accurately and efficiently detect the change points (where in the network the changes take place and how big the respective changes in the diffusion speed are) from the observed information diffusion data.

There are substantial number of studies on change detection in information diffusion process. Most of them treat change detection along the time axis alone in which all the diffusion information is projected on a single time line and the detection is formulated as a search problem in this time axis. These include [1–3, 7–9]. We have also approached this problem by directly dealing with the change of time interval between occurrences of a target event [6], and showed that our method outperformed Kleinberg's method [3] which is considered to be the state of the art. However, in reality information diffusion takes place along a diffusion path. Each path has multiple descendants (child nodes) and new paths start only from the children that are in the observed data. Thus, change in diffusion is both spatial and temporal. The above traditional sequence-based (time axis) approaches may be good enough to know a global trend over a long period of time, but is definitely not good enough to detect the correct change points. Information diffuses differently within different communities just as the sound velocity changes within different substances. Thus it is important to take both spatial and temporal factors into account in detecting changes, *i.e.*, where and when the change takes place.

We model these changes as changes in the time-delay parameter, where the delay is assumed to follow an exponential distribution. More precisely, we assume that the parameter changes are approximated by a step function along each diffusion path and propose an optimization algorithm that maximizes the likelihood of generating the observed diffusion sequence, and the number of change points are determined by MDL principle. The time complexity of the algorithm is almost proportional to the number of observed data points (candidates of possible change points).

We first demonstrate that the proposed method can detect the bursts using a real Twitter data quite efficiently. The results were very different from Kleinberg's method [3] which is considered to be the state of the art for burst detection along the time axis. This confirmed the need to explicitly use the network structure. Since we do not know the ground truth for the Twitter data, we generated synthetic data and embedded the change points of varying number using the same network structure with the Twitter data. The proposed method could successfully detect the correct change points for all cases with one very minor mis-detection, while Kleinberg's method again performed very poorly and the detected many incorrect change points.

2   Proposed Method

We consider information diffusion over a social network whose structure is defined as a directed graph $G = (V, E)$, where V and E $(\subset V \times V)$ represent a set of all nodes and a set of all links, respectively. Suppose that we observe a sequence of information diffusion $\mathcal{C} = \{(v_0, t_0), (v_1, t_1), \cdots, (v_N, t_N)\}$ that arose from the information released at the source node v_0 at time t_0. Here, v_n is an *active* node where the information has been propagated and t_n is its time. We assume, as a standard setting, that the actual information diffusion paths of a sequence $\mathcal{C}$ correspond to a tree $T_\mathcal{C}$ that is embedded in the directed graph G representing the social network [5], i.e., the parent node which passed the information to a node v_n is uniquely identified to be $v_{p(n)}$ if $n > 0$. Here, $p(n)$ is a function that returns the node identification number of the parent of v_n in the range of $\{0, \cdots, n-1\}$.

By setting that the time delay of information diffusion is represented as the simple exponential distribution $p(t_n - t_{p(n)}; r) = r \exp(-r(t_n - t_{p(n)}))$, we mathematically define the change point detection problem. For the actual information diffusion paths of a sequence $\mathcal{C}$, we consider the corresponding set of integers defined by $\mathcal{D} = \{0, 1, \cdots, N\}$. Let the node of the j-th change point be $n(j) \in \mathcal{D}$, then we assume that the delay parameter switches from r_j to r_{j+1} for the descendant nodes of $v_{n(j)}$ until another change took place. Namely, we are assuming a step function as a shape of parameter changes. Let the set comprising J change points be $\mathcal{S}_J = \{n(1), \cdots, n(J)\}$, and we set $n(0) = 0$ for the sake of convenience $(t_{n(j-1)} < t_{n(j)})$. Let the division of $\mathcal{D}$ by $\mathcal{S}_J$ be $\mathcal{D}_j$, i.e., $\mathcal{D} = \mathcal{D}_0 \cup \mathcal{D}_1 \cup \cdots \cup \mathcal{D}_J$, where $\mathcal{D}_j$ is a set of the descendant nodes of $v_{n(j)}$ until another change happens, and $|\mathcal{D}_j|$ represents the number of observed points in $\mathcal{D}_j$. Here, we request that $|\mathcal{D}_j| \neq 0$ for any $j \in \{0, \cdots, J\}$.

We consider the problem of detecting change points as a problem of finding a subset $\mathcal{S}_J \subset \mathcal{D}$ when the set of nodes of information diffusion result $\mathcal{C}$ is given. For this purpose, we consider maximizing the following objective function.

$$L(\mathcal{C}; \hat{r}_{J+1}, \mathcal{S}_J) = -N - \sum_{j=0}^{J} |\mathcal{D}_j| \log \left(\frac{1}{|\mathcal{D}_j|} \sum_{n \in \mathcal{D}_j} (t_n - t_{p(n)}) \right). \tag{1}$$

Here, as shown in [6], we can obtain this objective function by substituting the maximum likelihood estimate of the parameter $\hat{r}_{J+1}$ to the log-likelihood for $\mathcal{C}$ for a given set of change points $\mathcal{S}_J$. We first describe the simple method which is applicable when the number of change points J is large. This is a progressive binary splitting without backtracking. Below we describe the details of this algorithm **A**: after initializing $j \leftarrow 1$ and $\mathcal{S}_0 \leftarrow \emptyset$ (step **A1**), we fix the already selected set of $(j - 1)$ change points $\mathcal{S}_{j-1}$ and search for the optimal j-th change point $n(j)$ (step **A2**), and add it to $\mathcal{S}_{j-1}$ (step **A3**). We repeat this procedure from $j = 1$ to J. Here note that in the step **A3** elements of the change point set S_j are reindexed to satisfy $t_{n(i-1)} < t_{n(i)}$ for $i = 2, \cdots, j$. Clearly, the time complexity of the simple method is $O(NJ)$ which is fast. Thus, it is possible to obtain the result within a reasonable computation time for a large N. However, since this is a greedy algorithm, it can be trapped easily to a poor local optimal.

By inheriting the basic idea of our previous method [6], we propose a method which is computationally almost equivalent to the simple method but gives a solution of much better quality. Below we describe the details of this algorithm **B**: We start with the solution obtained by the simple method S_J (step **B1**), pick up a change point $n(j)$ from the already selected points, fix the rest $S_J \setminus \{n(j)\}$ and search for a better value $n(j)'$ (step **B2**), where $\cdot \setminus \cdot$ represents set difference. We repeat this from $j = 1$ to J. If no replacement is possible for all j ($j = 1, \cdots J$), i.e. $n(j)' = n(j)$ for all j, then no better solution is expected and the iteration stops.

So far, we have fixed the number of change points J, and proposed a method of finding the optimal parameter vector $\hat{\mathbf{r}}_{J+1}$ and inferring the change points S_J for the observed data C. Now, we present a method of estimating the value of J from C for solving the change points detection problem. To this end, we employ MDL (Rissanen's Minimum Description Length) [4]. More specifically, in order to describe the information diffusion model based on the obtained result S_J, we need the set of $J + 1$ time-delay parameters $\hat{\mathbf{r}}_{J+1}$, as well as the set of J change points S_J, which amounts to $2J + 1$ parameters. Thus we can consider the following MDL formula for the case of J change points:

$$ MDL(J) = -L(C; \hat{\mathbf{r}}_J, S_{J+1}) + \frac{1}{2}(2J + 1)\log(N). \tag{2} $$

Below we describe the details of this algorithm **C**: after initializing $J \leftarrow 0$ and $S_0 \leftarrow \emptyset$ (step **C1**), we compute S_{J+1} by the proposed algorithms **A** and **B**, and Calculate $MDL(J + 1)$ from Eq. (2) (step **C2**). We repeat this procedure from $J = 0$ by setting $J \leftarrow J + 1$ while $MDL(J + 1) \leq MDL(J)$. Here, we note that for model selection, we can consider employing various methods other than the MDL criterion and the likelihood ratio test, although we used the MDL criterion as a candidate.

3   Experiments

We applied the proposed method to the real-world information diffusion sequence which takes a form of tree and investigated how it can detect reasonable change points on the tree by visualizing the resulting change points and corresponding time-delay parameters estimated by it. To this end, we used a sequence of retweets extracted from Twitter[3], and formed a corresponding diffusion tree that has 477 nodes (tweets) and 476 edges (retweet actions). We refer to this dataset as the Retweet dataset.

---

[3] https://twitter.com/.

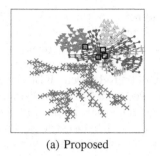
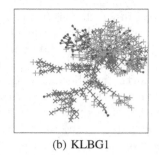

(a) Proposed   (b) KLBG1

Fig. 1. Visualization of changes of diffusion time on the information diffusion tree by the proposed and KLBG1 methods (Color figure online).

3.1   Results for Real Data

We applied our proposed method to the Retweet dataset, and obtained the result that the number of change points underlying in the tree is 4. Actually, the log-likelihoods for $J = 4$ and 5 are -3359.6 and -3353.9, respectively, and the corresponding MDL values are 3387.4 and 3387.9, respectively. We can observe that those values do not change significantly between $J = 4$ and 5, but it does not hold if J is smaller. Figure 1(a) visualizes the result for $J = 4$, in which nodes of the diffusion tree are denoted by different colors and different markers according to the estimated time-delay parameter values associated to them and the four change points detected are indicated with squares.

From these results, we can find that the given diffusion tree is clearly divided into 5 subtrees which have a certain number of nodes and whose root nodes are either the root node of the whole tree or change points detected by the proposed method. In addition, it can be observed that the diffusion speed clearly changes between different subtrees. Thus, these subtrees are likely to be considered as different communities in which information diffusion speed of a certain topic is different. Analyzing these subtrees more in depth is one of the future directions of this work.

Next, we compared the proposed method with conventional sequence-based methods [3,6] that detect change points by considering only a time series diffusion sequence without using any structural information of the network behind the diffusion. In this paper, we chose Kleinberg's method [3] as a representative one among them. It is based on hidden Markov model and has two parameters, γ and s. The parameter γ is used in its cost function, and we employed $\gamma = 1$ in this experiment. The parameter s is a scaling parameter and determines the delay parameter at the state j by $r_j = s^j r_0$ where the parameter r_0 is estimated by $r_0 = N/t_N$ as described in [3]. We set the scaling parameter s to 5 based on the observations obtained by applying our proposed method to the original dataset. Hereafter, we refer to Kleinberg's method with this setting as the KLBG1 method. In addition, we consider an alternative Kleinberg's method with another setting in which r_0 is fixed to 1.0, and refer to it as the KLBG2 method, which is used only for the experiments on the synthetic datasets discussed below.

Figure 1(b) shows, in the same manner as in Fig. 1(a), the result obtained by applying the KLBG1 method to the Retweet dataset. Comparing to Fig. 1(a), it is found that the number of change points detected by the KLBG1 method is substantially larger than the one by the proposed method. In addition, there are multiple small subtrees with an identical time delay parameter and they spread across a wide range of the diffusion tree. This is because the sequence-based methods use only a sequence of time stamps projected on a single time axis and do not take into account any structural information behind the diffusion process. Consequently, we cannot utilize this result to extract meaningful node groups or communities that could affect the information diffusion speed, which is possible by the proposed method.

3.2 Results for Synthetic Data

We constructed a synthetic sequence of information diffusion by utilizing the Retweet dataset. More specifically, to systematically regenerate the observation time points in which J change points are embedded, we divided $\mathcal{D}$ of the Retweet dataset into $J+1$ subsets $\mathcal{D}_0, \cdots, \mathcal{D}_J$ so that the original diffusion tree is decomposed into $J + 1$ subtrees each of which has at least 20 nodes. Then, we set the time-delay parameter r_j to 1.0 for $j = 0$ and $5 \times r_{pt(j)}$ for $j > 0$, where $pt(j)$ means the index such that $p(n(j)) \in \mathcal{D}_{pt(j)}$. It is noted that this coefficient of 5 is equivalent to the value of the scaling parameter of the KLBG1 and KLBG2 methods. After that, we generated observation time of nodes in each $\mathcal{D}_j$ according to the exponential distribution with the parameter r_j, varying J from 1 to 5, and generated 10 different datasets for each value of J.

To quantitatively evaluate the proposed method, we applied the proposed, KLBG1, and KLBG2 methods to the synthetic datasets, and compared their learning performance in terms of two criteria: the number of detected change points and the estimation error of the time-delay parameter. For each value of J, we applied each method to the 10 different synthetic datasets, each embedded with J change points, and computed an average over these 10 trials for each criterion. Figure 2(a) shows the number of change points detected by each method. It is obvious that the proposed method can almost exactly detect the number of embedded change points regardless of the value of J. In contrast, both the KLBG1 and KLBG2 methods overestimated the number of embedded change points. The KLBG2 method detected much more change points than the KLBG1 method did although the KLBG2 method used the true value of r_0 in addition to the true scale parameter s that was available for the KLBG1 method.

Next, we investigated the error E between the estimated time-delay parameter and the true one, defined as $E = N^{-1} \sum_{n=1}^{N} |\hat{r}(n) - r(n)|$, where $\hat{r}(n)$ is a parameter value that is estimated to have generated the time delay $t_n - t_{p(n)}$, and $r(n)$ is its true value. Since both the KLBG1 and KLBG2 methods do not use any structural information of the diffusion tree, we defined $p(n)$ as $p(n) = \arg\max_{m \in D}\{t_m | t_m < t_n\}$ for these two methods so that E gets to a

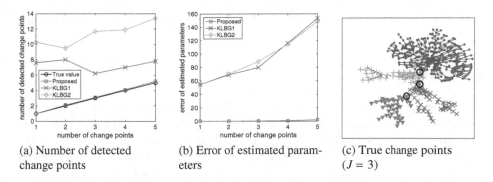

(a) Number of detected change points

(b) Error of estimated parameters

(c) True change points ($J = 3$)

Fig. 2. Learning performance by the proposed, KLBG1, and KLBG2 methods for $J = 1$ to 5.

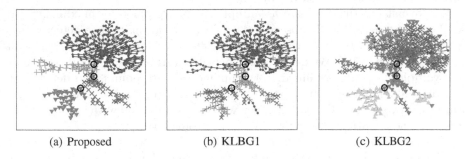

(a) Proposed

(b) KLBG1

(c) KLBG2

Fig. 3. Visualization results using the true change points and those estimated by the proposed, KLBG1, and KLBG2 methods for a synthetic dataset having $J = 3$ change points.

small value if they exactly detect change points and estimate the corresponding parameter values within small deviations. The results for each value of J are shown in Fig. 2(b), from which it is clear that the proposed method achieved extremely small errors, and thus can accurately estimate the parameter value for any value of J. On the other hand, the errors for the KLBG1 and KLBG2 methods are extremely large and increase in proportion to the number of embedded change points J.

Figure 3 visualizes, in the way similar to the case of Fig. 1(a), results for a synthetic dataset in which $J = 3$ change points were embedded. Figures 3(a) to (c) show the results of the proposed, KLBG1, and KLBG2 methods, respectively. The same 3 true change points illustrated by circles in Fig. 2(c) were used for the three methods. Comparing Figs. 2(c) and 3(a), we can see that the proposed method almost exactly detected the 3 true change points in the tree. In contrast, from Figs. 3(b) and (c), we see that they are much different from Fig. 2(c), and the diffusion speed changes at many nodes other than the true change points. The KLBG1 method is slightly better than the KLBG2 method, but the number of states it detected is 3 that is one less than the true value 4. The KLBG2 method that uses the true value of r_0 detected 5 states and there are many more change points than the true ones.

4   Conclusion

We addressed the problem of detecting the points at which the speed of information diffusion changed from a single observed diffusion sequence under the assumption that the delay of the information propagation follows the exponential distribution. Most of the existing change detection methods focus on changes in the time axis, ignoring the path along which information diffuses within the network. The proposed method is different and unique in that it explicitly takes the underlying network structure into account. It can deal with both spatial and temporal changes in information diffusion.

We formulated this problem as an optimization problem of maximizing the likelihood of generating the observed data. In doing so the change detected at a node is passed only to its descendants, and different information diffusion paths are handled in parallel. We devised an efficient iterative search algorithm whose time complexity is almost linear to the number of data points, and determined the optimal number of change points using MDL criterion. We tested the algorithm against the real Twitter data for which we do not know the ground truth and a synthetic data for which we know the ground truth.

The results for the real Twitter data revealed that the proposed method can detect change points efficiently. We also tested the other method that does not use the network structure data, choosing Kleinberg's burst detection method as one of the representative methods of this kind. The results were very different, which confirmed the need to explicitly use the network structure. The results for the synthetic data reveled that the proposed method could successfully detect the correct change points for all cases with one very minor mis-detection, while Kleinberg's method again performed very poorly and the detected many incorrect change points.

Acknowledgments. This work was partly supported by Asian Office of Aerospace Research and Development, Air Force Office of Scientific Research under Grant No. AOARD-13-4042, and JSPS Grant-in-Aid for Scientific Research (C) (No. 26330261).

References

1. Araujo, L., Cuesta, J.A., Merelo, J.J.: Genetic algorithm for burst detection and activity tracking in event streams. In: Runarsson, T.P., Beyer, H.-G., Burke, E.K., Merelo-Guervós, J.J., Whitley, L.D., Yao, X. (eds.) PPSN 2006. LNCS, vol. 4193, pp. 302–311. Springer, Heidelberg (2006)
2. Ebina, R., Nakamura, K., Oyanagi, S.: A real-time burst detection method. In: Proceedings of the 23rd IEEE International Conference on Tools with Artificial Intelligence (ICTAI), pp. 1040–1046 (2011)

3. Kleinberg, J.: Bursty and hierarchical structure in streams. In: Proceedings of the 8th ACM SIGKDD International Conference on Knowledge Discovery and Data Mining (KDD-2002), pp. 91–101 (2002)
4. Rissanen, J.: Stochastic Complexity in Statistical Inquiry. World Scientific, Singapore (1989)
5. Sadikov, E., Medina, M., Leskovec, J., Garcia-Molina, H.: Correcting for missing data in information cascades. In: Proceedings of the 4th ACM International Conference on Web Search and Data Mining (WSDM 2011), pp. 55–64 (2011)
6. Saito, K., Ohara, K., Kimura, M., Motoda, H.: Change point detection for burst analysis from an observed information diffusion sequence of tweets. J. Intel. Inf. Syst. (JIIS) **44**, 243–269 (2015)
7. Sun, A., Zeng, D., Chen, H.: Burst detection from multiple data streams: a network-based approach. In: IEEE Transactions on Systems, Man, and Cybernetics Society, Part C, pp. 258–267 (2010)
8. Zhang, X.: Fast algorithms for burst detection. Ph.D. Dissertation (New York University) (2006)
9. Zhu, Y., Shasha, D.: Efficient elastic burst detection in data streams. In: Proceedings of the 9th ACM SIGKDD International Conference on Knowledge Discovery and Data Mining (KDD-2003), pp. 336–345 (2003)

Multi-label Classification via Multi-target Regression on Data Streams

Aljaž Osojnik[1,2]([⊠]), Panče Panov[1], and Sašo Džeroski[1,2,3]

[1] Jožef Stefan Institute, Jamova cesta 39, Ljubljana, Slovenia
{aljaz.osojnik,pance.panov,saso.dzeroski}@ijs.si
[2] Jožef Stefan IPS, Jamova cesta 39, Ljubljana, Slovenia
[3] CIPKeBiP, Jamova Cesta 39, Ljubljana, Slovenia

Abstract. Multi-label classification is becoming more and more critical in data mining applications. Many efficient methods exist in the classical batch setting, however, in the streaming setting, comparatively few methods exist. In this paper, we propose a new methodology for multi-label classification via multi-target regression in a streaming setting and develop a streaming multi-target regressor iSOUP-Tree, which uses this approach. We experimentally evaluated two variants of the iSOUP-Tree algorithm, and determined that the use of regression trees is advisable over the use model trees. Furthermore, we compared our results to the state-of-the-art and found that the iSOUP-Tree method is comparable to the other streaming multi-label learners. This is a motivation for the potential use of iSOUP-Tree in an ensemble setting as a base learner.

1 Introduction

In recent years, the task of multi-label classification has been very prominent in the data mining research community [8]. It can be seen as a generalization of the ubiquitous multi-class classification task, where instead of a single label, each example is associated with multiple labels. This is one of the reasons why multi-label classification is the go-to approach when it comes to automatic annotation of media, such as images, texts or videos, with tags or genres.

Most research into multi-label classification has been in the batch context, however, strides have also been made to explore multi-label classification in the streaming setting [4,14,16]. The tendency of big data is clear and present in the research community, as well as in the real world. With an appropriate method, the streaming context allows for real-time analysis of large amounts of data, e.g., emails, blogs, RSS feeds, social networks, etc.

However, due to the nature of the streaming setting, there are several constraints that need to be considered. A data stream is potentially infinite sequence of examples, which needs to be analyzed with finite resources, in particular, in finite time and memory. The largest point of divergence from the batch setting is the fact that the underlying concept we are trying to learn, can change at any point. Therefore, the algorithm design is often divided into two parts: (1) learning the stationary concept, and (2) detecting and adapting to it's changes.

© Springer International Publishing Switzerland 2015
N. Japkowicz and S. Matwin (Eds.): DS 2015, LNAI 9356, pp. 170–185, 2015.
DOI: 10.1007/978-3-319-24282-8_15

In this paper, we focus on a method for multi-label classification in the streaming context that learns the stationary concept.

Many algorithms in the literature take the problem transformation approach to multi-label classification, both in the batch and the streaming setting. They transform the multi-label classification problem into several problems that can be solved with off-the-shelf methods, e.g., transformation into an array of binary classification problems. With this transformation, the label inter-correlations can be lost, and, consequently, the predictive performance can decrease.

In this paper, we take a different transformation approach and transform the multi-label classification problem into a multi-target regression problem. Multi-target regression is a generalization of single-target regression, i.e., it is used to predict multiple continuous variables. Many facets of the multi-label classification are also expressed in multi-target regression, e.g., the correlation between labels/variables, which motivated us to experiment with multi-label classification by using multi-target regression methods.

To address the multi-label classification task, we have developed a straightforward multi-label classification via multi-target regression methodology, and used it in a combination with a streaming multi-target regressor (iSOUP-Tree). The generality of this approach is paramount as it allows us to address multiple types of structured output prediction problems, such as multi-label classification and hierarchical multi-label classification, in the streaming setting. In this paper, we show that this approach is a viable candidate for the multi-label classification task on data streams. Furthermore, we explore the multi-target regressor in detail to determine which internal methodology is most appropriate for the task at hand. Finally, we perform comparisons with state-of-the-art methods for multi-label classification in the streaming setting.

The structure of the paper is as follows. First, we present the background and related work (Sect. 2). Next, we present the task of multi-label classification via multi-target regression on data streams (Sect. 3). Furthermore, we present the research questions and the experimental design (Sect. 4). Finally, we conclude with the discussion of the results (Sect. 5), conclusions, and further work (Sect. 6).

2   Background and Related Work

In this section, we review the state-of-the art in multi-label classification, both in the batch and the streaming context. In addition, we present the background of the multi-target regression task, which we use as a foundation for defining the multi-label classification via multi target regression approach.

2.1   Multi-label Classification Task

Stemming from the usual multi-class classification, where only one of the possible labels needs to be predicted, the task of *multi-label classification* (MLC) requires a model to predict a combination of the possible labels. Formally, this means that for each data instance x from an input space X a model needs to provide a prediction $\hat{y}$ from an output space Y, which is constructed as a powerset of the labelset $\mathcal{L}$, i.e., $Y = 2^{\mathcal{L}}$. This is in contrast to the multi-class classification task, where the output space is simply the labelset $Y = \mathcal{L}$. We denote the real labels of an instance x by y, and a prediction made by a model for x by $\hat{y}(x)$ (or $\hat{y}$).

In the batch setting, the problem transformation approach is commonly used to tackle the task of multi-label classification. Problem transformation methods are usually used as basic methods to compare to, and are used in a combination with off-the-shelf base algorithms. The most common approach, called *binary relevance* (BR), transforms a multi-label task into several binary classification tasks, one for each of the possible labels [17]. Binary relevance models have been often overlooked due to their inability to account for label correlations, though some BR methods are capable of modeling label correlations during classification.

Another common problem transformation approach is the *label combination* or *label powerset* (LC), where each subset of the labelset is considered as an atomic label for a multi-class classification problem [18,26]. If we start with a multi-label classification task with a labelset of $\mathcal{L}$, we transform this into a multi-class classification with a labelset $\mathcal{L}' = 2^{\mathcal{L}}$.

Third most common problem transformation approach is the *pairwise clasification*, where we have a binary model for each possible pair of labels [7]. This method performs well in some contexts, but for larger problems the method becomes intractable because of model complexity.

In addition to problem transformation methods, there are also adaptations of the well known algorithms that handle the task of multi-label classification directly. Examples of such algorithms are the adaptation of the decision tree learning algorithm for MLC [27], support-vector machines for MLC [9], k-nearest neighbours for MLC [28], instance based learning for MLC [5], and others.

2.2 Multi-label Classification on Data Streams

Many of the problem transformation methods for the multi-label classification task have also been used in the streaming context. Unlike the batch context, where a fixed and complete dataset is given as input to a learning algorithm, the streaming context presents several limitations that the stream learning algorithm must take into account. The most relevant are [2]: (1) the examples arrive sequentially; (2) there can potentially be infinitely many examples; (3) the distribution of examples need not be stationary; and (4) after an example is processed it is discarded or archived. The fact that the distribution of examples is not presumed to be stationary means that algorithms should be able to detect and adapt to changes in the distribution (*concept drift*). This sub-problem is called *drift detection*.

The first approach to MLC in data streams was a batch-incremental method that trains stacked BR classifiers [14]. Some methods for multi-class classification, such as Hoeffding Trees (HT) [6], have also been adapted to the multi-label classification task [16]. Hoeffding trees are incremental anytime decision trees for learning from data streams that use the notion that a small sample is usually sufficient for choosing an optimal splitting attribute, i.e., the use of the Hoeffding bound. Bifet et al. [3] also introduced the Java-based Massive Online Analysis (MOA)[1] framework, which also allows for the analysis of concept drift [2] and has become one of the main frameworks for data stream mining. Read et al. [16] proposed the use of multi-label Hoeffding trees with prunned sets (PS) at the leaves (HT$_{PS}$) and Bifet et al. [4] proposed the use of ensemble methods in this context (e.g., ADWIN Bagging).

Recently, Spyromitros et al. [24] introduced a parameterized windowing technique for dealing with the concept drift in multi-label data in a data stream context. Next, Shi et al. [21] proposed an efficient and effective method to detect concept drift based on label grouping and entropy for multi-label data. Finally, Shi et al. [22] proposed an efficient class incremental learning algorithm, which dynamically recognizes some new frequent label combinations.

2.3 Multi-target Regression

In the same way as was multi-label classification adapted from regular classification, we can look at the multi-target regression task as an extension of the single-target regression task. Multi-target regression (MTR) is the task of predicting multiple numeric variables simultaneously, or, formally, the task of making a prediction $\hat{y}$ from $\mathbb{R}^n$, where n is the number of targets for a given instance x from an input space X.

As in multi-label classification, there is a common problem transformation method that transforms the multi-target regression problem into multiple single-target regression problems. In this case, we consider each numeric target separately and train a single-target regressor for each of them. However, this *local* approach suffers from similar problems as the problem transformation approaches to multi-label classification, e.g., in this case the models do not consider the inter-correlations of the target variables. The task of simultaneous prediction of all target variables at the same time (the *global* approach) has been considered in the batch setting by Struyf and Džeroski [25]. In addition, Appice and Džeroski [1] proposed an algorithm for stepwise induction of multi-target model trees.

In the streaming context, some work has been done on multi-target regression. Ikonomovska et al. [13] introduced an instance-incremental streaming tree-based single-target regressor (FIMT-DD), which utilized the Hoeffding bound. This work was later extended to the multi-target regression setting [12] (FIMT-MT). There has been theoretical debate whether the use of the Hoeffding bound is appropriate [19], however, a recent study by Ikonomovska et al. [11] has shown that in practice the use of the Hoeffding bound produces good results. However, these algorithms had the drawback of ignoring nominal input attributes. Additionally, Shaker et al. [20] introduced an instance-based system for classification and regression (IBLStreams), which can be used for multi-target regression.

---

[1] http://moa.cms.waikato.ac.nz/, accessed on 2015/05/25.

3   Multi-label Classification via Multi-target Regression

In this section, we present the task of multi-label classification that is solved by transforming the problem into a multi-target regression setting. First, we present the problem formulation that describes the transformation procedure. Second, we describe the implementation of the proposed approach.

3.1   Problem Formulation

The problem transformation methods (see Sect. 2.1) generally transform a multi-label classification task into one, or several, binary or multi-class classification tasks. In this work, we take a different approach and transform a classification task into a regression task. The simplest example of a transformation of this type is to transform a binary classification task into a regression task. For example, if we have a binary target with labels *yes* and *no*, by transforming to the regression setting, we would consider a numeric target to which we would assign a numeric value of 0 if the binary label is *no* and 1 if the binary label is *yes*.

In the same way, we can approach the multi-class classification task. Specifically, if the multi-class target variable is ordinal, i.e., the class labels have a meaningful ordering, we can assign the numeric values from 0 to n to each of the corresponding n labels. This makes sense, since if the labels are ordered, a missclassification of a label into a "nearby" label is better than into a "distant" label. However, if the variable is not ordinal this makes less sense, as any given label is not in a strict relationship with other labels.

To address the multi-label classification task using regression, we transform it into a multi-target regression task (see Fig. 1). This procedure is done in two steps: first we transform the multi-label classification target variable into several binary classification variables, similar as in the BR method. However, instead of training one classifier for each of the binary variables, we further transform the values of the binary variable into numbers. A numeric target corresponding to a given label has a value 1 if the label is present in a given instance, and a value 0 if the label is not present.

For example, if we have a multi-label task with target labels $\mathcal{L} = \{red, blue, green\}$, we transform it into a multi-target regression task with three numeric target variables $y_{red}, y_{blue}, y_{green} \in \mathbb{R}$. If an instance is labeled with

	Target space		**Instance**
MLC	$y \subseteq \mathcal{L} = \{\lambda_1, \ldots, \lambda_n\}$		$y = \{\lambda_1, \lambda_3, \lambda_4\}$
	$\downarrow$	*transformation*	$\downarrow$
MTR	$y \in \mathbb{R}^n$		$y = (1, 0, 1, 1, \ldots)$

Fig. 1. Transformation from MLC to MTR. Used when the multi-target regressor is learning.

Fig. 2. Transformation from MTR to MLC. Used when transforming a multi-target regression prediction into a mulit-label classification one.

red and green, but not blue, the corresponding numeric targets will have values $y_{red} = 1, y_{blue} = 0$, and $y_{green} = 1$.

Since we are using a regressor, it is possible that a prediction for a given instance will not result in 0 or 1 for each of the targets. For this purpose, we use thresholding to transform back a multi-target regression prediction into a multi-label one (see Fig. 2). Namely, we construct the multi-label prediction in such a way that it contains labels with numeric values over a certain threshold, i.e., in our case, the labels selected are those with a numeric value over 0.5. It is clear, however, that a different choice of threshold leads to different predictions.

In the batch setting, thresholding could be done in the pre- and postprocessing phases, however, in the streaming setting it needs to be done in real time. Specifically, the process of thresholding occurs at two times. The first thresholding occurs when the multi-target regressor has produced a multi-target prediction, which must then be converted into a multi-label prediction. The second thresholding occurs when we are updating the regressor, i.e., when the regressor is learning. Most streaming regressors are heavily dependent on the values of the target variables in the learning process, so the instances must be converted into the numeric representation that the multi-target regressor can utilize.

The problem of thresholding is not only problematic in the MLC via MTR setting, but also when considering the MLC task with other approaches. In general, MLC models produce results which are interpreted as probability estimations for each of the labels, thus the thresholding problem is a fundamental part of multi-label classifcation.

3.2 Implementation

For the purpose of this work, we have reimplemented the FIMT and FIMT-MT algorithms [12] in the MOA framework to facilitate usability and visibility, as the original implementation was a standalone extension of the C-based VFML library [10] and was not available as part of a larger data stream mining framework. We have also extended the algorithm to consider nominal attributes in the input space when considering splitting decisions. This allows us to use the algorithm on a wider array of datasets, some of which are considered herein.

In this paper, we combined the multi-label classification via multi-target regression methodology with the extended version of FIMT-MT, reimplemented in MOA. We named this method the incremental Structured OUtput Prediction Tree (iSOUP-Tree), since it is capable of addressing multiple structured output prediction tasks, i.e., multi-label classification and multi-target regression.

Ikonomovska et al. [13] have considered the performance of FIMT-DD when a simple predictive model is placed in each of the leaves, i.e., in this case a single linear unit (a perceptron). Opposed to regular *regression trees* where the prediction in a given leaf for an instance x is made as the average value of the recorded target values, a *model tree* produces the prediction as a linear combination of input attribute values, i.e., $\hat{y}(x) = \frac{1}{|S|} \sum_{y \in S} y$, where S is the set of observed examples in a given leaf, and $\hat{y}(x) = \sum_{i=1}^{m} x_i w_i + b$, where m is the number of input attributes and w_i, b are the perceptron weights, respectively. It was shown that the performance was increased when using model trees, however, this was only experimentally confirmed for regression tasks, where the targets generally exhibit larger variations than in classification tasks.

Specifically, even when considering a classification task through the lens of regression, the actual target variables can only take values of 0 and 1. If we use a linear unit to predict one of the targets, we have no guarantee that the predicted value will land in the [0,1] interval, where as the regression tree will produce an average of zeroes and ones, which will always land in this interval. Additionally, the perceptrons in the leaves are trained in real-time according to the *Widrow-Hoff rule*, which consumes a non-negligible amount of time. This motivated us to consider the use of multi-target regression trees when addressing the task of multi-label classification via multi-target regression. We denote this algorithm variant iSOUP-RT and the model tree variant iSOUP-MT.

4   Experimental Design

In this section, we first present the experimental questions that we want to answer in this paper. Next, we describe the datasets and algorithms used in the experiments. Furthermore, we discuss the evaluation measures used in the experiments. Finally, we conclude with the employed experimental methodology.

Experimental Questions. Our experimental design is constructed in such a way to answer several lines of inquiry. First, we want to explore if the use of model trees improves predictive performance, as it was shown in the regular multi-target regression scenario [13]. Second, we want to compare the introduced methods to other state-of-the-art methods. In this case, we will limit ourselves to comparisons with basic multi-label classification methods. Specifically, this means that we will not be comparing to ensemble or other meta-learning methods, as these methods could potentially utilize the iSOUP-Tree models as base models. Finally, and most crucially, we will consider whether addressing the task of multi-label classification via multi-target regression is a viable approach. For this question, we will use the results from the experiments addressing the previous questions, since they are sufficient.

Table 1. Datasets used in the experiments. N – number of instances, L – number of labels, $\phi_{LC}(D)$ – average number of labels per instance.

Dataset	Enron	IMDB	MediaMill	Ohsumed	Slashdot	TMC
Domain	text	text	video	text	text	text
N	1702	120919	43907	13929	3782	28596
Attribs.	1001 binary	1001 binary	120 numeric	1002 binary	1079 binary	500 binary
L	53	28	101	23	22	22
$\phi_{LC}(D)$	3.4	2.0	4.4	1.7	1.2	2.2

Datasets. In the experiments, we use a subset of datasets listed in [16, Table 3] (see Table 1). The *Enron*[2] dataset [15] is a collection of labelled emails, which, though small by the data stream standards, exhibits some data stream properties, such as time-orderedness and evolution over time. The *IMDB*[3] dataset [16] is constructed from text summaries of movie plots from the Internet Movie Data-Base and is labelled with the relevant genres. The *MediaMill* (See footnote 2) dataset [23] consists of video data annotated with various concepts which was used in the TRECVID challenge. The *Ohsumed*[4] dataset [16] was constructed from a collection of peer-reviewed medical articles and labelled with the appropriate disease categories. The *Slashdot* (See footnote 3) dataset [16] was mined from http://slashdot.org web page and consists of article blurbs and is labelled with subject categories. The *TMC* (See footnote 2) dataset was used in the SIAM 2007 Text Mining Competition and consists of human generated aviation safety reports, labelled with the problems being described (we are using the version of the dataset specified in [26]).

Algorithms. To address our experimental questions, we performed experiments using our implementations of algorithms for learning multi-target model trees (**iSOUP-MT**) and multi-target regression trees (**iSOUP-RT**). In addition, to preform comparison with other state-of-the-art algorithms we reuse results of experiments [16], performed under the same experimental settings. These include the following basic algorithms: binary relevance classifier (**BR**), classifier chains (**CC**), multi-label Hoeffding Trees (**HT**) and pruned sets (**PS**).

Evaluation Measures. In the evaluation, we use a set of measures used in recent surveys and experimental comparisons of different multi-label algorithms in the batch setting [8]. These include the following measures: accuracy, Hamming loss, exact match, and ranking loss. Aside from ranking loss, we selected these measures based on the available results for other basic multi-label methods in [16], since we were unable to rerun the experiments with the code made available by the authors. The differences in implementation also disallow for the

---

[2] http://mulan.sourceforge.net/datasets-mlc.html, accessed on 2015/05/25.
[3] http://meka.sourceforge.net/, accessed on 2015/05/25.
[4] Provided on request by authors of [16].

comparison of running times. However, we will briefly consider the running times of the iSOUP-Tree variants.

In the following definitions, N is the number of examples in the evaluation sample, i.e., the size of one window w, while Q is the number of labels in the provided MLC setting. Accuracy for an example with a prediction set $\hat{y}_i$ and a real labelset y_i is defined as the Jaccard similarity coefficient between them, i.e., $\frac{|\hat{y}_i \cap y_i|}{|\hat{y}_i \cup y_i|}$. The *accuracy* over a sample is the averaged accuracy for each example: Accuracy $= \frac{1}{N} \sum_{i=1}^{N} \frac{|\hat{y}_i \cap y_i|}{|\hat{y}_i \cup y_i|}$. The higher the accuracy of a model the better its predictive performance.

The *Hamming loss* measures how many times an example-label pair is misclassified. Specifically, each label that is either predicted but not real, or vice versa, carries a penalty to the score. The Hamming loss of a single example is the number of such misclassified labels divided by the number of all labels, i.e., $\frac{1}{Q}|\hat{y}_i \triangle y_i|$ where $\hat{y}_i \triangle y_i$ is the symmetric difference of the $\hat{y}_i$ and y_i sets. The Hamming loss of a sample is the averaged Hamming loss over all examples: HL $= \frac{1}{N} \sum_{i=1}^{N} \frac{1}{Q}|\hat{y}_i \triangle y_i|$. The Hamming loss of a perfect model, which makes completely correct predictions, is 0 and the lower the Hamming loss the better the predictive performance of a model. Note, that the Hamming loss will generally be reported as the *Hamming score*, i.e., HS $= 1 -$ HL.

The *exact match* measure (also known as subset accuracy or 0/1-loss) is a very strict evaluation measure as it requires the predicted labelset to be identical to the real labelset. Formally, the exact match measure is defined as EM $= \frac{1}{N} \sum_{i=1}^{N} \mathbf{I}(\hat{y}_i, y_i)$, where $\mathbf{I}(\hat{y}_i, y_i) = 1$, iff $\hat{y}_i$ and y_i are identical. The higher the exact match, the better the predictive performance.

Since thresholding can have a large impact on performance measures and determining the optimal threshold is non-trivial, we are also interested in measures that are independent of the chosen threshold. One such measure is *ranking loss*, defined as RL $= \frac{1}{N} \sum_{i=1}^{N} \frac{|D_i|}{|y_i||\overline{y}_i|}$, where $\overline{y}_i = \mathcal{L} \setminus y_i$ is the complement of y_i in $\mathcal{L}$, $D_i = \{(\lambda_k, \lambda_l) \mid s(\hat{y}_i, k) \leq s(\hat{y}_i, l), (\lambda_k, \lambda_l) \in y_i \times \overline{y}_i\}$ and $s(\hat{y}_i, k)$ is the numeric score (probability) for the label λ_k in the prediction $\hat{y}_i$, before applying the threshold. Essentially, it measures how well the labels are ordered by score, i.e., the loss is low when the labels that aren't present have lower scores than the present labels. Consequently, lower values of ranking loss indicate better performance.

Experimental Setup. Throughout our experiments we use the holdout evaluation approach for data streams. This means that a *holdout set* (or a *window*) of fixed size is constructed once enough examples accumulate, after which the predictions on the holdout set are used to calculate and report the evaluation metrics. Following that, the model is then updated with the collected examples and the process is repeated until all of the examples have been used.

To answer the proposed experimental questions, we constructed the following experimental setup. To compare the predictive performance of iSOUP-MT and iSOUP-RT, we have decided to observe the evolution of the ranking loss over time. Ranking loss was selected as the measure of choice, as it is independent of a chosen threshold and, as discussed earlier, thresholding is a non-trivial problem to solve in the streaming context. In this case, the desired properties are low ranking loss and/or a strongly declining tendency of the ranking loss, indicating an improvement over time.

For our experiments, we used a window size of $w = \frac{N}{20}$, i.e., each of the streams was divided into 20 windows, and the measures were recorded at each window. This not only allows us to look at the time evolution of the selected measures, but is also identical to the experimental setup from Read et al. [16]. Since we wish to directly compare our results to the results provided therein, we averaged the selected measures over all 20 of the windows.

5   Results and Discussion

In this section, we present the results of the performed experiments that answer our experimental questions. First, we compare the performance of the iSOUP-MT and iSOUP-RT methods on several datasets using a set of evaluation measures. Next, we provide a comparison of our methods with different basic incremental ML methods using results from previous studies. Finally, we provide a discussion of results with a focus on possible improvements to our methodology.

Comparison of iSOUP-MT and iSOUP-RT. In Table 2, we show the comparison of iSOUP-MT and iSOUP-RT on a set of evaluation measures. The results show that with the exception of accuracy on the Slashdot dataset, iSOUP-RT generally achieves better or at least comparable results than iSOUP-MT and clearly uses less time. This indicates that model trees are generally worse than regression trees when using the MLC via MTR methodology. The implementation of iSOUP-MT that uses percetrons in the leaves of the trees should be adapted to capture the dependencies of labels on the input attributes more accurately or a different type of model should be used in their place.

Table 2. Comparison of iSOUP-MT and iSOUP-RT. The best result per dataset is shown in bold. Other than time, the results are an average over 20 windows.

Dataset evaluation measure		Enron	IMDB	MediaMill	Ohsumed	Slashdot	TMC
Exact match	iSOUP-MT	0.165	0.000	0.000	0.000	0.000	0.000
	iSOUP-RT	**0.194**	**0.001**	**0.044**	**0.072**	**0.001**	**0.103**
Hamming score	iSOUP-MT	0.740	0.903	0.560	0.765	0.620	0.516
	iSOUP-RT	**0.945**	**0.929**	**0.966**	**0.979**	**0.947**	**0.912**
Accuracy	iSOUP-MT	0.273	**0.005**	0.047	0.036	**0.065**	0.089
	iSOUP-RT	**0.276**	0.002	**0.346**	**0.114**	0.001	**0.322**
Ranking loss	iSOUP-MT	0.311	0.625	0.483	0.518	0.486	0.465
	iSOUP-RT	**0.105**	**0.180**	**0.058**	**0.250**	**0.220**	**0.126**
Time [s]	iSOUP-MT	15.02	549.97	363.83	96.63	17.60	53.67
	iSOUP-RT	**9.81**	**295.84**	**307.54**	**68.66**	**9.02**	**29.32**

In Fig. 3, we show the ranking loss diagrams, which show the comparison of the iSOUP-MT and iSOUP-RT methods on all 6 datasets used in our experiments. The figures clearly show that the evolution of the ranking loss measure is considerably better for the iSOUP-RT over all datasets. The only dataset where the ranking losses of iSOUP-MT and iSOUP-RT are comparable is the Enron dataset. However, it is a small dataset in terms of data streams, so the windows are small enough that the trees do not have time to significantly grow.

Comparison of Different Incremental Multi-label Methods. In this section, we present the results of the comparison of our methods (iSOUP-MT and

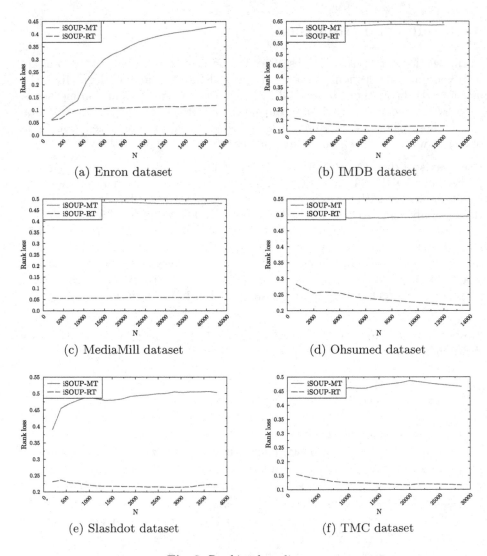

(a) Enron dataset

(b) IMDB dataset

(c) MediaMill dataset

(d) Ohsumed dataset

(e) Slashdot dataset

(f) TMC dataset

Fig. 3. Ranking loss diagrams

Table 3. Exact match measure. The best result per dataset is shown in bold. * marks results reused from [16, Table 6].

Dataset	iSOUP-MT	iSOUP-RT	BR*	CC*	HT*	PS*
Enron	0.168	**0.194**	0.006	0.007	0.058	0.086
IMDB	0.000	0.001	0.031	0.014	**0.108**	0.027
MediaMill	0.000	0.044	0.008	0.000	**0.050**	0.017
Ohsumed	0.000	0.072	0.115	0.054	0.083	**0.212**
Slashdot	0.000	0.001	0.000	0.000	**0.137**	0.113
TMC	0.000	0.076	0.149	0.123	0.087	**0.298**

iSOUP-RT) with other basic incremental multi-label methods. These include: binary relevance classifier (**BR**), classifier chains (**CC**), multi-label Hoeffding Trees (**HT**) and pruned sets (**PS**). Here, we note the results for these methods were reused from Read et al. [16, Tables 5, 6 and 7], because of inability to reproduce the experiments from the software links provided in [16].

In terms of the exact match measure our methods did not often score the best among compared algorithms (see Table 3). In this case, HT performed best on three of the datasets and was followed by PS with best results on two datasets. iSOUP-RT performed best on the Enron dataset. Notably, the results of iSOUP-RT are generally close to those of HT, except for a case where exact match is considerably higher for iSOUP-RT and a case where the opposite holds.

When considering the Hamming loss (presented in Table 4 as the Hamming score), however, iSOUP-RT out matched all other algorithms, except for the TMC dataset. Interestingly, the iSOUP-RT's results here are better aligned with PS's results, and not HT's, as in the case of exact match.

The results for the accuracy measure are less clear (see Table 5). PS performed the best with best results on three of the datasets, iSOUP-RT outperformed other algorithms in two cases and HT performed best on the IMDB dataset.

Table 4. Hamming loss (displayed as $1.0 - loss$). The best result per dataset is shown in bold. * marks results reused from [16, Table 7].

Dataset	iSOUP-MT	iSOUP-RT	BR*	CC*	HT*	PS*
Enron	0.740	**0.945**	0.524	0.503	0.926	0.934
IMDB	0.903	**0.929**	0.884	0.834	0.918	0.875
MediaMill	0.560	**0.966**	0.897	0.634	0.958	0.947
Ohsumed	0.765	**0.979**	0.913	0.866	0.900	0.947
Slashdot	0.620	**0.947**	0.055	0.054	0.915	0.912
TMC	0.516	0.912	0.907	0.871	0.884	**0.935**

Table 5. Accuracy. The best result per dataset is shown in bold. * marks results reused from [16, Table 5].

Dataset	iSOUP-MT	iSOUP-RT	BR*	CC*	HT*	PS*
Enron	0.273	**0.276**	0.144	0.142	0.134	0.241
IMDB	0.005	0.002	0.139	0.170	**0.210**	0.146
MediaMill	0.047	**0.346**	0.119	0.080	0.301	0.297
Ohsumed	0.036	0.114	0.297	0.292	0.125	**0.372**
Slashdot	0.065	0.001	0.054	0.054	0.145	**0.200**
TMC	0.089	0.322	0.415	0.446	0.171	**0.562**

Discussion. The results clearly indicate that the regression tree variant iSOUP-RT is a more appropriate method for the task of MLC via MTR than the model tree variant iSOUP-MT. This indicates that the perceptrons placed in the leaves significantly reduce the method's performance. This may be due to the mechanism of the perceptron, which does not guarantee that the result will land in the [0, 1] interval. Other types of leaf models should be considered and evaluated in the future, similar to [16] where the pruned sets (PS) method was used in the leaves of the Hoeffding trees.

A cursory glance makes it clear that there is a lot of variation in the majority of the results reported in the comparison of different methods. The exact match measure and accuracy fluctuate to a large extent and only the results of Hamming loss are consistent. However, with respect to the Hamming loss, the iSOUP-RT method consistently outperformed other methods, which possibly indicates that the learning mechanism is biased toward optimization of a similar measure.

Given the relatively small selection of evaluation measures and the observed variation among them, it would be prudent to consider other evaluation measures in a more in-depth experimental evaluation. This variation in the results is something that would be out of place in a more classical machine learning setting, however, there are many partially unexplored variables in the MLC context, e.g., drift-detection, thresholding, etc. Looking at the selected datasets also does not give us sufficient data to determine and analyze the effect of data set size on the performance of the various methods. Overall, we have shown that the MLC via MTR methodology is a valid approach for MLC. However, the use of perceptrons as models in the tree leaves is not advisable and other types of models should be considered. We have determined that iSOUP-RT's performance is similar to the other basic incremental multi-label learners. Therefore, iSOUP-RT is a suitable candidate for further experimentation, e.g., as a base model in ensemble methods explored in [16].

6   Conclusion and Future Work

In this paper, we have introduced the multi-label classification via multi-target regression methodology and introduced the iSOUP-Tree algorithm that is used to address the multi-label classification task.

We performed two sets of experiments, the first of which was designed to evaluate whether the use of model trees over regression trees increases the predictive performance as it was shown for the streaming multi-target regression task [13]. We observed the time evolution of the ranking loss, as well as the average ranking loss, exact match, Hamming loss and accuracy measures over the considered datasets. From these experiments, it was made clear that regression trees outperform model trees for the task of MLC via MTR.

The second set of experiments were designed to compare the introduced methods to other multi-label learners. To this end, the experimental design was equal to the one in [16]. While we were not able to establish clear superiority of one method over the other, we were able to determine that the introduced iSOUP-Tree method is a promising candidate for further experimentation, e.g., as a base model in state-of-the-art ensemble or other meta-learning techniques.

Additionally, due to the relatively unexplored nature of the streaming multi-label classification task, we plan to perform a more extensive experimental evaluation on more datasets and with respect to a wider set of evaluation measures. Specifically, we also wish to address the problems of drift detection and thresholding for the iSOUP-Tree method.

We also propose two other avenues of further work, in regards to extending the introduced methodology. The first one focuses on the model and the aim is to extend the iSOUP-Tree method using the option tree paradigm [11], used in the single-target regression setting, to the multi-target regression setting. This approach has been shown to outperform the regression tree methodology. The second extension is specific to the MLC via MTR methodology. In classical (batch) data mining, the task of hierarchical multi-label classification (HMC) is becoming more and more prevalent. In HMC, the labels are ordered in a hierachy and adhere to the hierarchy constraint, i.e., if an example is labeled with a label it also has to be labelled with the label's ancestors. We plan to extend the MLC via MTR methodology to be applicable to HMC tasks in the streaming setting.

Acknowledgements. We would like to acknowledge the support of the EC through the projects: MAESTRA (FP7-ICT-612944) and HBP (FP7-ICT-604102), and the Slovenian Research Agency through a young researcher grant and the program Knowledge Technologies (P2-0103).

References

1. Appice, A., Džeroski, S.: Stepwise Induction of Multi-target Model Trees. In: Kok, J.N., Koronacki, J., Lopez de Mantaras, R., Matwin, S., Mladenič, D., Skowron, A. (eds.) ECML 2007. LNCS (LNAI), vol. 4701, pp. 502–509. Springer, Heidelberg (2007)
2. Bifet, A., Gavaldà, R.: Adaptive Learning from Evolving Data Streams. In: Adams, N.M., Robardet, C., Siebes, A., Boulicaut, J.-F. (eds.) IDA 2009. LNCS, vol. 5772, pp. 249–260. Springer, Heidelberg (2009)
3. Bifet, A., Holmes, G., Kirkby, R., Pfahringer, B.: Moa: massive online analysis. J. Mach. Learn. Res. **11**, 1601–1604 (2010)

4. Bifet, A., Holmes, G., Pfahringer, B., Kirkby, R., Gavaldà, R.: New ensemble methods for evolving data streams. In: Proceedings of the 15th ACM SIGKDD International Conference on Knowledge Discovery and Data Mining, pp. 139–148. ACM (2009)
5. Cheng, W., Hüllermeier, E.: Combining instance-based learning and logistic regression for multilabel classification. Mach. Learn. **76**(2–3), 211–225 (2009)
6. Domingos, P., Hulten, G.: Mining high-speed data streams. In: Proceedings of the Sixth ACM SIGKDD International Conference on Knowledge Discovery and Data Mining, pp. 71–80. ACM (2000)
7. Fürnkranz, J., Hüllermeier, E., Mencía, E.L., Brinker, K.: Multilabel classification via calibrated label ranking. Mach. Learn. **73**(2), 133–153 (2008)
8. Gibaja, E., Ventura, S.: A tutorial on multilabel learning. ACM Comput. Surv. (CSUR) **47**(3), 52 (2015)
9. Gonçalves, E.C., Plastino, A., Freitas, A.A.: A genetic algorithm for optimizing the label ordering in multi-label classifier chains. In: IEEE 25th International Conference on Tools with Artificial Intelligence (ICTAI), 2013, pp. 469–476. IEEE (2013)
10. Hulten, G., Domingos, P.: VFML - a toolkit for mining high-speed time-changing data streams (2003). http://www.cs.washington.edu/dm/vfml/
11. Ikonomovska, E., Gama, J., Džeroski, S.: Online tree-based ensembles and option trees for regression on evolving data streams. Neurocomputing **150**, 458–470 (2015)
12. Ikonomovska, E., Gama, J., Džeroski, S.: Incremental multi-target model trees for data streams. In: Proceedings of the 2011 ACM Symposium on Applied Computing, pp. 988–993. ACM (2011)
13. Ikonomovska, E., Gama, J., Džeroski, S.: Learning model trees from evolving data streams. Data Min. Knowl. Disc. **23**(1), 128–168 (2011)
14. Qu, W., Zhang, Y., Zhu, J., Qiu, Q.: Mining Multi-label Concept-Drifting Data Streams Using Dynamic Classifier Ensemble. In: Zhou, Z.-H., Washio, T. (eds.) ACML 2009. LNCS, vol. 5828, pp. 308–321. Springer, Heidelberg (2009)
15. Read, J.: A pruned problem transformation method for multi-label classification. In: Proceedings of the 2008 New Zealand Computer Science Research Student Conference (NZCSRS 2008), pp. 143–150 (2008)
16. Read, J., Bifet, A., Holmes, G., Pfahringer, B.: Scalable and efficient multi-label classification for evolving data streams. Mach. Learn. **88**(1–2), 243–272 (2012)
17. Read, J., Pfahringer, B., Holmes, G., Frank, E.: Classifier chains for multi-label classification. Mach. Learn. **85**(3), 333–359 (2011)
18. Read, J., Pfahringer, B., Holmes, G.: Multi-label classification using ensembles of pruned sets. In: Eighth IEEE International Conference on Data Mining, 2008, ICDM 2008, pp. 995–1000. IEEE (2008)
19. Rutkowski, L., Pietruczuk, L., Duda, P., Jaworski, M.: Decision trees for mining data streams based on the McDiarmid's bound. IEEE Trans. Knowl. Data Eng. **25**(6), 1272–1279 (2013)
20. Shaker, A., Hüllermeier, E.: IBLStreams: a system for instance-based classification and regression on data streams. Evolving Syst. **3**(4), 235–249 (2012)
21. Shi, Z., Wen, Y., Feng, C., Zhao, H.: Drift detection for multi-label data streams based on label grouping and entropy. In: 2014 IEEE Data Mining Workshop (ICDMW), pp. 724–731. IEEE (2014)
22. Shi, Z., Xue, Y., Wen, Y., Cai, G.: Efficient class incremental learning for multi-label classification of evolving data streams. In: International Joint Conference on Neural Networks (IJCNN), 2014, pp. 2093–2099. IEEE (2014)

23. Snoek, C.G., Worring, M., Van Gemert, J.C., Geusebroek, J.M., Smeulders, A.W.: The challenge problem for automated detection of 101 semantic concepts in multimedia. In: Proceedings of the 14th Annual ACM International Conference on Multimedia, pp. 421–430. ACM (2006)
24. Spyromitros-Xioufis, E.: Dealing with concept drift and class imbalance in multi-label stream classification. Ph.D. thesis, Aristotle University of Thessaloniki (2011)
25. Struyf, J., Džeroski, S.: Constraint Based Induction of Multi-objective Regression Trees. In: Bonchi, F., Boulicaut, J.-F. (eds.) KDID 2005. LNCS, vol. 3933, pp. 222–233. Springer, Heidelberg (2006)
26. Tsoumakas, G., Vlahavas, I.P.: Random k-Labelsets: An Ensemble Method for Multilabel Classification. In: Kok, J.N., Koronacki, J., Lopez de Mantaras, R., Matwin, S., Mladenič, D., Skowron, A. (eds.) ECML 2007. LNCS (LNAI), vol. 4701, pp. 406–417. Springer, Heidelberg (2007)
27. Vens, C., Struyf, J., Schietgat, L., Džeroski, S., Blockeel, H.: Decision trees for hierarchical multi-label classification. Mach. Learn. **73**(2), 185–214 (2008)
28. Zhang, M.L., Zhou, Z.H.: A k-nearest neighbor based algorithm for multi-label classification. In: 2005 IEEE Granular Computing, vol. 2, pp. 718–721. IEEE (2005)

Periodical Skeletonization for Partially Periodic Pattern Mining

Keisuke Otaki[1,2]([⊠]) and Akihiro Yamamoto[1]

[1] Department of Intelligence Science and Technology, Graduate School
of Informatics, Kyoto University, Kyoto, Japan
ootaki@iip.ist.i.kyoto-u.ac.jp, akihiro@i.kyoto-u.ac.jp
[2] Research Fellow of the Japan Society for the Promotion of Science, Tokyo, Japan

Abstract. Finding periodical regularities in sequential databases is an important topic in Knowledge Discovery. In pattern mining such regularity is modeled as partially periodic patterns, where typical periods (e.g., daily or weekly) can be considered. Although efficient algorithms have been studied, applying them to real databases is still challenging because they are noisy and most transactions are not extremely frequent in practice. They cause a combinatorial explosion of patterns and the difficulty of tuning a threshold parameter. To overcome these issues we investigate a pre-processing method called skeletonization, which was recently introduced for finding sequential patterns. It tries to find clusters of symbols in patterns, aiming at shrinking the space of all possible patterns in order to avoid the combinatorial explosion and to provide comprehensive patterns. The key idea is to compute similarities within symbols in patterns from a given database based on the definition of patterns we would like to mine, and to use clustering methods based on the similarities computed. Although the original method cannot allow for periods, we generalize it by using the periodicity. We give experimental results using both synthetic and real datasets, and compare results of mining with and without the skeletonization, to see that our method helps us to obtain comprehensive partially periodic patterns.

Keywords: Sequential pattern mining · Partially periodic pattern · Skeletonization · Spectral clustering · Data preprocessing

1 Introduction

Finding patterns frequently appearing in databases is one of the important problems in data mining. Transactions of such databases naturally have timestamps as their auxiliary attributes. They are often ordered with their timestamps. A typical sorting using the order is the chronological order. For example sequences of transactions describing purchases of products in electronic commerce sites are sorted in the chronological order, from old transactions to new ones. If such chronological attributes are important to a database, such database is called a sequential database. As the order is directly related to time, typical periods

© Springer International Publishing Switzerland 2015
N. Japkowicz and S. Matwin (Eds.): DS 2015, LNAI 9356, pp. 186–200, 2015.
DOI: 10.1007/978-3-319-24282-8_16

related to clocks or calendars (e.g., hour, day, etc.) may contribute to regularities in the data. Thus assuming that such periodical behaviors may appear in various sequential databases (e.g., trajectory or life-log) is natural in data mining.

To get insights from sequential databases by capturing periodical regularities, *periodical pattern mining* have been studied [4,5,12]. The fundamental patterns are *full periodic patterns* and *partial periodic patterns* [4]. Note that, for the sake of simplicity, we here assume that patterns are sequences of symbols drawn from an alphabet Σ. For example, let $\Sigma = \{\text{sns}, \text{news}, \text{blog}, \text{shops}\}$. We consider a sequence $s = (\text{sns}, \text{news}, \text{blog}, \text{sns}, \text{news}, \text{blog}, \text{sns}, \text{shop}, \text{blog})$ representing some access log of Web pages. A pattern $(\text{sns}, \text{news})$ appears twice in s, and this is called *full periodic pattern*. As full periodic patterns require that all symbols be fully specified in them, they are not flexible and it is difficult to handle various periodical behaviors. As more flexible patterns, *partial periodic patterns* have been studied [5]. For example, a partial periodic pattern $(\text{sns}, \star, \text{blog})$ appears 3 times, where $\star$ is the wildcard symbol of length 1 representing any symbol in Σ. As partial periodic patterns can contain the symbol $\star$, they are more flexible than full periodic patterns to capture periodical behaviors in databases. In mining these periodic patterns, we assume that a given sequence s is divided into $\lceil \frac{|s|}{P} \rceil$ fragments, where P is a period of users' interest. In the example above, the pattern $(\text{sns}, \star, \text{blog})$ appears 3 times in fragments $(\text{sns}, \text{news}, \text{blog})$, $(\text{sns}, \text{news}, \text{blog})$, and $(\text{sns}, \text{shop}, \text{blog})$ of s. By listing such patterns, users can find frequently and periodically appearing combinations of symbols. A common way in the mining is to obtain a parameter θ from users, enumerate all candidates, and filter them by their frequency, called *support*, based on the user-specific parameter θ.

Although many efficient algorithms have been developed [4,12], it is still challenging to use them in practice because the number of enumerated patterns highly depends on the number $|\Sigma|$ of symbols we use. When databases get large, the size of Σ increases as well. This fact consequently makes evaluating patterns by their supports difficult because most patterns have similar and relatively small supports. That is, the space of (frequent) patterns on Σ get *sparse* with respect to the space of all possible patterns[1]. In addition, the number of patterns in the output become exponential to the size of Σ in the worst-case, and we essentially need to decrease the size Σ to get comprehensive patterns.

We resolve these issues by a recently developed novel pre-processing method, named *temporal skeletonization* [6], which aims at reducing the number of symbols. Since the temporal skeletonization cannot be applied to periodical settings, we generalize it by using the idea of periodical extensions of functions. Before going into the details, here we give two motivating examples of our study.

Motivating Example. For both numerical (e.g., price, temperature) and symbolic (e.g., item, product) sequences, preparing a large set Σ of symbols is inevitable if we deal with various databases. For example, Fig. 1(a) shows a sequence of electric power demand per day in UK, 2013. We discretize the sequence with

---

[1] Consider to find all partially periodic patterns up to the length k on Σ. Let $\Sigma_\star = \Sigma \cup \{\star\}$. All possible combinations are in $\Sigma_\star \cup \Sigma_\star^2 \cup \cdots \cup \Sigma_\star^k$, which can become much larger than that of all patterns appearing in databases in practice.

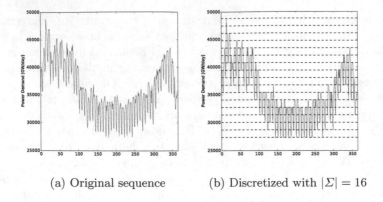

(a) Original sequence   (b) Discretized with $|\Sigma| = 16$

Fig. 1. A numeric sequence of power demand in UK, 2013 (Figure 1(a)), and its discretization with 16 symbols (corresponding to dashed lines) in Fig. 1(b).

dividing values into $|\Sigma|$ bins uniformly[2] as seen in Fig. 1(b) with $|\Sigma| = 16$ bins. It is obvious that we can represent a sequence as a symbolic sequence with a smaller loss with a larger set Σ. In Fig. 1(b), however, only a few combinations of Σ appear consecutively. It is difficult to tune the set Σ while taking a balance among the expressiveness and the sparseness. Now a typical periodical behavior is that *the demand gets higher every weekend*, which could be obtained by frequent patterns, where symbols corresponding to low values are followed by those doing to high values. We believe that such comprehensive and high-level patterns are more informative and useful to analyzing databases.

Similar tendencies can be seen in symbolic data as well. As an example we take into account those stored from LAST.FM (http://last.fm/), which are sequences of songs logged by users. Currently the site has 640 million songs as symbols. It is obviously intractable for both enumerating and analyzing patterns. We take some logs of users from an open dataset[3] (See Sect. 3 of [2]), where a sequence $s = (S_1, S_2, \ldots)$ is a log of a user and each S_i is the set of songs heard in the index i, where i corresponds to a 1 h interval of the log (e.g., the set S_4 shows the listened songs during 0 A.M. to 1 A.M.). For example, the sequence for user ID 808 is length 16, 913 log, where the user listened to 24, 310 songs and 1, 340 out of 16, 913 intervals are not empty (i.e., in other intervals the user did not listen to any songs). Then if we would like to analyze some daily behaviors (i.e., $P = 24$) including the empty situation, $24, 310^{24}$ is the upper bound[4] of the size of all combinations. Again, this is an intractable situation and data are obviously sparse and hard to analyze with pattern mining methods.

---

[2] For example, if the range of values $[0, 10)$ and $|\Sigma| = 4$, values in $[0, 10]$ would be categorized into either $[0, 2.5), [2.5, 5.0), [5.0, 7.5),$ or $[7.5, 10)$, and symbolic alphabets are assigned into those bins to encode the sequence into a symbolic sequence.

[3] http://www.dtic.upf.edu/~ocelma/MusicRecommendationDataset/.

[4] Of course most of them are infrequent patterns.

In pattern mining, therefore, Liu et al. [6] and others insisted that users carefully need to tune the set Σ and proposed the temporal skeletonization for symbolic sequences. Their idea is to construct clusters of symbols and assign each cluster a label. Then a sequence can be translated into a high-level and potentially comprehensive sequences of cluster labels, which roughly characterize the given sequence. By grouping symbols into clusters, we reduce the size of Σ. We develop such method for periodical analyses by generalizing the idea of temporal skeletonization [6], and discuss frequently occurring high-level patterns.

The rest of this paper is organized as follows. We give preliminaries in Sect. 2. Our method is formally described in Sect. 3. We provide our experimental results and discuss them in Sect. 4 and conclude our study in Sect. 5.

2 Periodic Pattern Mining and Temporal Skeletonization

Let Σ be the alphabet. The set $\Sigma^\star$ denotes the Kleene closure of Σ. We use $\Sigma^+ \equiv \Sigma^\star \setminus \{\epsilon\}$, where ϵ is the empty string. For a sequence s in $\Sigma^\star$, $|s|$ denotes the length of s, and we define $|\epsilon| = 0$. In addition, s_i and $s_{i,j}$ represent i-th element and the continuous subsequence from i to j of s ($i < j$), respectively. Let P be an fixed integer representing the period of users' interest.

2.1 Frequent Partially Periodic Pattern Mining

Periodical behaviors of databases can be modeled as *partially periodic patterns*. Transaction databases in the literature can be defined as (singleton) sets of sequences on Σ or those of subsets of Σ. Then patterns are intuitively sequences of (sets of) symbols. An important concept used is *periodical segments* of sequences.

Definition 1 (Event Sequences and Segments). For an *event sequence* $s \in \Sigma^+$ and a period P, s can be divided into $m(= \lceil \frac{|s|}{P} \rceil)$ *mutually disjoint segments*. We denote it by $s = \langle ps_1, ps_2, \ldots, ps_m \rangle$, where for $1 \le i \le m$, $ps_i = s_{im,(i+1)m-1}$. For example with a sequence $s = abcabdabb$ of symbols $\{a, b, c, d\}$ and $P = 3$, the sequence s is divided into $ps_1 = abc$, $ps_2 = abd$, and $ps_3 = abb$.

Definition 2 (Partial Patterns). A sequence from $\Sigma \cup \{\star\}$ is a *(partial) pattern*, where the special character $\star \notin \Sigma$ represents any event of length 1.

Periodical patterns we want to obtain are those appearing in periodical segments frequently. For a sequence s and a pattern p, we need to evaluate whether or not p is interesting. To estimate interest, we adopt *support*, as defined below.

Definition 3 (Support). For a sequence s and a pattern p of the same length, we say that s *is covered by* the pattern p if and only if $p_i = \star$ or $p_i = s_i$ for all $1 \le i \le |s|$, denoted by $s \preceq p$. The *support* of a partial pattern p, denoted by $\mathrm{Supp}_P(p)$, is defined as $\mathrm{Supp}_P(p, s) = |\{ps_i \mid s = \langle ps_1, \ldots, ps_m \rangle, ps_i \preceq p\}|$. For $s_2 = abcabdabb = \langle abc, abd, abb \rangle$ and $P = 3$, $\mathrm{Supp}_P(ab\star, s_2) = \frac{3}{3} = 1.0$.

If a partial pattern p satisfies $\mathrm{Sup}_P(p) \geq \theta$ with a threshold θ and a period P, then p is *frequent* and we call it a *partially periodic pattern* (PPP). We have the problem of finding frequent partially periodic patterns below.

Problem 1. Let θ be a user-specific threshold and P be a given period length. For a sequence s, the *Frequent Partially Periodic Pattern Mining* (*FPPPM*) problem is to list all partially periodic patterns p from s satisfying $\mathrm{Sup}_P(p) \geq \theta$.

Note that in this case a sequential database DB is a single sequence s from which we try to list all frequent partially periodic patterns. Without loss of generality, we can extend it to deal with a set $\mathrm{DB} = \{s_1, s_2, \ldots, s_M\}$ of sequences.

Several efficient algorithms have been developed for the frequent partially periodic pattern mining problem. For examples, Han *et al.* showed a fundamental algorithm using max sub-pattern trees [5] and Yang *et al.* proposed to use tuple representations for periodic patterns and a depth-first search algorithm based on the PREFIXSPAN [9] used in sequential pattern mining [12].

2.2 Temporal Skeletonization

We refer to the original definition of *temporal graphs* to explain the idea of the temporal skeletonization in [6], which tries to build a *similarity graph*[5] from a given database DB for capturing the similarity within symbols in Σ. Note that in this literature, a transaction in DB is defined as a sequence on Σ.

Definition 4 (Temporal Graphs [6]). Let $\mathrm{DB} = \{s_1, \ldots, s_N\}$ be the set of sequences of symbols from Σ. We define a weighted undirected graph $G = (V, E)$, where V corresponds to Σ. For two symbols $x, y \in \Sigma$, the weight $W_{x,y}$ of the edge corresponding to $\{x, y\}$ is defined as

$$W_{x,y} = \frac{1}{N} \sum_{n=1}^{N} \sum_{\substack{1 \leq i,j \leq |s_n|, |i-j| \leq r, i < j \\ e_i = s_{n,i}, e_j = s_{n,j}}} \mathbf{1}_{e_i = x \wedge e_j = y} \tag{1}$$

where N is the number of sequences, r be the *window width*, $\mathbf{1}_f$ is the indicator function that returns 1 if and only if the predicate f is true. Intuitively we count the number of *co-occurrences* of symbols x and y in the window of width r.

The right-hand side of Eq. 1 can be computed by checking the given database $\mathrm{DB} = \{s_1, \ldots, s_n\}$, where the indicator function can be replaced with other similarity measures. For example in [6], authors used the exponential function $\exp(-\frac{\|i-j\|^2}{k})$ with a parameter k to compute weights.

In constructing a temporal graph G, all indices of sequences in DB are taken into account several times. In each time we focus on some index i, we check all neighbors within width r. This computation is based on an implicit assumption;

---

[5] A *similarity graph* is a weighted graph in which vertices represent data points and edges represent the similarity between two points with their weights.

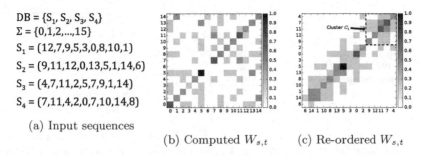

DB = {S_1, S_2, S_3, S_4}

Σ = {0,1,2,...,15}

S_1 = (12,7,9,5,3,0,8,10,1)

S_2 = (9,11,12,0,13,5,1,14,6)

S_3 = (4,7,11,2,5,7,9,1,14)

S_4 = (7,11,4,2,0,7,10,14,8)

(a) Input sequences

(b) Computed $W_{s,t}$ (c) Re-ordered $W_{s,t}$

Fig. 2. A toy example in [6] and two heatmaps: Fig. 2(b) shows the original W computed from DB, and Fig. 2(c) is the re-ordered one.

symbols x, y appearing closely and temporally (i.e., within the window of width *r*) *may belong to some meaningful temporal clusters.* After constructing G, users try to find clusters of symbols by applying clustering methods to G (such as *spectral clustering* [7,10]). The problem of finding clusters can be formulated as a standard graph-based optimization problem with some constraints as shown in [6], where an important step of clustering is to compute eigenvalues and eigenvectors of similarity matrices corresponding to a graph G. With the spectral clustering, a graph G can be represented as a heatmap, as shown in Fig. 2.

Example 1. The input is shown in Fig. 2(a). We compute the weights from {s_1, s_2, s_3, s_4} as seen in Fig. 2(b) and represent them by a heatmap, where both the *x*-axis and *y*-axis correspond to the order of the alphabet Σ. That is, on some (i, j), the thickness in the heatmap corresponds the value W_{Σ_i, Σ_j}. After applying the spectral clustering, we can re-order indices of W as shown in Fig. 2(c). For example, we can find a cluster of symbols such as $C_1 = \{4, 7, 9, 11, 12\}$, which is the upper right area in Fig. 2(c). Note that C_1 appears in prefixes of sequences in DB. With thanks to the temporal skeletonization, we can conjecture that all sequences in DB are in the form $(C_1, C_1, C_1, \dots)$. This prefix consisting of cluster labels can be regarded as a *high-level* (or, more abstract and comprehensive) pattern of the sequences given in DB.

3   Periodical Skeletonization

The key idea for taking into account periodical information is simple: Extending functions representing areas that we check in computing weights to some periodic functions of the periodicity P of our interest. If we would like to analyze some daily behaviors, we set $P = 31$ (or 30 days, for example).

The *sliding window* of width r used in the temporal skeletonization can be modeled by a *rectangular function* with width r and the origin i [6]. By modifying this function in a periodical manner, we can deal with the periodicity of

---

[6] The function is defined as $\text{Rect}_{i,r}(t) = 0$ if $|t - i| > r$, 1 otherwise.

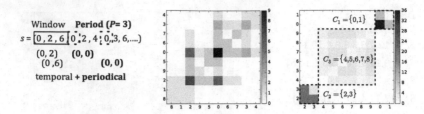

Fig. 3. An idea of the periodical skeletonization in Fig. 3(a). Figure 3(b) is a result only using the temporal information and Fig. 3(c) is that adopting the periodical information only, where rectangles are the discovered clusters.

occurrences of symbols. We can easily imagine such techniques on the analogy of Fourier series and Fourier transforms. Please recall the toy examples used in Sect. 2.1. For an input sequence $s = abcabdabb$ and $P = 3$, a frequent partially periodic pattern $ab\star$ appears 3 times in every segment abc, abd, and abb. This means that not only neighbors according to the sliding window $\text{Rect}_{i,r}(\cdot)$, but also periodical information from i, that is, $i+P, i+2P, i+3P, \dots$ could be used to search for similar intervals.

This observation inspires our modification for *periodical skeletonization*. Because now a target database in the FPPPM problem is a long sequence s as we defined in Problem 1, we make use of the same setting in the following.

Definition 5 (Periodical Graphs). In a *periodical graph* G, the weights from an input sequence s and a period P for two symbols x, y are computed as follows:

$$W_{x,y} = \sum_{\substack{1 \leq i,j \leq |s|, e_i = s_i, e_j = s_j \\ \text{if } e_i = x \wedge e_j = y}} \mathbf{1}_{|i-j| \leq r} + \mathbf{1}_{i \equiv j} \pmod{P} \tag{2}$$

The second term is newly introduced based on the periodical information. We can also replace the right-hand side of Eq. 2 with similarity functions by adapting them in a similar fashion based on the temporal skeletonization.

Example 2. Figure 3 illustrates examples of computing Eq. 2 from $s = (0, 2, 6, 0, 2, 4, \dots)$ with $\Sigma = \mathbb{N}$. Figure 3(b) is computed by the temporal skeletonization, while Fig. 3(c) adopts the periodical term only in Eq. 2. We can see 3 clusters as rectangles: $C_1 = \{0, 1\}, C_2 = \{2, 3\}$ and $C_3 = \{4, 5, 6, 7, 8\}$ in Fig. 3(c), and they are clear than those in Fig. 3(b).

We obtain this example by using an HMM in Fig. 4. We build an HMM $\mathcal{H}$ of three states T_1, T_2, T_3 with a cyclic relation (*i.e.*, it has transitions from T_1 to T_2, from T_2 to T_3, and T_3 to T_1, respectively with the probability 1). To simulate a partially periodic pattern $02\star$, in T_1 and T_2, $\mathcal{H}$ outputs 0 and 2, respectively

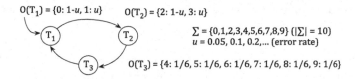

$O(T_1) = \{0: 1-u, 1: u\}$ $O(T_2) = \{2: 1-u, 3: u\}$

T_1 T_2

$\Sigma = \{0,1,2,3,4,5,6,7,8,9\}$ ($|\Sigma| = 10$)
$u = 0.05, 0.1, 0.2,...$ (error rate)

T_3 $O(T_3) = \{4: 1/6, 5: 1/6, 6: 1/6, 7: 1/6, 8: 1/6, 9: 1/6\}$

Fig. 4. Settings of HMMs used for generating synthetic data.

with high probability $100 \times (1 - u)\%$ and outputs 1 and 3 with low probability $100 \times u\%$. On the other hand in T_3, $\mathcal{H}$ generates $\{4, 5, 6, 7, 8, 9\}$ uniformly. By generating sequences of length N from such $\mathcal{H}$, we can obtain a sequence s containing the partially periodic pattern $02\star$ very frequently. Compared with the result in Fig. 3(b), we can see that C_1, C_2, C_3 correspond to T_1, T_2, T_3 more clearly as blocks in the heatmaps if we consider the periodicity in Fig. 3(c).

In summary, our periodical skeletonization is a method to make clusters of symbols using the co-occurrences and the *periodical co-occurrences* of symbols.

4   Experiments

We first report experiments with synthetic and real datasets which should have simple periodical behaviors to observe the effect of our proposal and to discuss the difference between two skeletonization methods. Then we apply both methods to analyze Last.fm datasets. With the results by skeletonization, we discuss the FPPPM problem for encoded sequences using cluster labels.

4.1   Preliminary Experiments

Datasets and Environment. Here we use both synthetic and real datasets. The summary of these datasets is shown in Table 1. A synthetic dataset is generated by using the HMM shown in Fig. 4. A real dataset, named PowerDemand, is a set of sequences of electric power demand in 2013, extracted from the GRIDWATCH system[7], which were previously used in Fig. 1. The original sequence records power demand in UK 12 times per hour, that is, roughly 300 times per day. We take the simple average (not *moving* average) of them to construct a hourly sequence of power demand in 2013, named PowerDemand-32. Because an yearly record may contain many periodical behaviors (e.g., daily, weekly, monthly, etc.), we extract a small subset, named PowerDemand-128F, of PowerDemand-32 and make the resolution of Σ more clear by increasing the size Σ from 32 to 128 and taking a part roughly from summer to autumn. For PowerDemand-128F, we expect that the sequence have the period $P = 7$. As another real dataset, we use Kyoto, a sequence of the daily temperatures in Kyoto from 1880 to 2014 with $P = 365$ and $|\Sigma| = 359$.

---

[7] http://www.gridwatch.templar.co.uk/.

Table 1. Summary of datasets and parameters used in experiments.

Name	Length	$|\Sigma|$	Period P	Note
HMM-600-u	600	10	3	with error rate $u = 0.25$
PowerDemand-32	365	32	7	Discretized with level 32
PowerDemand-128F	100	128	7	Subset of PowerDemand with level 128
Kyoto	43,833	359	365	Discretized with the resolution $0.1°C$

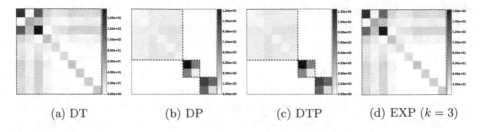

(a) DT   (b) DP   (c) DTP   (d) EXP ($k = 3$)

Fig. 5. Heatmaps representing similarity matrices of graphs from HMM-600-u with the parameter $P = 3$, $r = 2$, and $k = 3$. Figure 5(b) and (c) successfully show clear clusters as rectangles (surrounded by dashed lines).

We would like to show computed graphs and the discovered clusters by the spectral clustering algorithm via temporal/periodical graphs. To illustrate the discovered clusters, we additionally adopt the k-means algorithm in the spectral clustering. We set k by using the heuristic of the spectral clustering (Please see [11]), and use the normalized graph laplacian. In the following, we prepare the following labels to represent methods: (1) DT means the temporal skeletonization, (2) DP users the periodical information only, (3) DTP adopts both (1) and (2), and (4) EXP replaces the function $\delta(\cdot)$ with $\exp(\cdot)$ in the DTP setting.

We implemented the periodic skeletonization part in C++[8], and apply the spectral clustering algorithm (and k-means algorithm in it) by using a built-in implementation provided by SCIKIT-LEARN [8] based on Python 2.7.8. All experiments are run on a machine of Mac OS X 10.9 with $2 \times 2.26\,GHz$ Quad-Core Intel Xeon processors and 64 GB memory.

Results and Discussions. Out of several parameter settings we tried, we took a part of results to compare our periodical skeletonization with the temporal skeletonization. We showed results of synthetic data in Fig. 5, and those of real datasets in Fig. 6 with varying methods of computing weights slightly, where the labels $\{DT, DP, DTP, EXP\}$ represent the corresponding methods above used for computing similarity graphs of the temporal/periodical graphs.

From results using synthetic data, we can conjecture that periodical information of temporal graphs are helpful to find clusters of symbols compared with Fig. 5(a), (b) and (c), where we would like to extract periodical clusters, that

---

[8] GCC 4.7 with -STD=C++11 without any parallelization techniques.

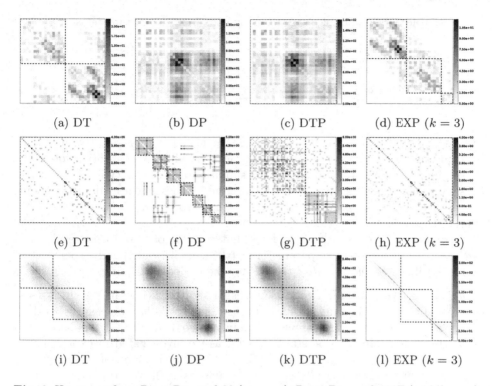

(a) DT   (b) DP   (c) DTP   (d) EXP ($k = 3$)

(e) DT   (f) DP   (g) DTP   (h) EXP ($k = 3$)

(i) DT   (j) DP   (k) DTP   (l) EXP ($k = 3$)

Fig. 6. Heatmaps from PowerDemand-32 (top row), PowerDemand-128F (middle row) and Kyoto (bottom row) with varying DT, DP, DTP, and EXP.

is, clusters representing $\{0, 1\}$ and $\{2, 3\}$, which corresponds to T_1 and T_2 in the HMM in Fig. 4, respectively. From the result using only temporal information in Fig. 5(a), however, we *cannot* find them. On the another hand, results using periodical information seen in Fig. 5(b) and both of them in Fig. 5(c) show two clusters $\{0, 1\}$ and $\{2, 3\}$ much clearly. The $\exp(\cdot)$ computation in Fig. 5(d) disturbs them.

Results from real datasets should be affected by properties of sequences and the periodicity of them. In two cases with PowerDemand-32 and PowerDemand-128F, for example, results were symmetric with respect to methods: If we use the periodical information in Fig. 6(b) and (c), we cannot find any clusters but in Fig. 6(f) and (g), we can find a few clusters of symbols, which are similar to results of synthetic data. We guessed that the difference between PowerDemand-32 and PowerDemand-128F is whether or not there exists many periodical behaviors in sequences. Because we selected a subsequence from PowerDemand-32 as PowerDemand-128F to remove multiple periodical information, the periodical skeletonization with a fixed period parameter $P = 7$ seemed to work well.

In results from Kyoto, we can see that there exist roughly 3 clusters in all outputs from Fig. 6(k) to (l). If we adopt the periodical information, those clusters are also emphasized on visualized temporal graphs. For example, by comparing

Table 2. Statistics of parts of logs by users from the Last.fm dataset.

User ID	$|L_{\text{all}}|$	$|L_{\text{ne}}|$ (non-empty intervals)	$|\Sigma|$	$\mathcal{S}$
User 672	384	147	247	2,329
User 808	529	147	578	2,108

results in Fig. 6(i) and (k), we confirmed that two dense clusters (top left and bottom right) in Fig. 6(k) are much more clear than in those Fig. 6(i). Note that these clusters are related to winter and summer, respectively. We inferred that these visualized results are helpful to analyze given sequential databases and enumerated patterns, particularly when we need to run methods iteratively to tune parameters.

Note that in both synthetic and real datasets, we observed that results are *not sensitive* with respect to parameters r (the window width) and k (parameter of $\exp(\cdot)$) if r is *enough smaller* than P for the periodical setting.

Conclusions. We conclude experiments using sequences containing clear periodical behaviors. Originally, results of clustering symbols are sensitive to definitions of similarities. The previous study reported in [6] that results of the skeletonization seemed to be stable. As far as we investigated in experiments, with respect to the parameters r and k, which control a kind of smoothing of symbols sequences, the results could be stable as well. We also see that our method could be helpful to *highlight periodical behaviors of sequences*. We guess that this result is also affected from the multiple periodicity, and conclude that the periodical skeletonization help us to find underlying structures. Although the method sometimes (as seen in PowerDemand-32) disturbs results, it seems to work as we expected particularly when the periodicity is clear.

4.2   A Case Study Using Last.fm Datasets

This section deals with datasets collected from the LAST.FM site. Because properties of data vary according to users, we would like to investigate how results get for real datasets. Datasets are obtained from [2] by gathering and ordering the logs of songs listened by users based on focusing the granularity "hour". One database corresponds to one sequence of sets of symbols (i.e., songs) by one user. For experiments, we take randomly users from the whole dataset, obtain sequences of sets of symbols, and use small parts of such sequences. We provide statistics of selected parts of sequences chosen by our method in Table 2. For a sequence $s = S_1, S_2, \ldots, S_M$ of $S_i \subseteq \Sigma$, L_{all} means $|s| = M$, L_{ne} shows $|\{S_i \mid S_i \neq \emptyset, S_i \text{ is in } s\}|$, Σ means the size of the set $|S_1 \cup \cdots \cup S_M|$, and $|\mathcal{S}| = \sum_i |S_i|$, respectively. We set $P = 24$ to analyze hourly behaviors.

We show results in Fig. 7. Here we do not want to say which clustering results are good (or bad). From experiments by periodical information in the skeletonization we confirmed two kind of results: A type increases the number of clusters compared with the ordinal temporal skeletonization (e.g., from Fig. 7(a) to (c)).

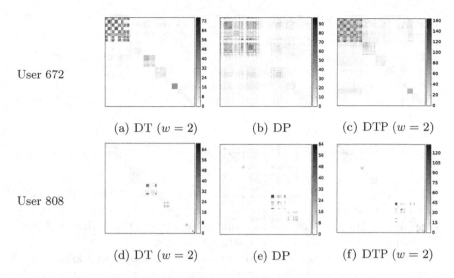

User 672

(a) DT ($w = 2$)   (b) DP   (c) DTP ($w = 2$)

User 808

(d) DT ($w = 2$)   (e) DP   (f) DTP ($w = 2$)

Fig. 7. Heatmaps from Last.fm datasets with varying DT, DP, and DTP.

Another type, in contrast, decreases the number of clusters (e.g., from Fig. 7(a) to (b), Fig. 7(d) to (e) and (f)). As the periodical information help us to consider periodical co-occurrences of symbols, if there exist some periodical behaviors of sequences, then applying the periodicity skeletonization should be helpful. We can only confirm that in some cases the clustering work for our purpose. We conjecture that for some databases our method does not work as they contain no periodical regularities.

Discussions. Discussing the quality of clusters is fundamentally impossible as we do not have any labels. Conceptually, the skeletonization does *not* use any semantic information of symbols, and results only depend on the co-occurrences of symbols. In our method, we intend that adding more computations by the periodicity have increased information we can use in the pre-processing step. Introducing additional resources for computing the similarities such as background knowledge or taxonomy is one of interesting future work. However, such knowledge resources are in general *high cost* compared with the skeletonization. Therefore, we guess that combining both methods is much effective for solving the sparseness problem. In addition, we also expect that introducing sophisticated clustering algorithms is important: For example, hierarchical spectral clustering [1], Non-negative Matrix Factorization (NMF) for clustering (e.g., [3]) should be helpful (e.g., considering multiple periods with hierarchy).

4.3 Partially Periodic Pattern Mining and Skeletonization

We now apply the periodical skeletonization into the FPPPM problem. Our purpose here is to obtain comprehensive and high-level patterns in mining by reducing the size $|\Sigma|$. We use clustering results obtained by the above experiments. For enumeration of patterns, we use the algorithm proposed by Yang

---

Algorithm 1. FPPPM with The Periodical Skeletonization

1: **procedure** PS-FPPPM(s (input), θ (threshold))
2:     Construct a periodical graph G from s ▷ The Periodical Skeletonization
3:     Estimate the number k of clusters with heuristics
4:     Compute assignments of symbols to clusters   ▷ $\Sigma \rightarrow \{1, 2, \ldots, k\}$
5:     Sort clusters $C_1, \cdots C_k$ with their cardinality ($|C_1| > |C_2| > \ldots |C_k|$)
6:     **for** $j = 1$ to k **do**
7:         Replace symbols in s using the cluster C_j and get $s^{(\geq j)}$ ▷ Re-encoding
8:         Apply PPPMINER for re-encoded $s^{(\geq j)}$ with the threshold θ

---

et al. [12], and call it PPPMINER. For the purpose, we re-implemented PPP-MINER in Python 2.7.8. Experimental settings are the same to those in Sect. 4.1. To examine how the periodical skeletonization affects enumerating patterns by the PPPMINER, we use Kyoto and Last.fm datasets.

Before mining, the number k of clusters is estimated by a well-known heuristics as we mentioned [11]. With this k, we propose an *incremental method*, where users replace symbols with cluster labels incrementally. The overview of this process is shown in Algorithm 1. We first sort clusters by the size. In descending order of the size, we incrementally re-encode an original sequence s with cluster labels as follows: First, we replace symbols in s belonging to the largest cluster labeled by C_1 (this new sequence is denoted by $s^{(\geq 1)}$ in Algorithm 1), second we do the same with the second largest cluster labeled by C_2, and so on (corresponding to Line 6 and Line 7). For each step in Line 8, we apply the PPPMINER to mine frequent partially periodic patterns.

Results. For the Kyoto dataset, let $k = 3$. We prepared four cases: Kyoto (original), Kyoto$^{(\geq 1)}$, Kyoto$^{(\geq 2)}$, and Kyoto$^{(\geq 3)}$. We show the number of enumerated patterns with $P = 365$ and with varying the threshold θ in Table 3(a).

For the User 672 dataset from the Last.fm dataset, we first apply both skeletonization methods as shown in Fig. 7(c) as well. We use the number $k = 10$ of clusters to pre-process. Out of $k = 10$ clusters illustrated in Fig. 7(c), for the integer j in Line 6, we use the largest cluster C_1 and get the re-encoded sequence User 672$^{(\geq 1)}$ corresponding to $j = 1$. In the same manner, we adopt the top three largest clusters C_1, C_2, and C_3 (i.e., $j = 3$) and get the sequence User 672$^{(\geq 3)}$. We finally use all clusters ($j = 10$) and label the obtained sequence as User 672$^{(\geq 10)}$. We show in Table 3(b) the numbers of enumerated patterns.

Discussions. In both cases we cannot find any frequent patterns *without* the periodical skeletonization. That is, without any pre-processing, databases are sparse and we cannot evaluate the support count well to get insights from datasets in the standard manner of partially periodic patterns mining. However, with thanks to the periodical skeletonization, we can discover many frequent patterns in other cases. Because the periodical skeletonization help us to find rough, characteristic patterns by clustering, we are now able to discover abstract but readable and high-level frequent partially periodic patterns.

Table 3. Numbers of enumerated patterns with (i.e., re-encoding labeled as $^{(\geq j)}$) and without (i.e., original data) the skeletonization.

(a) For the Kyoto dataset ($P = 365$)

Datasets	$\theta = 0.9$	0.7	0.5	$|\Sigma|$
Kyoto$^{(\geq 1)}$	9,065	57,596	133,027	224
Kyoto$^{(\geq 2)}$	28,134	210,806	523,021	97
Kyoto$^{(\geq 3)}$	54,354	349,648	917,403	3
Kyoto	0	0	0	359

(b) For User 672 dataset ($P = 24$)

Datasets	$\theta = 0.3$	0.2	0.1	$|\Sigma|$
User 672$^{(\geq 1)}$	128	318	51,304	177
User 672$^{(\geq 3)}$	128	319	22,540	144
User 672$^{(\geq 10)}$	127	260	5,718	10
User 672	0	0	0	247

For example in the Kyoto$^{(\geq 1)}$ setting, we found 9,065 patterns which characterize 90 % of segments in the given sequence. In addition, in the settings of User 672$^{(\geq 1)}$ and User 672$^{(\geq 3)}$, we *successfully* discover roughly 300 frequent patterns that characterize 20 % of segments, and this number is relatively small and easy to analyze. We also glanced at patterns listed with the cluster C_1 (the largest one), combinations of C_1 and songs ids $20, 22, 23, 27$ appeared frequently, i.e., patterns containing $(C_1, 20)$, $(C_1, 20, 22)$, etc. are mined many times.

We conjectured that these comprehensive and high-level patterns containing cluster labels would be helpful to analyze periodical behaviors in databases, particularly when we have some background knowledge about symbols in Σ. In addition, we confirmed that clustering using the skeletonization as a pre-processing of pattern mining work well to get more frequent patterns than those obtained from raw sequences. In our experiments, many patterns constructed by shifting symbols are also mined. For example, we often have the case a frequent pattern p occurring at 8 a.m. again becomes frequent at 9 a.m., 10 a.m., and so on. That is, the primitive definition of partially periodic patterns have some redundancy. It is our important future work to overcome such a redundancy problem. Tuning hyper-parameters including the number k of clusters and the width r of sliding windows is also our future work to enrich partially periodic pattern mining with pre-processing using the skeletonization. Further studies for the multiple periodicity are also our interesting future direction.

5   Concluding Remarks and Future Work

In this paper we have provided a new skeletonization method for dealing with partially periodic patterns based on the temporal skeletonization and periodical information. Our experiments with synthetic and real datasets have shown that our method could help us to obtain clusters of symbols even for periodical settings, particularly for a case where sequences have only one fixed period. Pattern mining results with the skeletonization have indicated that our method is helpful to obtain readable results with a relatively small computational cost as Σ get small. Even we use a large threshold θ, we can find frequent patterns which cannot be listed without the skeletonization. Using Last.fm datasets in our case study, we have tested that our method give some insights on relation of symbols used for describing databases, and their analyses might be important and helpful for Knowledge Discovery.

In future work, we would like to develop algorithms to reduce the redundancy of patterns more, based on well-studied concepts (e.g., closed patterns) together with the skeletonization. Further discussion using other pre-processing methods using semantic information (i.e., hierarchy) are also our important future work.

Acknowledgments. The authors would like to thank anonymous reviewers for their valuable comments. This study was partially supported by Grant-in-Aid for JSPS Fellows (26-4555) and JSPS KAKENHI Grant Number 26280085.

References

1. Alzate, C., Suykens, J.A.: Hierarchical kernel spectral clustering. Neural Netw. **35**, 21–30 (2012)
2. Celma, O.: Music Recommendation and Discovery in the Long Tail. Springer, Heidelberg (2010)
3. Cichocki, A., Zdunek, R., Amari, S.I.: Nonnegative matrix and tensor factorization [lecture notes]. IEEE Sig. Process. Mag. **25**(1), 142–145 (2008)
4. Han, J., Dong, G., Yin, Y.: Efficient mining of partial periodic patterns in time series database. In: Proceedings of 15th ICDE, pp. 106–115 (1999)
5. Han, J., Gong, W., Yin, Y.: Mining segment-wise periodic patterns in time-related databases. In: Proceedings of 4th KDD, pp. 214–218 (1998)
6. Liu, C., Zhang, K., Xiong, H., Jiang, G., Yang, Q.: Temporal skeletonization on sequential data: patterns, categorization, and visualization. In: Proceedings of 20th KDD, pp. 1336–1345 (2014)
7. Ng, A.Y., Jordan, M.I., Weiss, Y.: On spectral clustering: analysis and an algorithm. Adv. Neural Inf. Process. Syst. **13**, 849–856 (2001)
8. Pedregosa, F., Varoquaux, G., Gramfort, A., Michel, V., Thirion, B., Grisel, O., Blondel, M., Prettenhofer, P., Weiss, R., Dubourg, V., Vanderplas, J., Passos, A., Cournapeau, D., Brucher, M., Perrot, M., Duchesnay, E.: Scikit-learn: machine learning in python. J. Mach. Learn. Res. **12**, 2825–2830 (2011)
9. Pei, J., Han, J., Mortazavi-Asl, B., Wang, J., Pinto, H., Chen, Q., Dayal, U., Hsu, M.C.: Mining sequential patterns by pattern-growth: the prefixspan approach. IEEE Trans. Knowl. Data Eng. **16**(11), 1424–1440 (2004)
10. Shi, J., Malik, J.: Normalized cuts and image segmentation. IEEE Trans. Pattern Anal. Mach. Intel. **22**, 888–905 (1997)
11. Von Luxburg, U.: A tutorial on spectral clustering. Stat. Comput. **17**(4), 395–416 (2007)
12. Yang, K.J., Hong, T.P., Chen, Y.M., Lan, G.C.: Projection-based partial periodic pattern mining for event sequences. Exp. Syst. Appl. **40**(10), 4232–4240 (2013)

Predicting Drugs Adverse Side-Effects Using a Recommender-System

Diogo Pinto[2](✉), Pedro Costa[2], Rui Camacho[1,2], and Vítor Santos Costa[1,3]

[1] INESC TEC, Porto, Portugal
[2] DEI and Faculty of Engineering, University of Porto, Porto, Portugal
diogojapinto@gmail.com
[3] DCC, Faculty of Sciences, University of Porto, Porto, Portugal

Abstract. Adverse Drug Events (ADEs) are a major health problem, and developing accurate prediction methods may have a significant impact in public health. Ideally, we would like to have predictive methods, that could pinpoint possible ADRs during the drug development process. Unfortunately, most relevant information on possible ADRs is only available after the drug is commercially available. As a first step, we propose using prior information on existing interactions through recommendation systems algorithms. We have evaluated our proposal using data from the ADReCS database with promising results.

Keywords: Adverse drug effects · Adverse drug reactions · Singular value decomposition · Recommender-systems · Pharmacovigilance

1 Introduction

Adverse Drug Events (ADEs)[1] are events that indicate a relationship between the treatment and a negative outcome. It is estimated that, in the United States alone, ADEs account for up to 28 % of all emergency department visits [8], and 5 % of hospital deaths [3]. As a consequence, between 30 and 150 billion dollars are spent annually in hospitals treating those adverse events [4]. There is thus, not only a moral obligation on pursuing safer medicines, but also strong economic impact.

Randomized Controlled Trials (RCTs) are the main tool used to ensure drug quality. They are conducted in standardized conditions, nonetheless, authors have noticed under-representation of women and elderly patients in those trials [6]. Alongside RCTs being conducted regardless of the specific features of the drug or the patient, they often use small samples and with very little statistical significance. Due to these limitations, only ADEs that are common and that develop over short periods of time can be detected with high-confidence.

In this work, we aim at taking advantage of the ability of Machine Learning to process large amounts of data in order to find hidden connections. Our method is as follows. First, we collect data that would be publicly available before the

---

[1] Also referred to as Adverse Drug Reactions (ADR).

© Springer International Publishing Switzerland 2015
N. Japkowicz and S. Matwin (Eds.): DS 2015, LNAI 9356, pp. 201–208, 2015.
DOI: 10.1007/978-3-319-24282-8_17

drug enters the market. Second, we feed that information to a recommendation system. The output is the set of side-effects with higher estimated probability. We experimented our method, using two methods: Singular Value Decomposition (SVD) and Restricted Boltzmann Machines (RBM), and then combining them as an ensemble classifier.

The main contribution of the present work is the adaptation and evaluation of recommender systems to the problem of predicting ADEs. As the empirical evaluation shows, the technology is scalable and flexible, and enables ADE prediction at any stage of the drug's development (with special focus on the pre-marketing stage).

2   Related Work

Most ADR research is done on the post-marketing stage, where not only there is more information available, but also when large amounts of money were already invested and the cost of discovering a new ADE is considerably higher. Such research has relied on a variety of data sources. One major source has been electronic health records (EHRs), even though they pose challenges of their own [7].

Our approach is inspired on the excellent performance of recommender systems in sparse domains [5]. A significant boost to research in recommender systems was due to the NetFlix challenge. The winning entry of the competition was an ensemble of several algorithms, including various Singular Value models blended with RBM [2]. Our work applies and adapts these methods to the challenging task of ADE prediction.

3   Methods and Algorithms

3.1   Singular Value Decomposition

Formalization. The drug-ADE relationship is represented as a matrix $M \in R^{m \times n}$, where m is the number of drugs and n the number of ADEs. Whenever a drug d is known to cause ADE a, $M_{da} = 1$ This representation causes M to be sparse.

Matrix factorization allows not only the mapping of drugs and ADEs in factor-spaces but also the reduction of the matrix dimensionality. Consider that each drug is associated with a vector p_i and each ADE with a vector q_i such that:

$$M = PQ^\top \tag{1}$$

The Singular Value Decomposition (SVD) is a factorization of a real (in our case) or complex matrix. Let's consider the factorization of a real valued matrix. Formally, the singular value decomposition of an $m \times n$ matrix M is a factorization of the form $M = U \Sigma V^\top$, where U is an $m \times r$ orthonormal matrix, Σ is an $r \times r$ diagonal matrix with positive, non-zero, singular values in decreasing order, and $V^\top$ is an $r \times n$ orthonormal matrix, where r is the rank of matrix M. Then, it is possible to obtain matrices P and Q from Eq. 1 by:

$$P = U\sqrt{\Sigma} \qquad (2)$$

$$Q = \sqrt{\Sigma}V^\top \qquad (3)$$

We are interested in matrix Q of size $r \times n$, whose entries represent the "meta" relations between r pseudo-drugs and the n ADEs.

Dimensionality Reduction. The model generated by the method described above might suffer from over-fitting, since it would fit the noise present in M. One solution to generalize the model and reduce the effects of the noise is to find a matrix $\hat{M}$ which is the best rank k approximation of M, with $k < r$.

The problem to be solved is, then, to find the optimal value for k. The energy of the factorization of a matrix is defined by Rajaraman and Ullman [9] as the sum of the squares of all its singular values. The new reduced matrix is obtained by discarding a certain amount of that energy. k is the value that minimizes ϵ:

$$\left| \sum_{i=1}^{k} \Sigma_{ii}^2 - \alpha \sum_{i=1}^{r} \Sigma_{ii}^2 \right| = \epsilon, \qquad (4)$$

where $\alpha \in [0,1]$ is the amount of energy we wish to keep. Tests showed that the optimal value of alpha for this problem is 0.9, as greater values reduce the Recall and smaller values reduce the Precision.

Gradient Descent. After the dimensionality reduction step, it is possible to optimize P and Q by using gradient descent.

$$\min_{P,Q} \sum_{training} (m_{xi} - p_i q_x^\top)^2 + \lambda \left[\sum_x \|p_x\|^2 + \sum_x \|q_i\|^2 \right] \qquad (5)$$

The real goal is to find P and Q based on known drug-ADE relations so that we predict well the unseen values. This enables us to approximate missing drug-ADE relations as zeros.

3.2   Restricted Boltzmann Machines

Formalization. Restricted Boltzmann Machines (RBMs) [10] can be used to perform a binary factor analysis. An RBM is a stochastic neural network consisting on a layer of visible units, a layer of hidden units and a bias unit. The visible units represent, in this context, the drug's ADEs that we know. The hidden units are the latent factors that we want the model to learn. The visible units and the hidden units form a bipartite graph.

It is possible to reduce the dimensionality of a feature vector, in the case that the hidden layer has fewer units than the visible layer. By providing a drug d with size $1 \times n$ it is possible to obtain a vector f with its latent factors of size $1 \times l$ with l being the number of hidden units. On the other hand, vector l can also be used to obtain a vector $\hat{d}$. $\hat{d}_i$ represents the probability of drug d causing ADE i.

3.3   Ensembles

SVD and RBM are able to predict the probability of a drug causing a set of ADEs. More formally, given a drug d of size $1 \times n$, each method predicts a drug $\hat{d}$ of the same size:

$$\hat{d} = \begin{bmatrix} \hat{p}_1 & \hat{p}_2 & \cdots & \hat{p}_n \end{bmatrix} \tag{6}$$

where $\hat{p}_i$ represents the probability of drug d causing ADE i.

One can look at the problem of making an element-wise combination of $\hat{d}_{SVD}$ and $\hat{d}_{RBM}$ as a classification problem where, given two probabilities the model classifies the final probability as positive *(i.e. causing the ADE)* or negative *(i.e. not causing the ADE)*; or as a regression problem where, given two probabilities the model computes a new probability.

We have used a Support Vector Machine (SVM) with a Radial Basis Function (RBF) kernel for the classification problem. On the other hand, to solve the regression problem, a Support Vector Regression (SVR) algorithm was used, also with a RBF kernel. The SVR model was trained the same way as the SVM model, nonetheless, a threshold was required in order to be able to classify a drug as causing ADE i or not.

The Receiver Operating Characteristic (ROC) curve was computed and the threshold is obtained by using the Youden index [12], which maximizes *Sensitivity + Specificity* − 1. Graphically, the index is represented by the maximum height above the chance line.

4   Empirical Evaluation

4.1   Data

In these experiments we use the ADReCS[2] data-base as ground truth. This drug-ADE database is maintained by researchers at Xiamen University, and includes adverse drugs' reactions ontologies, that enable the standardization and hierarchization of ADE terms. The drug-ADE information of ADReCS was mainly sourced from the drug labels in the DailyMed, maintained by the U.S. National Library of Medicine (NLM) [1].

4.2   Methodology

The samples were randomly split into two different sets: 70 % into a training set and the remaining into a test set. In the case of the SVD and the regression ensemble, the elements of the resulting prediction vary between 0 and 1. A threshold is needed to distinguish between a positive and a negative example. To do that, the ROC curve is computed by using a validation set, and the Youden index [12] is used as the threshold, as described in Subsect. 3.3. The validation set is computed differently for each method and, therefore, it is explained in the corresponding Subsection. To test the model, the testing set is used. For each

---

[2] http://bioinf.xmu.edu.cn/ADReCS.

element of the testing set, 30 % of the ADEs are randomly removed and used as input to the model. The Precision, Recall and, whenever possible, the ROC area are computed.

SVD Experiments. To build the model, k-fold cross validation was applied to the training set, with $k = 10$. At each iteration, 9 folds are chosen as matrix M and matrix Q is computed, as described in Subsect. 3.1, leaving the remaining fold as the validation set.

After obtaining the model, the method was tested using the testing set. The computed metrics are presented on Table 1.

Table 1. Results of the SVD by removing 30 % of each drug's known ADEs present in the test set

	ROC area	Precision	Recall
Average	0.954	0.373	0.843
Standard deviation	0.054	0.120	0.136
Minimum	0.410	0.000	0.000
Maximum	1.000	1.000	1.000

The system performs well on the majority of the elements from the test set, as can be seen by the large ROC area and small standard deviation. With this approach, on average, the system is able to find 84 % of the ADEs of each drug. Nonetheless, about 37 % of the elements classified as positive are, indeed, positive, as concluded from the Precision.

A sensitivity analysis was performed, by varying the number of removed ADEs from the test set, in order to evaluate the system's performance under different conditions. As show in Fig. 1, the system's performance deteriorates as the level of information is reduced.

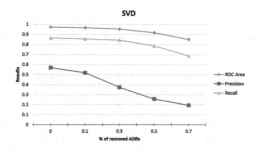

Fig. 1. Sensitivity Analysis of the different metrics by varying the number of removed ADEs from the test set

Table 2. Results of the RBM by removing 30 % of each drug's known ADEs present in the test set

	ROC area	Precision	Recall
Average	0.950	0.572	0.758
Standard deviation	0.051	0.276	0.196
Minimum	0.417	0.000	0.000
Maximum	1.000	1.000	1.000

RBM Experiments. The RBM model was built using 1000 hidden nodes, and was trained for 300 epochs. The results are presented on Table 2. Also, a sensitivity analysis was performed, the same way as for the SVD, and is presented on Fig. 2.

It is possible to conclude that the RBM provides better Precision but with lower Recall than the SVD. Also, this method deals better with the absence of information than the SVD.

Ensemble. In order to combine the two methods, 10 % of the training set is used as validation set. The results of combining SVD and RBM using the classification approach and the regression approach are compared on Table 3. Again, a sensitivity analysis was performed on the two approaches and presented on Fig. 3.

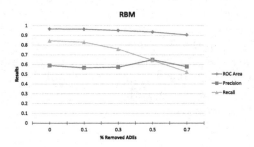

Fig. 2. Sensitivity Analysis of the different metrics by varying the number of removed ADEs from the test set

Table 3. Results of combining SVD and RBM, using SVM and SVR, by removing 30 % of each drug's known ADEs present in the test set

	Precision(SVM)	Recall(SVM)	ROC(SVR)	Precision(SVR)	Recall(SVR)
Average	0.974	0.718	0.990	0.687	0.909
SD	0.112	0.198	0.036	0.128	0.140
Minimum	0.000	0.000	0.509	0.000	0.000
Maximum	1.000	1.000	1.000	1.000	1.000

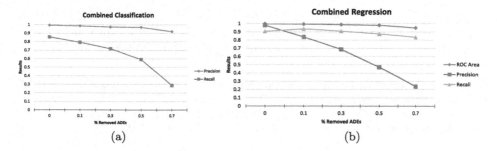

Fig. 3. Performance of all methods by varying the number of missing ADEs. (a) Precision. (b) Recall.

4.3 Discussion

By analyzing Fig. 4 it is possible to conclude that the combined approach is able to maximize one of the two metrics, but not both. The classification one is able to achieve high precision but has low recall, on the other hand, the regression approach achieves high recall but low precision.

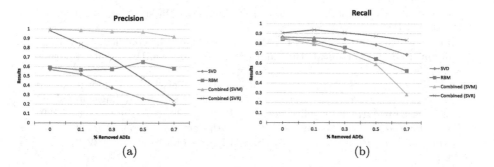

Fig. 4. Results of using the classification and regression approaches to combine the SVD and RBM models. (a) Classification. (b) Regression.

5   Conclusions and Future Work

The method presented here serves as a basis for further expansion. It is capable of taking other data to strengthen its results *e.g.*, molecular descriptors, molecular substructures, literature statistical analysis or even patients information.

Another connection particularly interesting is the comparison of the results with drug→side-effect reports that can be mined from a database such as the FDA Adverse Event Reporting System (FAERS), based on the approach of Rong Xu and QuanQiu Wang [11]. On the other hand, the comparison against other methods, such as different variations of SVM and Random Forests, could give more insight and even boost the precision and recall of the ensemble method.

In sum, there is still much work to be done based on this method, and most important, as a reminder to the research community of the importance on focusing on pre-marketing prediction (and consequent prevention) strategies for ADEs.

Acknowledgements. The authors gratefully acknowledge the financial support of Fundação para a Ciência e Tecnologia (FCT), through the research project "ADE - Adverse Drug Effects Detection" (PTDC/EIA-EIA/121686/2010), as well as the Master in Informatics and Computing Engineering (MIEIC) at FEUP.

References

1. Cai, M.-C., Xu, Q., Pan, Y.-J., Pan, W., Ji, N., Li, Y.-B., Jin, H.-J., Liu, K., Ji, Z.-L.: ADReCS: an ontology database for aiding standardization and hierarchical classification of adverse drug reaction terms. Nucleic Acids Res. **43**, 907–913 (2015)
2. Gower, S.: Netflix Prize and SVD, pp. 1–10 (2014)
3. Juntti-Patinen, L., Neuvonen, P.: Drug-related deaths in a university central hospital. Eur. J. Clin. Pharmacol. **58**(7), 479–482 (2002)
4. Lazarou, J., Pomeranz, B.H., Corey, P.N.: Incidence of adverse drug reactions in hospitalized patients: a meta-analysis of prospective studies. Jama **279**(15), 1200–1205 (1998)
5. Leskovec, J., Rajaraman, A., Ullman, J.D.: Mining of Massive Datasets. Cambridge University Press, UK (2014)
6. Martin, K., Bégaud, B., Latry, P., Miremont-Salamé, G., Fourrier, A., Moore, N.: Differences between clinical trials and postmarketing use. Br. J. Clin. Pharmacol. **57**(1), 86–92 (2004)
7. Page, D., Costa, V.S., Natarajan, S., Barnard, A., Peissig, P., Caldwell, M.: Identifying adverse drug events by relational learning. In: Proceedings of the AAAI Conference on Artificial Intelligence, vol. 2012, p. 790. NIH Public Access (2012)
8. Patel, P., Zed, P.J.: Drug-related visits to the emergency department: How big is the problem? Pharmacother. J. Hum. Pharmacol. Drug Ther. **22**(7), 915–923 (2002)
9. Rajaraman, A., Ullman, J.D.: Mining of Massive Datasets. Cambridge University Press, Cambridge (2011)
10. Smolensky, P.: Information processing in dynamical systems: Foundations of harmony theory (1986)
11. Xu, R., Wang, Q.: Large-scale combining signals from both biomedical literature and the fda adverse event reporting system (faers) to improve post-marketing drug safety signal detection. BMC Bioinform. **15**(1), 17 (2014)
12. Youden, W.J.: Index for rating diagnostic tests. Cancer **3**(1), 32–35 (1950)

Dr. Inventor Framework: Extracting Structured Information from Scientific Publications

Francesco Ronzano[✉] and Horacio Saggion

TALN Research Group, Universitat Pompeu Fabra,
C/Tanger 122, 08018 Barcelona, Spain
{francesco.ronzano,horacio.saggion}@upf.edu

Abstract. Even if research communities and publishing houses are putting increasing efforts in delivering scientific articles as structured texts, nowadays a considerable part of on-line scientific literature is still available in layout-oriented data formats, like PDF, lacking any explicit structural or semantic information. As a consequence the bootstrap of textual analysis of scientific papers is often a time-consuming activity. We present the first version of the Dr. Inventor Framework, a publicly available collection of scientific text mining components useful to prevent or at least mitigate this problem. Thanks to the integration and the customization of several text mining tools and on-line services, the Dr. Inventor Framework is able to analyze scientific publications both in plain text and PDF format, making explicit and easily accessible core aspects of their structure and semantics. The facilities implemented by the Framework include the extraction of structured textual contents, the discursive characterization of sentences, the identifications of the structural elements of both papers header and bibliographic entries and the generation of graph based representations of text excerpts. The Framework is distributed as a Java library. We describe in detail the scientific mining facilities included in the Framework and present two use cases where the Framework is respectively exploited to boost scientific creativity and to generate RDF graphs from scientific publications.

Keywords: Scientific text mining · Scientific information extraction · Software framework

1 Introduction: Mining Scientific Publications

Several studies concerning scientific text mining are built on the analysis of corpora populated with proper collections of scientific publications. The selection of the publications to include in these corpora is driven by the final goal of the study, but is also often **strongly influenced by the publishing format and the related ease of access to the contents of papers.** In order to perform text

The work described in this paper has been funded by the European Project Dr. Inventor (FP7-ICT-2013.8.1 - Grant no: 611383).

N. Japkowicz and S. Matwin (Eds.): DS 2015, LNAI 9356, pp. 209–220, 2015.
DOI: 10.1007/978-3-319-24282-8_18

mining, most of the tools need as input the plain text of a whole paper or parts of it. Even if several research communities and publishing houses are increasingly adopting structured, Web-friendly, textual formats to share papers, including XML dialects like JATS[1] [1], Elsevier Schemas[2] and RASH[3], layout-oriented data formats, like PDF, still represent one of the most exploited means to deliver publications and scientific results. Consequently, often the initial phases of scientific text mining studies suffer from the difficulty to access the information contained in papers because of both their unavailability as plain texts and the lack of any kind of structured way to browse their contents.

In order to mitigate this problem, we present the first version Dr. Inventor Framework, an integrated collection of text mining components useful to support the initial steps of scientific literature analysis. The Framework is delivered in the form of a publicly available self-contained Java library[4]. By integrating and customizing several text mining tools and on-line services, the Framework is able to process distinct facets of the contents of scientific publications. Papers, both plain texts and PDF documents, are analyzed so as to automatically identify and enrich their relevant structural and semantic elements. The Framework has been conceived and developed in the context of the European Project Dr. Inventor to support scientific creativity by automatically identifying analogies in document collections. Anyway, a wide range of scientific text mining analyses can be enabled and eased by relying on it.

In this paper we introduce the architecture of the first version of the Dr Inventor Framework. The set of Natural Language Processing components and external Web services that the Framework integrates and extends is described in detail in Sect. 2. Section 3 presents two use cases where the text analysis components of the Framework are exploited to support more complex tasks. In particular the rich structured representation of the contents of a paper that is generated by the Framework is used to extract knowledge graphs in order to foster scientific creativity and to represent the contents of the same paper as a RDF dataset. In Sect. 4 we provide our future plans to extend an evaluate the Dr. Inventor Framework.

2   Dr. Inventor Framework

The first version of the Dr. Inventor Framework enables a core set of structural, linguistic and semantic analyses over scientific papers. This Section provides, first of all, an overview of the general architecture of the Framework (Subsect. 2.1). In this way we introduce the high-level platform where the information provided by the text mining tools and Web services that are integrated in the Framework is merged thanks to the adoption of a shared data model of rich scientific

---

[1] http://jats.nlm.nih.gov/.

[2] http://www.elsevier.com/author-schemas/elsevier-xml-dtds-and-transport-schemas.

[3] http://cs.unibo.it/save-sd/rash/documentation/index.html.

[4] The last release of Framework together with the related documentation can be downloaded at: http://backingdata.org/dri/library/.

publication. Then, we describe in detail each one of the scientific text analysis components that are part of the first version of the Framework (Subsect. 2.2).

2.1 Architectural Overview

Considering the extensive adoption of Java as the programming language of choice in a multitude of contexts and the broad availability of both general-purpose and text mining libraries and tools implemented in Java, we chose to rely on Java to build Dr. Inventor Framework. In particular, we exploited and extended the Java-based text engineering architecture GATE [2]. We took advantage of GATE programming facilities to support the modeling, access and storage of textual annotations as well as the integration of external text mining tools and services. In fact, each component of the Framework is structured as a GATE Processing Resource[5].

An overview of the pipeline of text analysis components that are integrated in the first version of Dr. Inventor Framework is shown in Fig. 1. We can notice how each component relies on the output of previous ones and is responsible for the analysis of a defined aspect of a scientific publication. The Framework exposes the result of article analysis by specifying an enriched scientific publication data model: it consists of a set of Java classes that provides easy programmatic access to all the information extracted from a paper.

2.2 Scientific Text Components

This section introduces the set of text mining tools that have been integrated as components of the first version of the Dr. Inventor Framework.

PDF-to-text Converter. Objective: convert a PDF paper to a semi-structured text (XML) by spotting basic structural elements (title, abstract, sections, bibliographic entries, etc.), so as to enable further processing.

Currently there are several tools that support the conversion of PDF files to raw or semi-structured text. From the perspective of scientific text mining, these tools can be grouped into two broad categories: **general-purpose PDF to text converters** including Apache PDFbox[6], Pdf2xml[7] and Poppler[8] and **tools tailored to the conversion of PDF files of scientific articles** like LA-PDFText[9] [3], Cermine[10] [4], SectLabel[11] [5] or PDFX[12] [6]. After a comparative test of these tools, in the first version of the Dr. Inventor Framework,

---

[5] https://gate.ac.uk/sale/tao/splitch7.html.
[6] http://pdfbox.apache.org/.
[7] http://pdf2xml.sourceforge.net/ and http://sourceforge.net/projects/pdf2xml/.
[8] http://poppler.freedesktop.org/.
[9] https://code.google.com/p/lapdftext/.
[10] http://cermine.ceon.pl/.
[11] http://wing.comp.nus.edu.sg/parsCit/.
[12] http://pdfx.cs.man.ac.uk/.

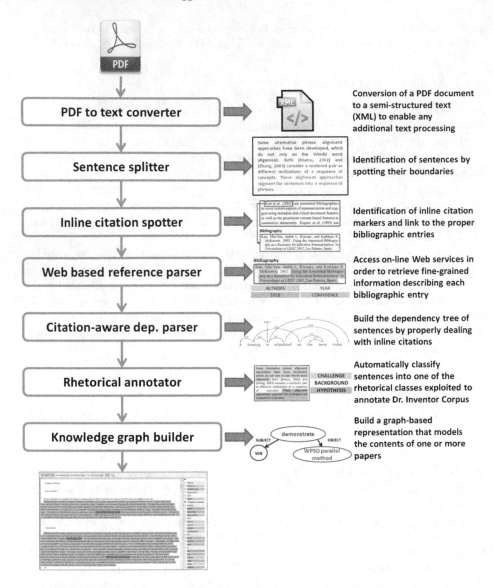

Fig. 1. Architectural overview of the components of Dr. Inventor Framework

we decided to face PDF-to-Text conversion by integrating the on-line Web API of PDFX. PDFX, thanks to its robust rule-based PDF analysis engine, manages to perform most of the times clean and consistent extractions of semi-structured textual contents from PDF files. The XML output of PDFX spots basic structural elements of the converted PDF publication: the title, the abstract, the sections and the bibliographic entries. PDFX has two main restrictions: the current PDF-to-Text conversion is available only as an on-line Web service and cannot deal with PDF files greater than 5 Mb. As a consequence, in future releases of the Framework we are planning to substitute PDFX with a customized version of a general purpose PDF-to-Text converter, like the Apache PDFbox Java library.

Sentence Splitter. Objective: spot sentences of scientific publications by identifying their boundaries.

We customized the rule-based sentence splitter integrated in ANNIE[13], the information extraction system bundled in GATE. To this purpose, we analyzed the sentence split errors performed on a set of 40 Computer Graphics papers (occurring with expressions like: i.e., et al., Fig., Tab.) and modified the sentence splitting rules of ANNIE in order to correctly deal with these situations.

Inline Citation Spotter. Objective: identify the inline citation markers in the text of the paper and link each inline citation marker to the proper reference in the bibliography section of the same paper.

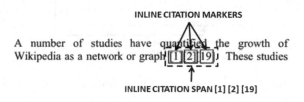

Fig. 2. Inline citation markers and spans

The inline citation spotter performs the following processing tasks (see Fig. 3):

- *task A*: identification of inline citation spans and markers (see Fig. 2) in the textual contents of the paper by means of a set of rules implemented in JAPE, an pattern matching formalism available in GATE[14] tailored to match widespread inline citations styles;
- *task B*: identification of the bibliographic entries, usually listed at the end of the paper. In the first release of the Framework, bibliographic entries are spotted by PDFX;
- *task C*: linking of each inline citation marker to the referenced bibliographic entry by means of a set of heuristics;
- *task D*: identification of the syntactic/non-syntactic role of each inline citation marker, in order to properly support the dependency parsing of the sentence in which the citation marker occurs (see following text analysis components). The first inline citation marker of Fig. 3 has a syntactic role in the sentence (subject), while the second one has no syntactic role. To verify the syntactic role of an inline citation span we exploit the approaches described in [7,8].

---

[13] https://gate.ac.uk/sale/tao/splitch6.html#chap:annie.
[14] https://gate.ac.uk/sale/tao/splitch8.html.

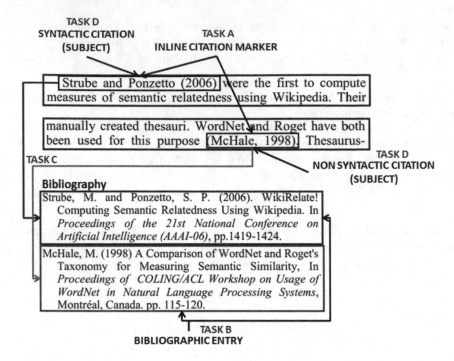

Fig. 3. The processing tasks of the inline citation spotter

Web-Based Reference Parser. Objective: access on-line Web services to parse bibliographic entries so as to retrieve fine-grained publication metadata (title, authors name, year of publication, etc. see Fig. 4).

In the first version of Dr. Inventor Framework we access and merge the results of the following on-line Web services in order to parse bibliographic entries:

- *FreeCite*[15]: this on-line tool analyzes citations by relying on a conditional random field sequence tagger trained on the CORA dataset, made of 1838 manually tagged bibliographic entries[16];
- *CrossRef*[17]: this Web service matches free form citations/bibliographic entries to the items of CrossRef publications metadata archive;
- *Bibsonomy*[18]: its Web API enables the retrieval of the BibTeX metadata of a publication from the Bibsonomy database by providing its title in the query.

We merge the results retrieved by querying the REST endpoints of these three Web services, trying to determine for each bibliographic entry the title of the paper, the year of publication, the list of authors, and the venue or journal of

---

[15] http://freecite.library.brown.edu/welcome.
[16] https://hpi.de/naumann/projects/repeatability/datasets/cora-dataset.html.
[17] http://search.crossref.org/help/api.
[18] http://www.bibsonomy.org/help/doc/api.html.

publication. We give precedence to Bibsonomy results over CrossRef and Freecite outputs, when the outputs of their responses disagree.

Citation-Aware Dependency Parser. Objective: execute the dependency parsing of the sentences of the paper by properly dealing with sentences that include inline citations.

In the first version of Dr. Inventor Framework we rely on the MATE[19] dependency parser [9] to determine the syntactic structure of the sentences of a paper. MATE analyzes a paper sentence-by-sentence by performing the following tasks: tokenization, lemmatization, POS tagging and dependency parsing. We modified the parser to correctly deal with inline citation spans when building the dependency tree of a sentence. In particular we exclude inline citation spans from the textual contents to parse if they have no syntactic role in the sentence to analyze (see the second citation shown in Fig. 3). On the contrary, if we determine that a citation has a syntactic role with respect to the sentence to parse (like the first citation of Fig. 3 where the inline citation span is the subject of the sentence), we replace the whole inline citation span with the word '*citation*' before processing it by MATE.

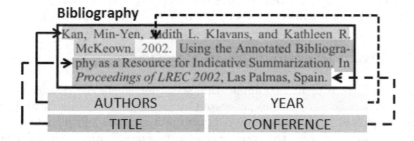

Fig. 4. Example of parsed bibliographic entry

Scientific Discourse Annotator. Objective: automatically classify each sentence of a paper as belonging to one of following scientific discourse categories: Approach, Challenge, Background, Outcomes and Future Work.

We trained and integrated in the Framework a classification model useful to determine the scientific discourse category of the sentences of a paper. The classifier was trained over the set of 8877 sentences belonging to 40 Computer Graphics papers that constitute the Dr. Inventor Rhetorically Annotated Corpus [10]. All the sentences of the Corpus have been manually characterized by three annotators by associating a scientific discourse category, obtaining a value of inter-annotator agreement (Cohen's k) equal to 0.6567. We identified the set of scientific discourse categories that we used to characterize the sentences of the corpus by relying on the scientific discourse annotation schemata proposed

---

[19] https://code.google.com/p/mate-tools/.

by [14,15]. This set of scientific discourse categories has been also determined with the purpose of correctly characterizing the contents of Computer Graphics papers. Alternative sets of scientific discourse categories have been proposed like the one referred to as IRMAD (Introduction, Results, Method, Abstract and Discussion) [16,17], that is tailored to characterize the discursive structure of articles concerning medicine and biology.

The sentence classifier integrated in the Framework exploits both lexical and syntactic features to model each sentence and is based on a Support Vector Machine with linear kernel [11]. We exploited the machine learning Java library Weka[20] to support all the tasks related to rhetorical sentence classification. This classifier obtains a F1 score equal to 0.764 as result of a 10-fold cross validation over the set of sentences of the Dr. Inventor Rhetorically Annotated Corpus.

Knowledge Graph Builder. Objective: build graph-based representations that models the contents of papers.

In the first version of Dr. Inventor Framework we implemented a set of rules to extract subject-verb-object triples from the dependency graph generated by the Citation-aware dependency parser. We merge the contents of each subject and object node with its eventual modifiers. The output of this process constitutes the knowledge-graph generated by the Knowledge graph builder. In Fig. 5 we show an example of a subject-verb-object triple of a knowledge-graph.

3   Dr. Inventor Framework in Practice

In this Section we present two application scenarios where the rich, structured versions of scientific publications generated by means of the Dr. Inventor Framework are exploited to support more complex data analysis and modeling tasks. The first scenario concerns the core aim of Dr. Inventor Project: the automated identification of analogies in scientific literature to support creativity. In the second scenario we deal with the creation of RDF graphs of scientific publications.

3.1   Scenario 1: Looking for Creative Analogies

The Dr. Inventor Framework is developed in the context of the European Project Dr. Inventor. The final aim of this Project is to provide a set of tools and computational approaches to support and stimulate scientific creativity. In particular, the Project focuses on the investigation of analogy-based creativity that is the creative process triggered by the identification of unforeseen analogies between a pair of artifacts (e.g. research papers). Since Dr. Inventor deals with the scientific domain, the core artifacts that are considered are Research Objects. A Research Object is any kind of output of research like publications, reports, patents,

---

[20] http://www.cs.waikato.ac.nz/ml/weka/.

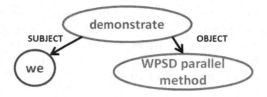

Fig. 5. Example of subject-verb-object triple extracted by the Knowledge-graph builder

datasets, workflows, etc. Dr. Inventor aims at supporting users to explore creative associations between Research Objects by browsing their latent structural similarities.

As explained in detail in [12], Research Objects and in particular the contents of scientific publications are modeled in Dr. Inventor by means of sets of subject-verb-object triples as required by Gentner's theoretical framework to model analogies [13]. By relying on the Knowledge graph builder component, the Dr. Inventor Framework enables the automated extraction and aggregation of such triples to represent the contents of research papers. Figure 6 sketches how, starting from a sentence extracted from a PDF paper, its content is parsed by the Citation-aware dependency parser and then a Knowledge graph is built by extracting subject-verb-object triples. Dr. Inventor is investigating measures

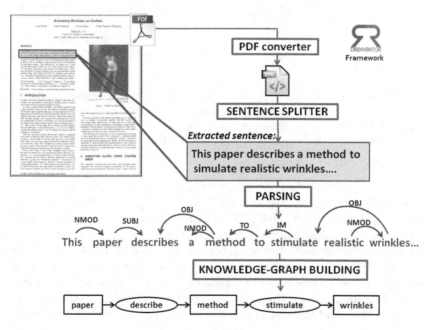

Fig. 6. Example of the steps to extract subject-verb-object triples from a PDF paper (verbs are into rectangular boxes)

of structural and semantic similarity among the triples extracted from scientific publications in order to identify and propose analogies among the corresponding contents of the original papers [12]. An evaluation of the quality of the extracted triples together with their usefulness to support scientific creativity is planned in the following stages of Dr. Inventor Project.

3.2   Scenario 2: Modeling Papers Contents as RDF Datasets

Dr. Inventor Framework has been exploited to enable the automated generation of RDF datasets that represent the contents of scientific articles thanks to the detailed, structured information that can be extracted from scientific publications. In particular, in the context of the Semantic Publishing Challenge[21], organized as part of the European Semantic Web Conference 2015, we have exploited the Framework to extract information from the header and the bibliography of papers as required by one of the three Tasks of the same Challenge. Considering the set of PDF papers that are accessible online on the CEUR-WS Proceedings Web Portal[22], we used the Framework to easily parse their contents. Then we modeled the information we extracted as RDF triples by using widespread Semantic Publishing Ontologies, as shown in Fig. 7.

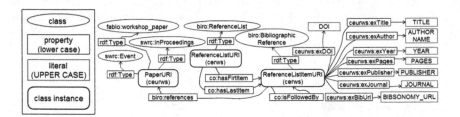

Fig. 7. RDF data model of bibliographic data extracted from scientific publications - Namespace prefixes: co: Component Ontology, fabio: FRBR-aligned Bibliographic Ontology, biro: Bibliographic Reference Ontology, swrc: Semantic Web for Research Communitie Ontology, ceurws: CEUR-WS pubblications Ontology

Thanks to the RDF data modeling of scientific publications, by means of simple SPARQL queries we are able to perform complex bibliographic searches among the papers published at CEUR-WS. For instance we can easily select all the papers that have less than a specific number of citations or that cited only works published before 2010. In the context of the Semantic Publishing Challenge 2015, 10 SPARQL queries where proposed so as to evaluate the information extracted from a set of test papers and modeled as RDF triples. Considering the subset of 5 SPARQL queries that deal with contents mined from the headers and bibliographic entries of the test paper, we achieved an average F-1 score of 0.55.

---

[21] https://github.com/ceurws/lod/wiki/SemPub2015.
[22] http://ceur-ws.org/.

4   Conclusions and Future Works

We presented the initial version of the Dr. Inventor Framework, a publicly available Framework that enables the extraction of several types of structural and semantic information from the plain text or PDF document of scientific publications. The Framework is intended as an integrated collection of utilities useful to support the execution of more complex text analysis tasks over scientific articles. After a detailed description of the components that are integrated in the Framework, we have presented two practical examples of its use in specific scientific literature mining tasks: the extraction of subject-verb-object triples from papers contents in order to foster creative analogies and the generation of RDF datasets describing scientific articles' contents. Dr. Inventor Framework is distributed as a Java library.

The version of the Dr. Inventor Framework presented in this paper is still in its early stages. As future work, we plan to improve the performances and perform both intrinsic and extrinsic evaluations of the components of the Framework by giving priority to the ones that support the information extraction needs of the Project Dr. Inventor. In particular, we aim at improving the Knowledge graph builder by both integrating a co-reference resolution module in order to be able to effectively merge duplicated nodes and by validating its output with respect to the effectiveness in supporting the discovery of scientific analogies. We will integrate scientific paper summarization capabilities by relying on the support of the SUMMA library [18]. Moreover, we plan to integrate in the Framework and customize an existing PDF-to-Text conversion library in order not to rely on external Web service like PDFX. We also plan to improve and further validate the performances of the Scientific discourse annotator by relying on sequence classifiers. To conclude, we would also explore the possibility to provide access to the text analysis components of the Framework by a set of RESTful Web services.

References

1. Huh, S.: Coding practice of the Journal Article Tag Suite extensible markup language. Sci. Editing **1**(2), 105–112 (2014)
2. Cunningham, H., Maynard, D., Bontcheva, K.: Text Processing with GATE. Gateway Press CA, Murphys (2011)
3. Ramakrishnan, C., Patnia, A., Hovy, E.H., Burns, G.A.: Layout-aware text extraction from full-text PDF of scientific articles. Source Code Biol. Med. **7**(1), 7 (2012)
4. Tkaczyk, D., Szostek, P., Dendek, P.J., Fedoryszak, M., Bolikowski, L.: CERMINE-automatic extraction of metadata and references from scientific literature. In: 11th IAPR International Workshop on Document Analysis Systems (DAS), pp. 217–221. IEEE (2014)
5. Councill, I.G., Giles, C.L., Kan, M.Y.: ParsCit: an open-source CRF reference string parsing package. In: LREC Proceedings (2008)
6. Constantin, A., Pettifer, S., Voronkov., A.: PDFX: fully-automated PDF-to-XML conversion of scientific literature. In: Proceedings of the 2013 ACM Symposium on Document Engineering. ACM (2013)

7. Abu-Jbara, A., Radev., D.: Coherent citation-based summarization of scientific papers. In: Meeting of the Association for Computational Linguistics: Human Language Technologies, vol. 1. Association for Computational Linguistics (2011)
8. Abu-Jbara, A., Radev., D.: Reference scope identification in citing sentences. In: North American Chapter of the Association for Computational Linguistics: Human Language Technologies. Association for Computational Linguistics (2012)
9. Bohnet, B.: Very high accuracy and fast dependency parsing is not a contradiction. In: Proceedings of the 23rd International Conference on Computational Linguistics. Association for Computational Linguistics (2010)
10. Fisas, B., Saggion, H., Ronzano, F.: On the discursive structure of computer graphics research papers. In: Proceedings of the Linguistic Annotation Workshop, NAACL (2015)
11. Schlkopf, B., Smola, A.J.: Learning with Kernels: Support Vector Machines, Regularization, Optimization, and Beyond. MIT press, Cambridge (2002)
12. O'Donoghue, D., Abgaz, Y., Hurley, D., Ronzano F., Saggion, H.: Stimulating and simulating creativity with Dr inventor. In: International Conference of Scientific Creativity (2015)
13. Gentner, D.: StructureMapping: a theoretical framework for analogy. Cogn. Sci. **7**(2), 155–170 (1983)
14. Teufel, S., Siddharthan, A., Batchelor, C.: Towards discipline-independent argumentative zoning: evidence from chemistry and computational linguistics. In: Proceedings of the Conference on Empirical Methods in Natural Language Processing, vol. 3, pp. 1493–1502. Association for Computational Linguistics (2009)
15. Liakata, M., Teufel, S., Siddharthan, A., Batchelor, C.R.: Corpora for the conceptualisation and zoning of scientific papers. In: LREC (2010)
16. Agarwal, S., Yu, H.: Automatically classifying sentences in full-text biomedical articles into Introduction, Methods Results and Discussion. Bioinformatics **25**(23), 3174–3180 (2009)
17. Guo, Y., Reichart, R., Korhonen, A.: Improved information structure analysis of scientific documents through discourse and lexical constraints. In: HLT-NAACL, pp. 928–937 (2013)
18. Saggion, H.: SUMMA a robust and adaptable summarization tool. Traitement Automatique des Langues **49**, 103–125 (2008)

Predicting Protein Function and Protein-Ligand Interaction with the 3D Neighborhood Kernel

Leander Schietgat$^{(\boxtimes)}$, Thomas Fannes, and Jan Ramon

Department of Computer Science, KU Leuven,
Celestijnenlaan 200A, 3001 Leuven, Belgium
{leander.schietgat,thomas.fannes,jan.ramon}@cs.kuleuven.be

Abstract. Kernels for structured data have gained a lot of attention in a world with an ever increasing amount of complex data, generated from domains such as biology, chemistry, or engineering. However, while many applications involve spatial aspects, up to now only few kernel methods have been designed to take 3D information into account. We introduce a novel kernel called the 3D Neighborhood Kernel. As a first step, we focus on 3D structures of proteins and ligands, in which the atoms are represented as points in 3D space. By comparing the Euclidean distances between selected sets of atoms, the kernel can select spatial features that are important for determining functions of proteins or interactions with other molecules. We evaluate the kernel on a number of benchmark datasets and show that it obtains a competitive performance w.r.t. the state-of-the-art methods. While we apply this kernel to proteins and ligands, it is applicable to any kind of 3D data where objects follow a common schema, such as RNA, cars, or standardized equipment.

1 Introduction

Over the past years, kernel functions for structured data have gained a lot of attention and were successfully applied to many real-world problems. Chemoinformatics is an area which is of particular interest. Since molecules are naturally represented by graphs, graph kernels have proven very suitable for this kind of problems and they have obtained excellent results [5,6,21]. However, until now, attention has mostly focused on the so-called small (mostly drug) molecules. In this context, even NP-hard problems can usually still be solved efficiently in practice. The ability to tackle proteins, which are two orders of magnitude larger, is a far bigger challenge.

Proteins are macromolecules that play a crucial role in a wide range of biological processes. They are responsible for, e.g., signaling responses between or within cells, the formation of structural elements, or catalyzing chemical reactions. In order to obtain more insights into these processes, many of them have been modeled by machine learning and data mining methods, with applications as predicting the function of proteins [24], the ligands they bind to [6,15], or drug resistance in HIV [8].

© Springer International Publishing Switzerland 2015
N. Japkowicz and S. Matwin (Eds.): DS 2015, LNAI 9356, pp. 221–235, 2015.
DOI: 10.1007/978-3-319-24282-8_19

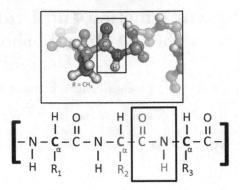

Fig. 1. Chemical structure of the peptide bond (bottom) and the three-dimensional structure of a peptide bond between an Alanine and an adjacent amino acid (top) (from: Wikipedia/Protein). The black box indicates a single peptide bond. We call all the atoms belonging to the side chains or residues (abbreviated as R_1, R_2, R_3 in the figure) side chain atoms. We call all other atoms backbone atoms.

A protein consists of a polypeptide chain of amino acids linked with peptide bonds (Fig. 1). The linked series of carbon, nitrogen and oxygen atoms is known as the protein backbone and the variable parts of the amino acids are the residues or side chains. Proteins can be represented as graphs using different approaches [22]. In some approaches, atoms are represented as vertices and bonds as edges whereas others use nodes to represent amino acids and edges to represent the strength of interaction between the side chains of two amino acids. The structural similarities can be measured with graph mining tools such as kernels [21] or distance measures [25]. Analyzing structural similarity is motivated by the fact that proteins having similar structures are more likely to exhibit common biochemical properties [2].

However, existing graph kernels have multiple limitations. First, they perform poorly on large labeled graphs [26]. Second, they do not take into account 3D information directly. By transforming the protein structures into graphs, information about angles and exact distances is lost. Third, the size of protein graphs have a large impact on their efficiency. In general, it is not clear yet whether learning on 3D structures directly results in accurate models and whether this approach performs better than state-of-the-art graph kernels applied to proteins.

In this paper, we propose a new kernel for 3D data, called the 3D Neighborhood Kernel (3DNK). In contrast to existing kernels, it takes spatial distances into account, focusing on geometry rather than relationships in a graph. As a first step, we focus on two biological applications involving 3D structures of proteins and ligands. The first task involves the classification of proteins into enzymes and Gene Ontology (GO) classes [20], while the second task involves the prediction of binding affinities, representing interaction strength between proteins and ligands [1]. We will compare our kernel to four state-of-the-art methods: two graph kernels (the Fast Subtree Kernel (FSTK) [26] and the Neighborhood

Subgraph Pairwise Distance Kernel (NSPDK) [6]) and two methods that were created specifically to solve the aforementioned biological tasks (the MAMMOTH kernel [20] and RF-Score [1]). Our results show that the 3DNK kernel is competitive with the state-of-the-art methods w.r.t. predictive performance, while it can be applied to a larger variety of tasks.

The remainder of the paper is structured as follows. Section 2 presents the necessary definitions and notation w.r.t. support vector machines and kernel methods. We describe the 3DNK kernel in Sect. 3. Section 4 performs an experimental evaluation of 3DNK and a comparison with the state-of-the-art methods. Section 5 discusses related work and Sect. 6 concludes.

2   Preliminaries

Kernel Functions. Given a set X and a function $K : X \times X \to \mathbb{R}$, we say that K is symmetric if for any x_i and $x_j \in X$ holds that $K(x_i, x_j) = K(x_j, x_i)$, and K is positive-semidefinite if for any $n \geq 1$ and any $x_1, \ldots, x_n \in X$, the matrix K defined by $K(x_i, x_j)$ is positive-semidefinite, that is, $\sum_{ij} c_i c_j K(x_i, x_j) \geq 0$ for all $c_1, \ldots, c_n \in \mathbb{R}$ or equivalently if all its eigenvalues are nonnegative. The function K is called a kernel function and $K(x_i, x_j)$ represents a measure of similarity between x_i and x_j, which can be for example vectors, strings, trees, graphs, or 3D structures.

Support Vector Machines (SVMs). Let X be a non-empty set of n training examples associated with class labels $\{\mathbf{x}_i, y_i\}_{i=1}^{n}$, $\mathbf{x}_i \in X = \mathbb{R}^d$, $d \in \mathbb{N}$ the dimension of input space, and $y_i \in \mathbb{R}$ the target value (discrete in the classification case, a numerical value in the regression case). The task is to learn a function $f : X \to y$ to predict the target value of unlabeled examples. An SVM gives the solution to, for example, the binary classification problem by introducing a hyperplane that separates the training data into positive and negative examples. An infinite number of such hyperplanes exists. Let $\mathbf{w} \in \mathbb{R}^d$ be the weight vector that determines the orientation of the hyperplane and $b \in \mathbb{R}$ be the threshold that determines the offset of the plane from the origin. The class of such hyperplanes is then given by $\langle \mathbf{w}, \mathbf{x} \rangle + b = 0$ and corresponds to the decision function $f(\mathbf{x}) = sgn(\langle \mathbf{w}, \mathbf{x} \rangle + b)$. The maximum margin classifier identifies the optimal hyperplane that is distinguished by the maximum distance from the nearest training objects in both the classes. The optimal solution is unique and sparse, and it is determined by data points close to the decision boundaries called *support vectors*. In the regression case, a similar method is applied, but the orientation of the hyperplane is determined by the ϵ-sensitive loss, which measures the deviation from the target values y_i [7].

In many real-world situations, the data are not easily separable in the input space. However, linear separation can be achieved if the input data are projected onto some higher dimensional dot product feature space F. Let $\phi : X \to F$ be a non-linear mapping from input space X to feature space F. Surprisingly, the explicit mapping of data from input space to feature space is not needed. A kernel function, mapping $K(x_i, x_j)$ to $\langle \phi(x_i), \phi(x_j) \rangle$ makes it possible to compute dot products in the feature space without explicitly knowing the mapping ϕ [7,28].

3   The 3D Neighborhood Kernel

In this section, we will first present a family of kernels that act on 3D point sets (Sect. 3.1). Then, we discuss how we instantiate different kernels from this family in order to solve two specific biological tasks (Sect. 3.2). Finally, we show how we can make the kernel more efficient (Sect. 3.3).

3.1   The 3D Neighborhood Kernel Family

Let $P = \mathbb{R}^3 \times \Sigma$ be the set of all 3D points, embedded in a Euclidian space and labeled over an alphabet Σ. For a point $p \in P$, let $\zeta(p)$ and $\lambda(p)$ represent its 3D coordinates, i.e., a tuple $\langle x_p, y_p, z_p \rangle$, and its label, respectively. Furthermore, let the input space $\mathcal{X} = 2^P$ represent the set of all possible 3D point sets. We will call a point set $X \in \mathcal{X}$ an example.

Let $n \in \mathbb{N}$ be a parameter. Let $\mathcal{F}_\Delta$ be the family of all functions $\Delta : \mathcal{X} \to \mathcal{X}$ for which $\forall X : \Delta(X) \subseteq X$, i.e., for any point set, Δ outputs a subset of that point set. Let $\mathcal{F}_\Phi$ be the family of all functions $\Phi : \mathcal{X} \times P \to P^n$ such that for every $X \in \mathcal{X}$ and $p \in X$, $\Phi(X, p) = \langle p_1, p_2, \ldots, p_n \rangle \in X^n$ is a tuple of n points with $p_i \in X$ for $1 \leq i \leq n$. For a function $\Phi \in \mathcal{F}_\Phi$, we define:

$$d_\Phi(X, p) = \langle \|\zeta(p) - \zeta(p_1)\|, \|\zeta(p) - \zeta(p_2)\|, \ldots, \|\zeta(p) - \zeta(p_n)\| \rangle,$$

where $\Phi(X, p) = \langle p_1, p_2, \ldots, p_n \rangle$. This function generates a tuple of Euclidian distances, where the distances are those from p to the corresponding point decided by $\Phi(X, p)$.

The idea of the 3D Neighborhood Kernel (3DNK) is to compare point sets based on their 3D structure. More specifically, the kernel performs the following steps on its two input point sets: *(i)* for each of both point sets, a subset of points is selected (called the *selected* points) according to a user-specified criterion Δ; *(ii)* for each selected point, its neighborhood is retrieved according to a user-specified neighborhood function Φ; and *(iii)* for each point in the sets of selected points, a feature vector is constructed describing the local spatial conformation of that point in its neighborhood. The distance between two point sets is then calculated by comparing the feature vectors of all pairs of identically labeled, selected points. The construction of a tuple of distances for a point $a \in \Delta X$ is shown in Fig. 2.

Definition 1 (3DNK family). *Let $n \in \mathbb{N}$ be a neighborhood size parameter, $\Delta \in \mathcal{F}_\Delta$ be a selection function, $\Phi \in \mathcal{F}_\Phi$ be a neighborhood function, and $\sigma \in \mathbb{R}^+$ be a parameter for the Gaussian RMS width. The 3DNK family, $K_{\Delta,\Phi} : \mathcal{X} \times \mathcal{X} \to \mathbb{R}$, is defined as follows:*

$$K_{\Delta,\Phi}(X, Y) = \sum_{a \in \Delta(X)} \sum_{b \in \Delta(Y)} K_G\left(d_\Phi(X, a), d_\Phi(Y, b)\right) \cdot I\left(\lambda(a) = \lambda(b)\right),$$

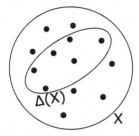
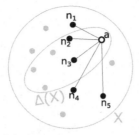
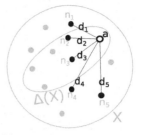

(a) An example X and its $\Delta(X)$.

(b) Neighborhood for a selected point: $\Phi(X,a) = \langle n_1, n_2, n_3, n_4, n_5 \rangle$.

(c) $d_\Phi(X,a)$ $=$ $\langle d_1, d_2, d_3, d_4, d_5 \rangle$,   where d_i is the distance between a and the corresponding neighborhood point n_i.

Fig. 2. Overview of the defined functions Δ, Φ and d_Φ.

where $K_G : \mathbb{R}^n \times \mathbb{R}^n \rightarrow \mathbb{R}$ is a Gaussian-based distance kernel, i.e.,

$$K_G(v_a, v_b) = exp\left(\frac{-\|v_a - v_b\|^2}{\sigma^2}\right),$$

and $I(x) = 1$ if x is true, and 0 otherwise.

It can be easily verified that instantiations of the 3DNK family will lead to positive-semidefinite kernels.

3.2 Instantiations of the 3DNK Kernel

In this section, we discuss how we instantiate the two function parameters of the 3DNK family in order to solve two biological tasks: predicting protein function and protein-ligand interaction. We denote with $\mathcal{C}$ the set of all chemical elements.

Predicting Protein Function. In this setting, an example is a protein 3D structure, consisting of atoms. For such a protein X and an atom $a \in X$, we define $\Delta(X)$ as the set of its side chain atoms, and the neighborhood function $\Phi(X,a)$ will only select backbone atoms of X in the neighborhood of a. The motivation for this is that the distances between atoms from the backbone and atoms from the side chains will influence binding pocket geometry, and hence determine protein function. For example, when predicting resistance in HIV proteins, resistant proteins will generally have similar backbones, but the acquired mutations (changing the side chains of the protein) will influence the distances between backbone and side chain atoms.

Let X be a protein, $a \in \Delta(X)$ a side chain atom, and $\Sigma = \mathcal{C}$. We present two approaches for mapping a side chain atom to a feature vector, i.e., two candidates for the function Φ as given in Definition 1, resulting into two kernels:

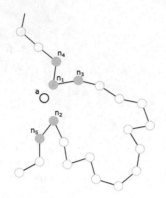 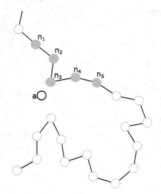

(a) Nearest neighbor neighborhood selection.

(b) Sequence window neighborhood selection centered around the nearest backbone atom (n_3).

Fig. 3. The backbone atoms n_1 to n_5 for an atom a (neighborhood size 5).

1. **Nearest Neighbor.** For each side chain atom a, $\Phi_{nn}(X, a)$ maps the atom to the n nearest backbone atoms, ordered in ascending order (Fig. 3(a)). We denote this kernel with 3DNK$_{nn}$.
2. **Sequence Window.** For each side chain atom a, $\Phi_{sw}(X, a)$ maps the atom to n backbone atoms, by using a window of size n over the backbone, centered around its nearest backbone atom. The tuple of size n consists of the backbone atoms in the window, ordered from left to right (Fig. 3(b)). Note that amino acids towards both ends of the sequence do not have a complete window, and therefore these side chain atoms are not used in the kernel. We denote this kernel with 3DNK$_{sw}$.

Predicting Protein-Ligand Interaction. In this setting, an example X is the set of ligand atoms and the protein binding pocket atoms to which the ligand is bound. We define $\Delta(X)$ as the set of ligand atoms, while the neighborhood function Φ will select atoms from the protein binding pocket. The motivation for this is that the distances between the ligand atoms and the atoms from the binding pocket will influence the binding affinity. We again present two approaches:

1. **Ligand Type.** For each ligand atom a, we construct the neighborhood by using Φ_{nn} to select the nearest protein atoms (Fig. 4(a)). We reuse the notation 3DNK$_{nn}$, as it is very similar to the one of the previous learning task.
2. **Ligand-protein Atom Type.** Contrary to 3DNK$_{nn}$, which does not take into account the atom types at the protein side, here we construct for each ligand atom a and each atom type t the neighborhood by selecting the nearest

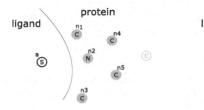 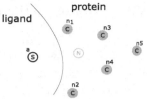

(a) Nearest neighbor neighborhood selection.

(b) Nearest neighbor, atom-type dependent neighborhood selection of atom a and neighbors with label C, $\lambda(a) = (S, C)$.

Fig. 4. The protein atoms n_1 to n_5 for an atom a (neighborhood size 5).

protein atoms having an atom type t. Constructing multiple neighborhoods per ligand atom (one for each atom type), can be done by applying the following procedure: *(i)* for each ligand atom $a \in \Delta(X)$ and for each atom type $t \in \Sigma$, add a new point p to $\Delta(X)$ with $\zeta(p) = \zeta(a)$ and $\lambda(p) = (\lambda(a), t) \in \Sigma^2$; *(ii)* remove the original ligand atom a from $\Delta(X)$. For a point p with label $(\lambda(a), t)$, $\Phi_{at}(X, p)$ selects the n nearest protein atoms with label t. The kernel value only depends on selected points with the same label, and hence two ligand atoms a and b will be compared if and only if *(i)* they have the same atom type, and *(ii)* both their neighborhoods are constructed with respect to the protein atoms with the same atom type. This process is shown in Fig. 4(b). We denote this kernel with 3DNK$_{at}$.

3.3 Implementation Optimizations

A trivial implementation of this kernel would give a complexity of

$$\mathcal{O}\left(n^2 \times |\Delta(X)| \times |\Delta(Y)| \times |X| \times |Y|\right).$$

As the neighborhood function Φ in Definition 1 only depends on a single example, calculating the feature vectors can be done as a preprocessing step. Furthermore, we can optimize both preprocessing and kernel value computation:

– **Preprocessing.** Our different versions of the neighborhood function Φ proposed in Sect. 3.2 depend on finding one or more nearest atoms, for which we use a k-d tree [11]. Constructing a k-d tree requires time $|X| \log |X|$, and finding n nearest neighbors requires time $n \log |X|$ for each atom. The total runtime for the preprocessing step can thus be upper bounded by

$$\mathcal{O}\left((|X| + n|S|) \log |X|\right),$$

subquadratic for values of $n < |X| / \log |X|$.

– **Kernel Value Approximation.** Computing the actual kernel value takes
quadratic time in $|S|$. However, for a point from $\Delta(X)$, only those points
from $\Delta(Y)$ with a similar feature vector will significantly influence the feature vector, i.e., their representations in feature space are near. For a point a
mapped to $\boldsymbol{v_a} \in \mathbb{R}^n$, we consider only those points $\boldsymbol{v_b}$ that lie within a hypercube with side $2r$ around $\boldsymbol{v_a}$. This limits the number of points but induces an
error ϵ in the kernel value. The value for r for a given ϵ is then

$$r = \sigma \sqrt{2 \log \left(\frac{|S|}{\epsilon} \right)}.$$

Finding those points inside the hypercube can be done efficiently by using
orthogonal range trees [17], which have a worst-case complexity of

$$\mathcal{O} \left(n|S|^{1-\frac{1}{n}} + k \right),$$

where k is number of points inside the hypercube. Note that k grows to $|S|$
as ϵ tends to zero. In our implementation we use a k-d tree as defined in [4].

The total complexity of both preprocessing and computing the kernel value can
be upper bounded by

$$\mathcal{O} \left(n \times |X| \times \left(|X|^{1-\frac{1}{n}} + k \right) \right),$$

which is at most quadratic in $|X|$.

The implementation of the 3DNK kernel can be downloaded at https://dtai.
cs.kuleuven.be/software/3DNK.

4   Experimental Evaluation

In this section, we perform an experimental analysis to assess the predictive performance of the different instantiations of 3DNK on the one hand, and w.r.t. the
state-of-the-art methods on the other hand. In order to do so, we conduct experiments on four benchmark datasets (defining 31 binary classification problems
and 1 regression problem) for four state-of-the-art methods.

4.1   Datasets

We assembled a first dataset (HIV) on our own, adopted two dataset collections
of protein 3D structures (EC and GO) from Qiu et al. [20], and adopted a fourth
dataset of 3D structures of protein-ligand pairs (PDBbind) from Ballester and
Mitchell [1].

HIV Resistance (HIV). This dataset contains 2048 protein structures belonging to HIV protease. The protein sequences were retrieved from the Stanford database (http://hivdb.stanford.edu). 1024 sequences from patients treated with the protease inhibitor indinavir were labeled as therapy-resistant while 1024 other sequences from patients who did not receive any treatment were labeled as therapy-naive. Since there were no 3D structures available, we generated each protein's 3D structure through homology modelling, using the tool MODELLER [23] and applying standard parameters.

Enzyme Classification (EC). The Enzyme Classification (EC) dataset [10] contains 998 protein structures derived from the EC hierarchy, of which the top level consists of six enzyme classes. The benchmark contains 498 PDB structures representing these classes, plus an additional 498 PDB structures of non-enzymes. The dataset defines 7 different binary classification tasks: one predicts whether a protein structure is an enzyme, and the six others predict the correct enzyme class within the set of enzymes, adopting a *one-vs-all* strategy. The average number of examples per dataset is 569.

Gene Ontology Classification (GO). The Gene Ontology (GO) dataset links 1024 proteins to 23 GO terms [20]. All GO terms are leaves of the GO hierarchy, while 11 of them belong to the *molecular function* branch, 8 to the *biological process* branch and 4 to the *cellular component* branch. The authors transform this multi-label problem into a set of binary classification problems in the following way. For each GO term T, they partition the set of proteins into three sets. First, all proteins that are annotated with T are labeled as positive. Next, all paths from T to the root of the GO hierarchy are traversed. Any protein having a child of the terms belonging to these paths is not taken into consideration, since the authors argue that such proteins might not be properly assigned. Finally, a randomized sample of proteins (ensuring a ratio of negatives to positives of 3 to 1) of that are not on the path from T to the root are labeled as negative. The average number of examples per dataset is 173.

Protein-Ligand Interactions (PDBbind). The PDBbind benchmark dataset was designed to assess the performance of scoring functions for molecular docking. The aim is to predict whether a small molecule (called ligand) will bind to a target protein. The strength of the binding is expressed as a numerical value representing the log-value of the measured binding affinity, constituting a regression problem. Here we use the 2007 version of the PDBbind database [29], which was divided by [1] into a training set of 1105 examples and a carefully selected test set of 195 examples, which has an equal number of representatives for each protein family.

4.2   State-of-the-art Methods

We compare our method against four state-of-the-art methods that take as input graphs or 3D structures of proteins, ligands or combinations thereof. The first two are graph kernels which have been applied to biological data before, while the last two methods were designed specifically to solve the tasks of predicting protein function and protein-ligand interaction.

Fast Subtree Kernel. The Fast Subtree Kernel (FSTK) is a graph kernel proposed by Shervashidze and Borgwardt [26] and is based on the Weisfeiler-Lehman test for graph isomorphism. FSTK iteratively looks at neighborhoods of nodes and unfolds the structure to get a tree-like pattern called a subtree. It then counts the matching subtree patterns of height up to h in two graphs G and G'. The authors show in their paper that FSTK outperforms four state-of-the-art graph kernels.

Fast Neighborhood Subgraph Pairwise Distance Kernel. Costa and De Grave [6] propose a fast graph kernel (NSPDK) based on the pairwise distance of neighborhood subgraphs and show that it outperforms four state-of-the-art graph kernels. Their decomposition kernel works as follows. First, pairs of so-called neighborhood subgraphs are generated, and then the kernel counts the number of identical pairs of neighboring graphs of radius r at distance d between two graphs.

Mammoth Kernel. Qiu et al. [20] propose a kernel that is based on the structural alignment between two proteins. However, this alignment score cannot be converted into a kernel function directly, because it is not positive-semidefinite. Instead, the authors employ an empirical kernel map as follows. For a given dataset of structures $X = x_1, \ldots, x_n$, a structure x_i is represented as an n-dimensional vector in which the jth entry is the Mammoth score between x_i and x_j. The resulting Mammoth kernel incorporates information about the alignability of a given pair of proteins. In their paper, the authors compare the Mammoth kernel to five other state-of-the-art kernels for protein structures and show that it outperforms them.

RF-Score. Ballester and Mitchell [1] introduce RF-Score as an alternative to traditional scoring functions for molecular docking. RF-Score uses random forests to make predictions based on 36 features they extract from the protein-ligand pairs. Each feature is an occurrence count of a particular atom type pair (one from the ligand and one from the protein) at a maximum distance of 12 Angström from each other. In their paper, the authors show that RF-Score outperforms 18 scoring functions on a testset of 195 examples. Since we use exactly the same training and test set, we adopt the results of the different scoring functions.

4.3   Methodology

To evaluate the kernels, we generate the kernel matrices, train support vector machines (SVMs) on them and evaluate their predictive performance. As SVM implementation we use SVM^{light} [14]. To evaluate the methods on the PDBbind dataset, we compare with published results [1].

Parameter Settings. We optimized the following parameters of the different methods on a separate tuning set. We tuned the regularization parameter c of the SVM out of the values $\{10^{-3}, 10^{-2}, 10^{-1}, 10^0, 10^{-1}, 10^2, 10^3\}$. For 3DNK, the neighborhood size (parameter n) was tuned out of the set $\{15, 21, 27, 33\}$, and the parameter σ out of the set $\{10^{-2}, 10^{-7/4}, 10^{-3/2}, 10^{-5/4}, 10^{-1}, 10^{-3/4}, 10^{-1/2}, -10^{-1/4}, 10^0, 10^{1/4}, 10^{1/2}, 10^{3/4}, 10^1\}$, while the precision parameter ϵ was set to 10^{-3}. Since FSTK and NSPDK are graph kernels, we could not give them the 3D data directly. Instead, we adopted the strategy of Borgwardt et al. [3] and constructed for each protein structure a graph in which every amino acid is a node. Next, we added an edge between two amino acids if the amino acids are less than a certain distance from each other. For the protein datasets (HIV, EC and GO), we tuned FSTK and NSPDK for thresholds of 6 and 8 Angström (values suggested by the authors). For the PDBbind dataset, there are no amino acids on the ligand side, so we generated graphs with atoms as nodes instead. Since these are graphs of much smaller granularity, we added a distance threshold of 4 Angström to create graphs. For FSTK, we tuned one additional parameter h (the number of iterations) out of $\{1, ..., 11\}$. For NSPDK, we tuned two additional parameters: the distance parameter out of $\{1, 2, 3, 4\}$ and the radius parameter out of $\{0, 1, 2\}$, as recommended by the authors. For the MAMMOTH kernel, no parameters had to be tuned.

Evaluation. To evaluate the classification models, we use the area under the ROC curve (AUROC) score [19]. To evaluate the regression models, we use Pearson's correlation coefficient (R), Spearman's correlation coefficient (R_S) and standard deviation of the difference between predicted and measured binding affinity (SD) in order to be able to compare with the published results of [1]. For HIV, EC and GO, a stratified 10-fold cross-validation is used. To optimize the above mentioned parameters, we constructed a tuning set through an internal 5-fold cross-validation over the training data. For PDBbind, we used the same training and test split as in [1]. Here, the parameters were optimized through a 10-fold cross-validation over the training data.

We compute the statistical significance of the different methods by computing standard deviations on the AUROC and regression scores. Method A then significantly outperforms method B at the 1 % level under a t-test if their confidence intervals do not overlap.

4.4 Results

Predicting Protein Function. In Table 1 we compare $3DNK_{sw}$ and $3DNK_{nn}$ to the state-of-the-art methods. Since RF-Score only works for protein-ligand interaction data, we could not run it for this task. First, the results show that there is no significant difference between $3DNK_{sw}$ and $3DNK_{nn}$, since their confidence intervals at 1 % around their AUROC scores overlap on the three datasets. Second, $3DNK_{sw}$ and $3DNK_{nn}$ perform significantly better than FSTK on HIV, but there are no significant differences with FSTK on EC and GO. Third, $3DNK_{sw}$ and $3DNK_{nn}$ perform significantly worse than NSPDK on HIV,

Table 1. AUROC of 3DNK and the state-of-the-art methods for the benchmark classification datasets. The best scoring method per dataset is indicated in bold. For EC and GO, averaged results are reported.

Dataset	$3DNK_{sw}$	$3DNK_{nn}$	FSTK	NSPDK	Mammoth
HIV	0.848 ± 0.008	0.853 ± 0.008	0.717 ± 0.010	$\mathbf{0.896 \pm 0.007}$	0.863 ± 0.008
EC	0.575 ± 0.021	$\mathbf{0.600 \pm 0.021}$	0.573 ± 0.021	0.535 ± 0.021	0.536 ± 0.021
GO	0.744 ± 0.033	0.710 ± 0.035	0.687 ± 0.035	0.660 ± 0.036	$\mathbf{0.859 \pm 0.026}$

while $3DNK_{nn}$ significantly outperforms NSPDK on EC. On GO, there are no significant differences between NSPDK and $3DNK_{sw}$/$3DNK_{nn}$. Fourth, Mammoth significantly outperforms $3DNK_{sw}$ and $3DNK_{nn}$ on the GO dataset, while $3DNK_{nn}$ significantly outperforms Mammoth on EC. On HIV, there are no significant differences between Mammoth and $3DNK_{sw}$/$3DNK_{nn}$. The Friedman test combined with the Nemenyi post-hoc test, a non-parametric test procedure for statistical comparisons of classifiers over multiple datasets [9], confirms that there are no significant differences between $3DNK_{sw}$/$3DNK_{nn}$ and the state-of-the-art methods.

Predicting Protein-Ligand Interaction. In Table 2 we compare $3DNK_{nn}$ and $3DNK_{at}$ to the state-of-the-art scoring functions. Since Mammoth only works on proteins, we could not apply it here. Furthermore, we could not run FSTK due to excessive memory requirements. For RF-Score (the top-scoring method), the confidence interval around its performance for R at 1% is $[0.691, 0.840]$. This shows that the performance of $3DNK_{at}$ is not significantly different than the one of RF-Score.

Table 2. Pearson's correlation coefficient (R), Spearman's correlation coefficient (R_S) and standard deviation of the difference between predicted and measured binding affinity (SD) of 3DNK and the state-of-the-art methods (including the results published in [1]) for the PDBbind benchmark regression dataset. FSTK and RF-Score could not be applied on this dataset. Methods are ordered by decreasing R.

Method	R	R_S	SD	Method	R	R_S	SD
1 RF-Score	0.776	0.762	1.58	12 DS::LigScore2	0.464	0.507	2.12
2 $3DNK_{at}$	0.730	0.75	1.67	13 GlideScore-XP	0.457	0.435	2.14
3 NSPDK	0.685	0.679	1.83	14 DS::PMF	0.445	0.448	2.14
4 $3DNK_{nn}$	0.652	0.688	1.81	15 GOLD::ChemScore	0.441	0.452	2.15
5 X-Score::HMScore	0.644	0.705	1.83	16 SYBYL::D-Score	0.392	0.447	2.19
6 DrugScore^{CSD}	0.569	0.627	1.96	17 DS::Jain	0.316	0.346	2.24
7 SYBYL::ChemScore	0.555	0.585	1.98	18 GOLD::GoldScore	0.295	0.322	2.29
8 DS::PLP1	0.545	0.588	2.00	19 SYBYL::PMF-Score	0.268	0.273	2.29
9 GOLD::ASP	0.534	0.577	2.02	20 SYBYL::F-Score	0.216	0.243	2.35
10 SYBYL::G-Score	0.492	0.536	2.08	21 FSTK	–	–	–
11 DS::LUDI3	0.487	0.478	2.09	22 Mammoth	–	–	–

Conclusion. The different instantiations of 3DNK perform competitively when compared to the state-of-the-art methods on the two tasks.

5   Related Work

A first group of methods that learn on geometrical data can be found in the field of pattern mining. Kuramochi and Karypis [16] present a framework in which the frequent pattern mining task is upgraded to the geometrical level. Their algorithm finds frequent geometric subgraphs (with 3D coordinates) which are rotation, scaling and translation invariant. Because noise is often present in these types of data, they perform an inexact matching based on a user-defined tolerance threshold. Nowozin and Tsuda [18] approach the task, which they call frequent subgraph retrieval, from a slightly different angle: they start from a database and a query graph and look for all subgraphs of this query graph in the database, given a geometric tolerance factor.

In the context of inductive logic programming, Srinivasan et al. [27] use a logical description of the 3D coordinates and chemical properties of molecules in order to learn structure-activity relationships. Borgwardt et al. [3] introduce graph kernels for proteins. They convert protein structures into a graph, with nodes representing secondary structure elements (integrated in the nodes are chemical properties) and propose a kernel based on random walks which uses appropriate kernels on the node level to take into account their continuous attributes. The authors also use a hyperkernel to select the best kernels and their weights for a specific dataset. Shervashidze and Borgwardt [26] describe a way to convert graphs and propose an efficient graph kernel on them. Costa and De Grave [6] propose a fast graph kernel based on the pairwise distance of neighborhood subgraphs and show that it outperforms four state-of-the-art graph kernels, including the one of [26]. Ceroni et al. [5] incorporate the 3D structure directly in their decomposition kernel, but is limited to small molecules. Hinselmann et al. [12] present a graph decomposition kernel for small molecules which also takes into account 3D information. The idea is to assign each atom the distance information to the remaining atoms and their corresponding atom type. This information is stored in a trie, which holds information on the shortest path and the geometrical environment. This leads to efficient computation of the local kernels.

Qiu et al. [20] proposed the MAMMOTH kernel. In [13], the authors convert their kernel into a paired variant in order to decide whether two proteins interact with each other. Ballester and Mitchell [1] proposed RF-Score. These methods were discussed in Sect. 4.2.

6   Conclusions and Further Work

In this paper, we introduced the 3DNK kernel, which acts on 3D structures. We applied 3DNK to two biological tasks and compared it to four state-of-the-art methods. Empirical evaluation showed that 3DNK performed competitively w.r.t. the state-of-the-art graph kernels and two methods that were designed specifically to solve the two respective biological tasks. Contrary to these application-specific methods, 3DNK is more broadly applicable and can solve both tasks equally well as those methods. The results suggest that the information in 3D structures can be exploited successfully and that the kernel can be deployed on a variety of problems.

In future work, we will explore various aspects of the 3DNK family further (such as the parameter space) and search for application domains on which new instantiations can be applied.

Acknowledgements. The authors would like to thank students Davy De Mits and Sunil Aryal for conducting preliminary experiments, Dr. Kurt De Grave and Dr. Fabrizio Costa for assistance with running NSPDK, and Jérôme Renaux for proof-reading. This research was supported by ERC-StG 240186 MiGraNT and IWT-SBO Nemoa.

References

1. Ballester, P.J., Mitchell, J.B.O.: A machine learning approach to predicting protein-ligand binding affinity with applications to molecular docking. Bioinformatics **26**(9), 1169–1175 (2010)
2. Borgwardt, K.: Graph Kernels. Ph.D. thesis, Computer Science, Ludwig-Maximilians-University Munich (2007)
3. Borgwardt, K., Ong, C., Schonauer, S., Vishwanathan, S., Smola, A., Kriegel, H.: Protein function prediction via graph kernels. Bioinformatics **21**(S1), i47–i56 (2005)
4. de Berg, M., Cheong, O., Kreveld, M., Overmars, M.: Computational Geometry: Algorithms and Applications. Springer, Heidelberg (2000)
5. Ceroni, A., Costa, F., Frasconi, P.: Classification of small molecules by two- and three-dimensional decomposition kernels. Bioinformatics **23**(16), 2038–2045 (2007)
6. Costa, F., De Grave, K.: Fast neighborhood subgraph pairwise distance kernel. In: Proceedings of the 27th International Conference on Machine Learning, pp. 255-262 (2010)
7. Cristianini, N., Shawe-Taylor, J.: An Introduction to Support Vector Machines and Other Kernel Based Methods. Cambridge University Press, UK (2000)
8. Deforche, K.: Modeling HIV resistance evolution under drug selective pressure. Ph.D. thesis, Katholieke Universiteit Leuven (2008)
9. Demšar, J.: Statistical comparisons of classifiers over multiple data sets. J. Mach. Learn. Res. **7**, 1–30 (2006)
10. Dobson, P.D., Doig, A.J.: Predicting enzyme class from protein structure without alignments. J. Mol. Biol. **345**, 187–199 (2005)
11. Friedman, J.H., Bentley, J.L., Finkel, R.A.: An algorithm for finding best matches in logarithmic expected time. ACM Trans. Math. Softw. **98**, 209–226 (1977)
12. Hinselmann, G., Fechner, N., Jahn, A., Eckert, M., Zell, A.: Graph kernels for chemical compounds using topological and three-dimensional local atom pair environments. Neurocomputing **74**, 219–229 (2010)
13. Hue, M., Riffle, M., Vert, J.-P., Stafford Noble, W.: Large-scale prediction of protein-protein interactions from structures. BMC Bioinform. **11**(144), 1–9 (2010)

14. Joachims, T.: Learning to Classify Text using Support Vector Machines: Methods, Theory, and Algorithms. Springer, US (2002)
15. King, R.D., Muggleton, S., Srinivasan, A., Sternberg, M.J.E.: Structure-activity relationships derived by machine learning: the use of atoms and their bond connectivities to predict mutagenicity by inductive logic programming. Proc. Natl. Acad. Sci. **93**, 438–442 (1996)
16. Kuramochi, M., Karypis, G.: Discovering frequent geometric subgraphs. In: Proceedings of the 2004 IEEE International Conference on Data Mining, pp. 258–265 (2004)
17. Lee, D.T., Wong, C.K.: Worst-case analysis for region and partial region searches in multidimensional binary search trees and balanced quad trees. Acta Informatica **9**, 23–29 (1977)
18. Nowozin, S., Tsuda, K.: Frequent subgraph retrieval in geometric graph databases. In: Proceedings of the 2008 IEEE International Conference on Data Mining, pp. 953–958 (2008)
19. Provost, F., Fawcett, T.: Analysis and visualization of classifier performance: comparison under imprecise class and cost distributions. In: Proceedings of the Third International Conference on Knowledge Discovery and Data Mining, pp. 43–48. AAAI Press (1998)
20. Qiu, J., Hue, M., Ben-Hur, A., Vert, J.-P., Stafford Noble, W.: A structural alignment kernel for protein structures. Bioinformatics **23**(9), 1090–1098 (2007)
21. Ramon, J., Gärtner, T.: Expressivity versus efficiency of graph kernels. In: Proceedings of the First International Workshop on Mining Graphs, Trees and Sequences (MGTS2003), pp. 65–74 (2003)
22. Saidi, R., Maddouri, M., Nguifo, E.M.: Comparing graph-based representations of protein for mining purposes. In: Proceedings of the KDD-09 Workshop on Statistical and Relational Learning in Bioinformatics, pp. 35–38 (2009)
23. Sali, A., Blundell, T.L.: Comparative protein modelling by satisfaction of spatial restraints. J. Mol. Biol. **234**, 779–815 (1993)
24. Schietgat, L., Vens, C., Struyf, J., Blockeel, H., Kocev, D., Džeroski, S.: Predicting gene function using hierarchical multi-label decision tree ensembles. BMC Bioinform. **11**(2), 1–14 (2010)
25. Schietgat, L., Ramon, J., Bruynooghe, M.: A polynomial-time maximum common subgraph algorithm for outerplanar graphs and its application to chemoinformatics. Ann. Math. Artif. Intell. **69**, 343–376 (2013)
26. Shervashidze, N., Borgwardt, K.: Fast subtree kernels on graphs. In: Bengio, Y., Schuurmans, D., Lafferty, J., Williams, C.K.I., Culotta, A. (eds.) Advances in Neural Information Processing Systems, vol. 22, pp. 1660–1668. Curran, USA (2009)
27. Srinivasan, A., Page, D., Camacho, R., King, R.D.: Quantitative pharmacophore models with inductive logic programming. Mach. Learn. **64**, 65–90 (2006)
28. Suykens, J., Van Gestel, T., De Brabanter, J., De Moor, B., Vandewalle, J.: Least Squares Support Vector Machines. World Scientific, Singapore (2005)
29. Wang, R., et al.: The PDBbind database: methodologies and updates. J. Med. Chem. **48**, 4111–4119 (2005)

Hierarchical Multidimensional Classification of Web Documents with MultiWebClass

Francesco Serafino$^{(\boxtimes)}$, Gianvito Pio, Michelangelo Ceci,
and Donato Malerba

Department of Computer Science, University of Bari Aldo Moro, Via Orabona 4,
70125 Bari, Italy
{francesco.serafino,gianvito.pio,michelangelo.ceci,
donato.malerba}@uniba.it

Abstract. Most of works on text categorization have focused on classifying documents into a set of categories with no relationships among
them (flat classification). However, due to the intrinsic structure that
can be found in many domains, recent works are focusing on more complex tasks, such as multi-label classification, hierarchical classification
and multidimensional classification. In this paper, we propose the hierarchical multidimensional classification task, where documents can be
classified according to different dimensions/viewpoints (e.g., topic, geographic area, time period, etc.), where in each dimension categories can
be organized hierarchically. In particular, we propose the system Multi-
WebClass, a multidimensional variant of the system WebClassIII, which
discovers correlations among categories belonging to different dimensions
and exploits them, according to two different strategies, to refine the set
of features used during the learning process. Experimental evaluation
performed on both synthetic and real datasets confirms that the exploitation of correlations among categories can lead to better results in terms
of classification accuracy, possibly reducing specialization error or generalization error, depending on the strategy adopted for the refinement
of the feature sets.

Keywords: Structured output prediction · Text categorization · Hierarchical classification · Multidimensional classification

1 Introduction

The number of web documents continuously and massively increases every day
and their automatic classification is considered an essential task. In recent years,
a plethora of classification algorithms has been developed. Some of them work in
the single label classification setting, where categories are not organized according to any specific schema. However, (web) documents can naturally be classified
into several hierarchically organized categories. For example, a blog article classified as Sport could, at the same time, be classified as Tennis, Roland Garros,

© Springer International Publishing Switzerland 2015
N. Japkowicz and S. Matwin (Eds.): DS 2015, LNAI 9356, pp. 236–250, 2015.
DOI: 10.1007/978-3-319-24282-8_20

and so on. For this reason, recent works have focused on the hierarchical classification task, where class labels are hierarchically organized and each object is associated to more than one class label (according to the hierarchy).

Moreover, documents can be classified according to different classification dimensions. For example, a web page could be classified according to its topic, the referenced geographical information or the publication date. By considering multiple dimensions of classification, each of which possibly hierarchically organized, it is possible to define the task of Hierarchical Multidimensional Classification. More formally, this task represents the combination of Multidimensional Classification and Hierarchical Classification, where: *(i)* more than one class attribute is associated to each document, each describing the document according to a different point of view and *(ii)* each class attribute is hierarchically organized.

In the literature, several works about multidimensional classification and hierarchical classification have been proposed. Some works consider the first task as a variant of (and, accordingly, convertible to) the latter, whereas other works consider these two tasks separately. For example, in [2] a multidimensional classification method based on Bayesian Network is proposed. The authors perform multidimensional classification on a flat set of labels by organizing class variables (dimensions), feature variables and bridges (from classes to features) as three distinct network subgraphs. In [14], the authors propose the application of multidimensional classification approaches to biomedical texts, in order to extract specific portions of text containing scientific content.

In [16] the authors propose a framework which implements three different classifiers (kNN, naïve Bayes and centroid-based) in order to evaluate three different techniques for multidimensional classification: flat-based, hierarchical-based and multidimensional-based. In the first two cases, they convert the multidimensional model into flat and hierarchical models, respectively, whereas in the last case they consider each dimension separately. Experiments performed on two datasets showed that the multidimensional-based and hierarchical-based approaches outperform the flat-based approach.

Moreover, in a recent work [7], the hierarchical multidimensional classification task is solved by considering it as a multi-label classification task. First, the system builds a set of probabilistic multi-class classifiers (one for each non-leaf node in the hierarchy) which are applied simultaneously to each test instance. Second, a probability is computed for each path in the hierarchy, by combining the output of the classifiers learned for the nodes involved in the path. Finally, the path with the highest probability is the output of the classification.

In this paper, we extend the system WebClassIII [5] (described in Sect. 2), which offers a hierarchical classification framework, with the more complex task of Hierarchical Multidimensional Classification. Moreover, we exploit the possible multi-dimensionality of the data in order to improve the classification with respect to each single dimension, which are generally classified independently by existing works. In order to exploit such possible dependencies, we propose the identification of correlations among categories belonging to different hierarchies.

Such correlations can be exploited to improve the classification accuracy with respect to a given hierarchy, even when the other hierarchies are not the main subject of the classification task. The discovery of correlations is motivated by the reasonable assumption that documents labeled with a given category along one dimension could be usually labeled with another given category belonging to another dimension. For example, we can consider documents organized according to the *Geographic* dimension and the *Topic* dimension. If many documents labeled with *Rome* in the *Geographic* hierarchy (structured as *Europe* → *Italy* → *Rome*), are also frequently labeled as *Traffic* in the *Topic* hierarchy (structured as *News* → *Accident* → *Traffic*), then it is possible that there is a correlation between the categories *Rome* and *Traffic*, possibly representing the fact that: "Rome is affected by an high number of accidents due to traffic".

The rest of the paper is organized as follows. In the next section, the classification framework implemented in WebClassIII, which represents the background of this work, is described. In Sect. 3 the extension of WebClass to perform hierarchical multidimensional classification is presented. Some experimental results on both synthetic and real datasets are reported and discussed in Sect. 4. Final conclusions and remarks are reported in Sect. 5.

2   Background: WebClassIII

WebClass is a classification framework for HTML pages. The last version of WebClass, i.e., WebClassIII, is an extension for the hierarchical text categorization [5] which exploits three different classification approaches: Naïve Bayes [10], centroid-based [6] and SVM [11].

The hierarchical organization of categories is exploited in all the phases of the document classification, namely feature selection, learning of the classification model, and categorization of a new document. Documents are represented as bag-of-words (where each term is associated to its frequency in the document). In general, two alternatives can be considered [1]: *(i)* the same feature space is used to represent documents belonging to all categories or *(ii)* several specific feature spaces are used to represent documents belonging to different categories. In WebClassIII an intermediate solution is adopted. In particular, for each category, a different document representation is used to decide which subcategory (temporary represented in the same feature space of its parent) is the most appropriate for a given document.

In the learning phase, starting from the root, the system builds a classification model for each category c. When c has only a subcategory, a dummy subcategory is introduced. The training documents associated to the dummy subcategory are those associated only to c (and not to the subcategory). Therefore, the sum of probabilities of all the direct subcategories of c is not necessarily 1.0, since the probability that the document does not belong to any subcategory should be taken into account.

The classification phase is performed in a top-down fashion from the root to the leaves, according to a greedy strategy. When the document reaches an internal category c, it is represented on the basis of the feature set associated to c and the system computes a score for each direct subcategory (dummy categories are not considered during the classification), according to the classification model learned on c. The document is associated to the subcategory with the highest score above a precomputed threshold (one for each category, see [4] for details). The search proceeds recursively from that subcategory, until no score is greater than the corresponding threshold or a leaf category is reached. The first case mainly happens when the document deals with a general rather than a specific topic or when the document belongs to a specific category which does not appear in the hierarchy. If the search stops at the root, then the document is marked as unclassified.

In the following, we report some details about the identification of an appropriate subset of terms (dictionary) for representing documents belonging to each category, which are then extended in order to exploit correlations among categories.

In particular, documents are initially tokenized, and the set of tokens is filtered in order to remove HTML tags, punctuation marks, numbers and tokens with less than three characters. After tokenization, two standard pre-processing methods are applied, that are stopword removal (based on words in Glimpse [9]) and stemming (using Porter algorithm [12]).

WebClassIII associates each category with a subset of words which best represent documents of that category. In particular, each word $w_{i,c'}$ that appears in at least a document of the category c' (direct subcategory of c), WebClassIII computes a weight $v_{i,c'}$ and builds a dictionary $Dict_{c'}$ containing n_{dict} words with the highest weight. The weight $v_{i,c'}$ is computed as follows:

$$v_{i,c'} = TF_{c'}(w_i) \times DF_{c'}^2(w_i) \times \frac{1}{CF_c(w_i)} \tag{1}$$

where:

- $TF_{c'}(w)$ is the *maximum term frequency* of w over the documents belonging to the category c';
- $DF_{c'}(w)$ is the *document frequency* computed as the percentage of documents of category c' in which w occurs;
- $CF_c(w)$ is the *category frequency* computed as the number of direct subcategories of c having at least a document in which w occurs.

It is noteworthy that positive examples for c' are sufficient for the computation of $TF_{c'}(w)$ and $DF_{c'}(w)$, while the computation of $CF_c(w)$ also requires the negative examples for c'.

The feature set associated to each category c, which is exploited for learning the corresponding classifier, consists of the union of the dictionaries associated to all the subcategories of c (called Hierarchical Feature Set in [5]).

3   Hierarchical Multidimensional Classification

In this Section, we describe MultiWebClass, an extension of WebClassIII which is able to perform Hierarchical Multidimensional Classification. The most

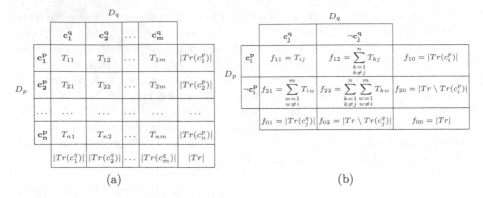

Fig. 1. (a) Contingency matrix between two dimensions D_p and D_q. (b) Contingency matrix between two categories $c_i^p \in D_p$ and $c_j^q \in D_q$ built using artificial categories.

straightforward solution would consist in learning a classification model for each dimension independently. However, as discussed in Sect. 1, this solution is not able to catch possible correlations among categories belonging to different dimensions. In MultiWebClass we adopt a different solution which first identifies possible correlations among categories belonging to different dimensions and then exploit such correlations in order to *extend* the feature sets used in the learning phase so to improve predictive performances. In the following subsections, we describe the proposed approach.

3.1 Discovery of Correlations Between Categories

The identification of correlations between two variables is a common task in statistics which is usually solved by means of a contingency matrix, where rows and columns represent the values of the first and the second variables, respectively. Inspired by this commonly used solution, we use the same strategy for categories. In particular, we build a contingency matrix as shown in Fig. 1(a), where:

- the variables on rows and columns represent two classification dimensions $D_p = \{c_1^p, c_2^p, \ldots, c_n^p\}$ and $D_q = \{c_1^q, c_2^q, \ldots, c_m^q\}$, where c_i^p is the i-th category of the p-th dimension;
- each cell value T_{ij} represents the number of documents labeled as both the categories c_i^p and c_j^q in the training set;
- $Tr(c_i^p)$ represents the set of training documents labeled as c_i^p in the hierarchy of the p-th dimension.

Starting from such contingency matrix, we construct a further 2×2 contingency matrix for each pair of categories c_i^p, c_j^q belonging to different dimensions ($p \neq q$) as shown in Fig. 1(b). In this matrix, we build two artificial categories $\neg c_i^p, \neg c_j^q$ which consist of all the documents not belonging to c_i^p and c_j^q, respectively. Obviously, we exclude root categories when computing such contingency matrices.

The correlation between two categories can be symmetric or not, on the basis of the considered correlation measure. In this work, we exploit a variant of the *Confidence* measure [15]. Such measure is asymmetric and is defined as $\gamma(c_i^p, c_j^q) = \frac{f_{11}}{f_{10}}$ and $\gamma(c_j^q, c_i^p) = \frac{f_{11}}{f_{01}}$, where f_{11}, f_{10} and f_{01} are the values of the contingency matrix between c_i^p and c_j^q (see Fig. 1(b)). This measure is actually a frequency-based estimation of the probability that documents belonging to c_i^p also belong to c_j^q (or vice versa). This measure, however, can lead to unreliable probabilities in the case that very few documents are used for its computation. To overcome this issue, we consider $\frac{f_{11}}{f_{10}}$ (and $\frac{f_{11}}{f_{01}}$) as a proportion in a statistical population and we use the Wilson confidence interval [17] in order to make conservative decisions about the presence of a correlation. We use the Wilson score interval since it directly derives from the Pearson's chi-squared test with two cases (here the two cases for $\frac{f_{11}}{f_{10}}$ are: a document that belongs to c_i^p also belongs to c_j^q or not). Formally, the Wilson score interval for $\frac{f_{11}}{f_{10}}$ is defined as:

$$\left[\frac{f_{11} + \frac{z_0^2}{2}}{f_{10} + z_0^2} - \frac{\sqrt{f_{10}} z_0}{f_{10} + z_0^2} \sqrt{\frac{f_{11} f_{12}}{f_{10}^2} + \frac{z_0^2}{4 f_{10}}} \quad , \quad \frac{f_{11} + \frac{z_0^2}{2}}{f_{10} + z_0^2} + \frac{\sqrt{f_{10}} z_0}{f_{10} + z_0^2} \sqrt{\frac{f_{11} f_{12}}{f_{10}^2} + \frac{z_0^2}{4 f_{10}}} \right] \tag{2}$$

where z_0 is the Z-score value (according to the normal distribution) for a given confidence $1 - \alpha$.

Since we are interested in making conservative decisions about the presence of a correlation, we consider the lower bound of this interval as the probability that c_i^p and c_j^q are correlated:

$$\gamma(c_i^p, c_j^q) = \frac{f_{11} + \frac{z_0^2}{2}}{f_{10} + z_0^2} - \frac{\sqrt{f_{10}} z_0}{f_{10} + z_0^2} \sqrt{\frac{f_{11} f_{12}}{f_{10}^2} + \frac{z_0^2}{4 f_{10}}} \tag{3}$$

Due to the asymmetry of the considered correlation measure, as shown in Algorithm 1, we pair-wisely search for correlations between two categories belonging to two different dimensions D_p and D_q in both the directions $D_p \to D_q$ and $D_q \to D_p$. Note that we are only interested in the discovery of positive correlations (i.e., a given category on a dimension possibly implies a category in another dimension). For this reason, we do not consider the proportions f_{21}/f_{01} and f_{12}/f_{10}. Moreover, since we are only interested to highly correlated pairs of categories, we consider as correlated two categories c_i^p and c_j^q only if $\gamma(c_i^p, c_j^q) > \beta$, where β is a user-defined threshold.

Finally, since we use the correlations to extend the feature sets and we use hierarchical feature sets (this aspect will be clarified in the next subsection), if the correlation $c_i^p \to c_j^q$ is identified, then it is possible to prove that, for each $c_k^q \in ancestors(c_j^q)$, there exists the correlation $c_i^p \to c_k^q$. This can be easily proved by observing Eq. (3). Indeed, for each $c_k^q \in ancestors(c_j^q)$ we have that $\gamma(c_i^p, c_k^q) \geq \gamma(c_i^p, c_j^q)$. This directly follows from the following observations:

- f_{01} has the same value in both $\gamma(c_i^p, c_k^q)$ and $\gamma(c_i^p, c_j^q)$, since it is the number of documents labeled as c_i^p;

Algorithm 1. Discovery of correlations among categories

---

input : The set of dimensions $\mathcal{D} = \{D_1, D_2, \ldots, D_s\}$;
a correlation measure $\gamma(\cdot, \cdot)$;
a threshold β to consider a discovered correlation as relevant.

output: $correlations = \{\langle c, CorrelatedSet_c \rangle\}_c$, where $CorrelatedSet_c$ is the set of categories d, s.t. exists the correlation $c \to d$.

1   $correlations \leftarrow \emptyset$;
2   **for** *all pairs of dimensions* D_p, D_q **do**
3      $correlations \leftarrow correlations \cup findCorrelations(D_p, D_q, \beta)$;
4      $correlations \leftarrow correlations \cup findCorrelations(D_q, D_p, \beta)$;

5   **return** $correlations$;

6   **findCorrelations**(D_p, D_q, β)
7   $correlations \leftarrow \emptyset$;
8   $exploredPairs \leftarrow \emptyset$;

9   **for** $c_i^p \in D_p$ *in pre-order* **do**
10      $correlatedSet \leftarrow \emptyset$;
11   **for** $c_j^q \in D_q$ *in post-order* **do**
12   **if** $\langle c_i^p, c_j^q \rangle \notin exploredPairs$ **and** $\gamma(c_i^p, c_j^q) \geq \beta$ **then**
13              $correlatedSet \leftarrow correlatedSet \cup \{c_j^q\}$;
             // Skip the exploration of ancestors of c_j^q
14   **for** $c_k^q \in ancestors(c_j^q)$ **and** $c_k^q \neq root(D_q)$ **do**
15                  $exploredPairs \leftarrow exploredPairs \cup \{\langle c_i^p, c_k^q \rangle\}$;
16                  $correlatedSet \leftarrow correlatedSet \cup \{c_k^q\}$;

17      $correlations \leftarrow correlations \cup \{\langle c_i^p, correlatedSet \rangle\}$;
18   **return** $correlations$;

- f_{12} has the same value in both $\gamma(c_i^p, c_k^q)$ and $\gamma(c_i^p, c_j^q)$, since it is the number of documents which are not labeled as c_i^p;
- f_{11} for $\gamma(c_i^p, c_k^q)$ is greater than or equal to f_{11} for $\gamma(c_i^p, c_j^q)$, since c_k^q contains documents in c_j^q.

In order to take into account this property, we visit the first hierarchy in pre-order and the second hierarchy in post-order. The effect is a reduction of the number of correlations between pairs of categories to be evaluated (see Algorithm 1, lines 14–16).

3.2   Exploiting Discovered Correlations

As shown in Sect. 2, in WebClassIII, the feature set associated to each category is the union of the dictionaries of its subcategories. In this work, we exploit the discovered correlations to extend the feature set of some categories. In particular, given a discovered correlation $c_i^p \to c_j'^q$ (where $c_j'^q$ is a subcategory of c_j^q), we

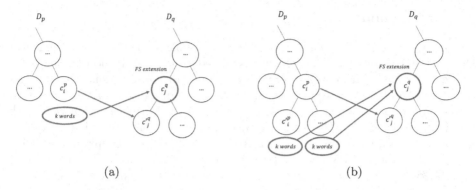

Fig. 2. Feature set extension FSE I, when (a) c_i^p is a leaf and when (b) c_i^p is not a leaf. Red arrows indicate correlations, while blue arrows indicate feature set extension (Color figure online).

extend the feature sets of the involved categories according to two different strategies:

- **FSE I: Category Dictionaries.** This strategy extends the feature set of the category c_j^q by including features in the dictionaries of categories in the dimension D_p:
  - When c_i^p is a leaf, then the feature set of the category c_j^q is extended by including the top-k ($k < n_{dict}$) terms of the dictionary associated c_i^p (see Fig. 2(a));
  - When c_i^p is not a leaf, the feature set of the category c_j^q is extended by including the top-k ($k < n_{dict}$) terms from the dictionaries of the subcategories of c_i^p (see Fig. 2(b)).

  The top-k terms are selected according to Eq. 1. The rationale behind this strategy is to use, in the classifier associated to c_j^q, some of the features that the classifier associated to c_i^p uses to discriminate among its child categories.
- **FSE II: Correlated Words.** This strategy works similarly to FSE I, but adds a different set of terms to the feature set of c_j^q. In particular, the top-k ($k < n_{dict}$) words, according to Eq. 1, are those that appear in the feature sets of both c_i^p and $c_j'^q$, but not to c_j^q. Note that the feature selection algorithm implemented in WebClassIII could have pruned some features of $c_j'^q$ when building the feature set of c_j^q. This strategy could restore such features because of the presence of the correlation.

FSE II is more conservative than FSE I since it uses, for feature extension, features extracted from the same dimension. On the contrary, since FSE I incorporates in one dimension features coming from a completely different dimension, it is more "daring", but can also incur into errors (possibly) coming from unrelated features.

4   Experiments

The experimental evaluation has been performed using, in WebClass, the Naïve Bayes classification algorithm which was proved to perform the best among those implemented in the framework [5]. For evaluation purposes, we consider both real and synthetically generated datasets. Since we are interested in the evaluation of the contribution given by the exploitation of the correlations, in this work the results obtained with both the proposed feature set extension strategies introduced in MultiWebClass (i.e., FSE I and FSE II) are compared only with the results obtained by WebClass III. This because WebClass III does not exploit correlations among different hierarchies and represents the non-multidimensional counterpart of the approach we propose.

4.1   Datasets

Real Dataset. The evaluation on real data has been performed on the dataset Reuters Corpora Volume 1 (RCV1) [8]. It contains more than 800,000 stories collected by the international news agency Reuters from 20th August 1996 to 19th August 1997. News are manually labeled according to the following three classification dimensions:

- *Topic*: the main subject of a story (hierarchically organized);
- *Industry*: the type of business discussed in a story (hierarchically organized);
- *Region*: a geographic location or an economic/political group (not hierarchically organized).

The training and testing sets are obtained according to [8], i.e., by selecting documents published from 20th August 1996 to 31st August 1996 as training set, and documents published from 1st September 1996 to 19th August 1997 as testing set. Coherently with [5], we considered only documents associated to a single category. The result is a set of 4,517 training documents and 146,248 testing documents.

Synthetic Datasets. The evaluation on synthetic data has been performed by means of the 5 fold cross validation approach on a set of datasets generated by simulating the presence of correlations among categories belonging to different dimensions. In particular, the generation of the datasets takes into account the following aspects:

- *Dimensions.* The set of dimensions $\mathcal{D}$ is created using as parameters: the number of dimensions s, the tree *depth* and the *degree* of nodes. Each dimension is represented as a full and perfectly-balanced tree.
- *Dictionaries.* Each leaf category is associated with a fixed number of terms (50 in our experiments) randomly selected from the Ispell American English[1] dictionary, whereas each internal category is associated with the dictionary obtained as the union of the terms selected for its subcategories.

---

[1] http://fmg-www.cs.ucla.edu/geoff/ispell-dictionaries.html.

– *Document Generation.* For each category c, a set of d_c documents is generated. Each document consists of a set of t terms, randomly selected from the dictionary associated to c, and a set of t_g general terms randomly taken from the Ispell American English dictionary. The complete set of documents belonging to the category c is then obtained as the union of the d_c documents (properly) belonging to c and the set of documents belonging to its subcategories.

– *Correlation Injection.* Correlations are injected between pairs of categories belonging to different dimensions. Injection is performed by labeling documents belonging to a category of a given dimension also with a given category of another dimension. The system takes an input parameter n_{corr} which represents the number of correlations to inject into leaf categories, whereas correlations between internal categories are injected in a bottom-up fashion. In particular, for each correlation $c_i^p \rightarrow c_j^q$ injected at l-th level, we introduce a correlation $father(c_i^p) \rightarrow father(c_j^q)$ at $(l-1)$-th level. In order to avoid the injection of several and redundant correlations involving the same internal categories, only the most frequent correlations are preserved. The percentage of correlations to preserve is given by the user-defined parameter $CorrRatio$.

We generated all the synthetic datasets with the following parameters: $s = 3$ (i.e., 3 hierarchically organized classification dimensions), $degree = 2$ (i.e., each internal node has 2 children), $n_{corr} = 20$ (i.e., for each pair of dimensions, we inject a correlation for 20 randomly selected pairs of leaf categories) and $CorrRatio = 0.40$ (i.e., we preserve the top-40 % most frequent correlations among parent categories).

For each category, all the documents are generated by randomly selecting 80 terms from the dictionary of the category (i.e., $t = 80$) and 20 terms from the Ispell American English dictionary (i.e., $t_g = 20$). The number of hierarchical levels (i.e., the parameter $depth$) and the number of documents d_c for each category are set to different values in order to generate different synthetic datasets. In particular, we generated five different datasets, that are:

– **4–20**, which has 4 hierarchical levels and 20 documents for each category;
– **4–30**, which has 4 hierarchical levels and 30 documents for each category;
– **4–40**, which has 4 hierarchical levels and 40 documents for each category;
– **4–50**, which has 4 hierarchical levels and 50 documents for each category;
– **3–40**, which has 3 hierarchical levels and 40 documents for each category.

While the results obtained on the datasets 4–20, 4–30, 4–40 and 4–50 are compared in order to analyze the performance by varying the number of documents for each category, results obtained on the datasets 3–40 and 4–40 are compared in order to obtain a preliminary analysis on the sensitiveness of the algorithm to the complexity of the hierarchical structure of each classification dimension.

4.2   Experimental Setting

In the following, we report some details about the parameter setting of MultiWebClass. In particular, we set the confidence value $1 - \alpha$ to 0.95. Consequently, the Z-score value is $z_0 = 1.96$. The threshold value β has been set to 0.3. n_{dict} (size of dictionaries) is set to 25, which provided good results in WebClassIII. The number of terms k, propagated by the proposed strategies for feature set extension is set to 20 (coherently with n_{dict}).

As regards the classification dimensions, we considered the *Topic* dimension for RCV1 (which is typically used for classification purposes [13]), exploiting the correlations with the other two dimensions (i.e., *Industry* and *Region*) and the first dimension for the synthetic datasets, exploiting the correlations with the other two dimensions[2].

The comparison between WebClassIII and MultiWebClass has been performed according to five evaluation measures, that are:

- *Accuracy*, which is the percentage of correctly classified documents;
- *Generalization Error*, which is the percentage of documents classified as a super-category of the correct category;
- *Specialization Error*, which is the percentage of documents classified into a subcategory of the correct category;
- *Misclassification Error*, which is the percentage of documents classified as a category which is in a different path with respect to the correct category in the hierarchy;
- *Unknown Ratio*, which is the percentage of documents that are not classified (actually classified in the root category of the hierarchy).

Intuitively, the sum of all the considered measures is always equal to 1.

4.3 Results

According to the experimental setting, in this section we report the results obtained with all the considered datasets and perform three different analyses on:

- synthetic datasets with a fixed depth of the hierarchies (i.e., $depth = 4$) and a different number of documents for each category (i.e., $d_c = \{20, 30, 40, 50\}$);
- synthetic datasets with a fixed number of documents for each category (i.e., $d_c = 40$) and different depths of the hierarchies (i.e., $depth = \{30, 40\}$);
- real data, i.e., on the RCV1 dataset.

Synthetic Datasets with Fixed Depth. Results for this analysis are reported in Table 1. As it can be observed from the table, all the considered approaches were able to make at least a decision in the root node, i.e., the Unknown Ratio is 0. Moreover, by observing the Misclassification Error, we can see that all the systems were almost always able to consider the correct path in the hierarchy. The main differences can be observed in the other three measures. In particular, FSE I always leads to better accuracy values when the number of documents

---

[2] In the case of synthetic datasets, results do not depend on the specific dimension.

per category increases (i.e., $d_c > 20$). This approach also obtains good results in terms of Generalization Error, sometimes at the cost of a slightly higher Specialization Error, which confirms that FSE I generally leads to less conservative decisions, if compared to FSE II. Overall, by comparing the results obtained by FSE I and FSE II with those obtained by WebClassIII, it is possible to see that the exploitation of the discovered correlations leads to better results.

Synthetic Datasets with Different Depth. This analysis aims at evaluating the performance with respect to the depth of the hierarchy. Results are reported in Table 2. As expected, the higher the complexity of the classification hierarchy, the lower the classification accuracy. However, the proposed approaches always lead to better results. A more detailed analysis reveals that, as expected, moving from 3 to 4 levels in the hierarchy leads to reduce the advantage of the MultiWebClass with respect to WebClassIII (percentage gain in accuracy changes from 9.3 % to 8.1 % in the case of FSE I). As regards the specific error measures, we can observe that all the systems generally prefer to make Generalization Errors with respect to Specialization Errors when the complexity of the hierarchy increases.

Table 1. Classification results of synthetic dataset 4–20, 4–30, 4–40 and 4–50 on the first dimension.

Dataset 4–20					
System	Accuracy	Gen. Error	Spec. Error	Misclass. Error	Unknown Ratio
MWC - FSE I	0.759	**0.168**	0.073	**0.000**	**0.000**
MWC - FSE II	**0.764**	0.217	**0.019**	**0.000**	**0.000**
WebClass III	0.685	0.246	0.069	**0.000**	**0.000**
Dataset 4–30					
System	Accuracy	Gen. Error	Spec. Error	Misclass. Error	Unknown Ratio
MWC - FSE I	**0.795**	**0.153**	**0.052**	**0.000**	**0.000**
MWC - FSE II	0.751	0.162	0.087	**0.000**	**0.000**
WebClass III	0.698	0.245	0.057	**0.000**	**0.000**
Dataset 4–40					
System	Accuracy	Gen. Error	Spec. Error	Misclass. Error	Unknown Ratio
MWC - FSE I	**0.731**	0.261	**0.008**	**0.000**	**0.000**
MWC - FSE II	0.711	**0.229**	0.030	0.030	**0.000**
WebClass III	0.676	0.245	0.079	**0.000**	**0.000**
Dataset 4–50					
System	Accuracy	Gen. Error	Spec. Error	Misclass. Error	Unknown Ratio
MWC - FSE I	**0.829**	**0.099**	0.072	**0.000**	**0.000**
MWC - FSE II	0.715	0.147	0.138	**0.000**	**0.000**
WebClass III	0.727	0.247	**0.026**	**0.000**	**0.000**

Table 2. Classification results of synthetic dataset 3–40 and 4–40 on the first dimension.

Dataset 3–40					
System	Accuracy	Gen. Error	Spec. Error	Misclass. Error	Unknown Ratio
MWC - FSE I	**0.834**	**0.129**	0.037	**0.000**	**0.000**
MWC - FSE II	0.833	0.131	**0.036**	**0.000**	**0.000**
WebClass III	0.763	0.169	0.068	**0.000**	**0.000**
Dataset 4–40					
System	Accuracy	Gen. Error	Spec. Error	Misclass. Error	Unknown Ratio
MWC - FSE I	**0.731**	0.261	**0.008**	**0.000**	**0.000**
MWC - FSE II	0.711	**0.229**	0.030	0.030	**0.000**
WebClass III	0.676	0.245	0.079	**0.000**	**0.000**

Real Data. Finally, we compare the results obtained with MultiWebClass and WebClassIII on RCV1. Results are reported in Table 3.

From the results we can see that, differently from what we observed for synthetic datasets, Unknown Ratio is nonzero. This because, contrary to the synthetic datasets, testing set contains documents that do not come from the same data distribution of training documents. However, despite higher Unknown Ratio, FSE II obtains better results in terms of Accuracy, Generalization Error, Specialization Error and Misclassification Error. This confirms the more conservative nature of FSE II, which makes more accurate predictions and avoids wrong decisions when the degree of uncertainty is high. This general behavior can suggest us the use of FSE II when we would like to obtain a more accurate classification, at the price of some unclassified instances, whereas FSE I (and, in some cases, the original WebClass III) is more appropriate when we want to force classification (reducing Unknown Ratio), at the price of a higher Specialization Error. This observation does not hold for synthetic datasets since the unknown ratio is always zero due to the considerations about data distribution reported before.

In Table 4, we show some correlations discovered by our approach and used for feature set extension. It is noteworthy that most of them appear reasonable. For example, some regions which are usually subject to political issues

Table 3. Classification results of RCV1 on the dimension *Topic*.

Reuters RCV1					
System	Accuracy	Gen. Error	Spec. Error	Misclass. Error	Unknown Ratio
MWC - FSE I	0.559	0.141	0.003	0.087	0.210
MWC - FSE II	**0.566**	**0.129**	**0.002**	**0.078**	0.225
WebClass III	0.561	0.188	0.003	0.101	**0.147**

Table 4. Correlations discovered on RCV1 with correlation strength greater than 0.3.

Source category	Target category	Correlation Strength
Region.Cyprus	Topic.Government/Social	0.51841
Region.EuropeanUnion	Topic.EuropeanCommunity	0.51322
Region.Macedonia	Topic.Government/Social	0.46769
Industry.PortsAndShippingServices	Topic.Corporate/Industrial	0.42569
Region.Ghana	Topic.EquityMarkets	0.42438
Region.Malawi	Topic.Government/Social	0.35930
Region.Syria	Topic.Government/Social	0.35479
Region.Bahrain	Topic.Government/Social	0.35027
Region.Jamaica	Topic.Corporate/Industrial	0.34238
Region.Malta	Topic.Government/Social	0.32404
Industry.PortsAndShippingServices	Topic.Capacity/Facilities	0.31651
Region.CzechRepublic	Topic.Markets	0.30804
Region.UnitedArabEmirates	Topic.Government/Social	0.30070

are correlated to the topic *Government/Social*. Moreover, some regions whose economy is based on some specific business activities are correlated to the topic *Corporate/Industrial*.

5 Conclusions and Future Work

In this paper we tackled the Hierarchical Multidimensional Classification task and presented, at this purpose, the system MultiWebClass. In particular, MultiWebClass discovers correlations between categories belonging to different hierarchies and exploits them by extending (according to two different strategies) the feature sets used for learning classifiers.

Results on both synthetic and real datasets show that the exploitation of the discovered correlations, which appear reasonable after a quick qualitative analysis, can lead to better classification performances in terms of accuracy. Moreover, the different strategies proposed for feature set extension appear appropriate for different goals (i.e., higher accuracy *vs* higher number of classified instances), since they have a different degree of conservativeness when making decisions.

For future work, we intend to deeply analyze the sensitiveness of MultiWebClass to different parameter settings. We will also consider additional strategies for exploiting the discovered correlations, possibly including negative correlations. Moreover, inspired by the work in [3], we will explore the task of multidimensional hierarchical classification in the transductive setting. Finally, we intend to perform experiments on additional real-world datasets, also related to different application domains (e.g., biological data).

Acknowledgements. We would like to acknowledge the support of the European Commission through the project MAESTRA - Learning from Massive, Incompletely annotated, and Structured Data (Grant number ICT-2013-612944).

References

1. Apté, C., Damerau, F., Weiss, S.M.: Automated learning of decision rules for text categorization. ACM Trans. Inf. Syst. (TOIS) **12**(3), 233–251 (1994)
2. Bielza, C., Li, G., Larranaga, P.: Multi-dimensional classification with bayesian networks. Int. J. Approximate Reasoning **52**(6), 705–727 (2011)
3. Ceci, M.: Hierarchical text categorization in a transductive setting. In: Workshops Proceedings of the 8th IEEE International Conference on Data Mining (ICDM 2008), Pisa, Italy, 15–19 December, 2008, pp. 184–191. IEEE Computer Society (2008)
4. Ceci, M., Malerba, D.: Hierarchical Classification of HTML Documents with Web-ClassII. In: Sebastiani, F. (ed.) ECIR 2003. LNCS, vol. 2633, pp. 57–72. Springer, Heidelberg (2003)
5. Ceci, M., Malerba, D.: Classifying web documents in a hierarchy of categories: a comprehensive study. J. Intell. Inf. Syst. **28**(1), 37–78 (2007)
6. Han, E.-H.S., Karypis, G.: Centroid-Based Document Classification: Analysis and Experimental Results. In: Zighed, D.A., Komorowski, J., Żytkow, J.M. (eds.) PKDD 2000. LNCS (LNAI), vol. 1910, pp. 424–431. Springer, Heidelberg (2000)
7. Hernández, J., Sucar, L.E., Morales, E.F.: Multidimensional hierarchical classification. Expert Syst. Appl. **41**(17), 7671–7677 (2014)
8. Lewis, D.D., Yang, Y., Rose, T.G., Li, F.: Rcv1: A new benchmark collection for text categorization research. J. Mach. Learn. Res. **5**, 361–397 (2004)
9. Manber, U., Wu, S., et al.: Glimpse: A tool to search through entire file systems. In: Usenix Winter, pp. 23–32 (1994)
10. Mitchell, T.M.: Machine Learning. McGraw Hill series in computer science. McGraw-Hill, Tom Mitchell (1997)
11. Platt, J., et al.: Fast training of support vector machines using sequential minimal optimization. In: Advances in Kernel Methods Support Vector Learning, 3 (1999)
12. Porter, M.F.: An algorithm for suffix stripping. Program **14**(3), 130–137 (1980)
13. Schapire, R.E., Singer, Y.: Boostexter: A boosting-based system for text categorization. Mach. Learn. **39**(2/3), 135–168 (2000)
14. Shatkay, H., Pan, F., Rzhetsky, A., Wilbur, W.J.: Multi-dimensional classification of biomedical text: Toward automated, practical provision of high-utility text to diverse users. Bioinformatics **24**(18), 2086–2093 (2008)
15. Tan, P.-N., Kumar, V., Srivastava, J.: Selecting the right interestingness measure for association patterns. In: Proceedings of the Eighth ACM SIGKDD International Conference on Knowledge Discovery and Data Mining, pp. 32–41. ACM (2002)
16. Theeramunkong, T., Lertnattee, V.: Multi-dimensional text classification. In: Proceedings of the 19th International Conference on Computational Linguistics, vol. 1, pp. 1–7. Association for Computational Linguistics (2002)
17. Wilson, E.B.: Probable inference, the law of succession, and statistical inference. J. Am. Stat. Assoc. **22**(158), 209–212 (1927)

Evaluating the Effectiveness of Hashtags as Predictors of the Sentiment of Tweets

Credell Simeon[(✉)] and Robert Hilderman

University of Regina, Regina, SK S4S 0A2, Canada
{simeon3c,Robert.Hilderman}@uregina.ca

Abstract. Recently, there has been growing research interest in the sentiment analysis of tweets. However, there is still a need to examine the contribution of Twitter-specific features to this task. One such feature is hashtags, which are user-defined topics. In our study, we compare the performance of sentiment and non-sentiment hashtags in classifying tweets as positive or negative. By combining subjective words from different lexical resources, we achieve accuracy scores of 83.58 % and 83.83 % in identifying sentiment hashtags and non-sentiment hashtags, respectively. Furthermore, our accuracy scores surpass those scores obtained using models that apply a single lexical resource. We apply derived properties of sentiment and non-sentiment hashtags, including their sentiment polarity to classify tweets. Our best classification models achieve accuracy scores of 81.14 % and 86.07 % using sentiment hashtags and non-sentiment hashtags, respectively. Additionally, our models perform comparably to supervised machine learning algorithms, and outperform a scoring algorithm developed in a previous study.

1 Introduction

Since its inception in 2006, Twitter, a microblogging application, has gained increasing popularity with approximately 302 million monthly users and an estimated 500 million new daily posts[1]. Twitter provides a platform whereby registered users can post short text messages called tweets. Tweets are opinionated statements, which convey sentiments about different topical issues. Therefore, we can apply sentiment analysis to determine whether the sentiment contained within the text is either positive or negative [5]. Positive tweets express favorability whereas negative tweets express unfavorability towards a subject. Thus, sentiment analysis is useful in assessing people's attitudes and emotions towards products and services offered by businesses [13], or political candidates in general elections [3].

For sentiment analysis, research studies have applied both machine learning techniques and lexicon-based methods. The lexicon-based approach depends entirely on using opinion lexicons, which are dictionaries of positive and negative words, to detect subjectivity in text [19]. By contrast, supervised machine learning applies learning algorithms to large number of labeled data. Unlike other

---

[1] https://about.twitter.com/company.

© Springer International Publishing Switzerland 2015
N. Japkowicz and S. Matwin (Eds.): DS 2015, LNAI 9356, pp. 251–265, 2015.
DOI: 10.1007/978-3-319-24282-8_21

text, tweets can contain a significant amount of information which can make the sentiment analysis task challenging [4]. Therefore, it is important to examine the unique nature of tweets, as this plays a significant role in determining their overall sentiment.

Tweets are highly informal text messages, which are restricted to 140 characters. They are conversational in nature and thus, they contain many features including: abbreviations, slangs, acronyms, repetitions of characters in words e.g., "yeaaaah", punctuation marks, and emoticons. In terms of Twitter-specific features, these are described below.

1. **Retweets** are copies of an original tweet that are posted by other users [7]. They are denoted by the letters, "RT".
2. **Mentions** are used for replying directly to others. They begin with the "@" symbol followed by the name of a Twitter user e.g., "@john".
3. **URL links** are used to direct users to interesting pictures, videos or websites for additional information.
4. **Hashtags** are user-defined topics, keywords or categories denoted by the hash symbol, "#". Hashtags can be a single word or a combination of consecutive words, e.g., "#believe" and "#wishfulthinking", respectively. A tweet can contain multiple hashtags, which can be located anywhere in the text.

Of all the features of tweets described previously, hashtags have been selected as the focus of our study. The significance of hashtags lies in their unique ability to simultaneously connect related tweets, topics, and communities of people who share similar interests. Each hashtag is a sharable link, which can be used to promote specific ideas, search for popular content, engage other users, and group related content. Most importantly, hashtags are useful for determining the popularity of topics, and the overall sentiment that is being expressed by groups of users. Consequently, hashtags are being used by many other platforms including photo-sharing applications such as Instagram[2] and social networks like Facebook[3], Tumblr[4] and Google+[5].

Additionally, hashtags contain sentiment and topic information. Hashtags that contain only topic information are considered to be non-sentiment bearing. However, hashtags that contain sentiment information, such as an emotion expressed by itself or directed towards an entity, are considered to be sentiment bearing. These two types of hashtags are similar to the sentiment, sentiment-topic and topic hashtags that were proposed in a previous study [17]. Examples of sentiment and non-sentiment hashtags are "#best" and "#football", respectively.

In this study, we hypothesize that hashtags can be used as accurate predictors of the overall sentiment of tweets. Based on this assumption, we can identify three major opportunities for improving the sentiment analysis of tweets. Firstly,

---

[2] http://instagram.com/.
[3] http://facebook.com/.
[4] https://www.tumblr.com/.
[5] https://plus.google.com/.

we might be able to accurately determine the sentiment of a large volume of tweets without having to examine individual tweets. Secondly, we can reduce dependency on manual annotation of tweets, which can be time-consuming and labor-intensive [4,6,19]. Thirdly, by focusing on a single feature, we reduce the effort required in determining the optimal combination of the various features in the tweets. Therefore, our study applies the derived properties of hashtags, including their sentiment polarity, in order to classify tweets as positive and negative. We describe these properties as "derived" because they are not part of the definition of a hashtag, but they are resulting because of it. For instance, a hashtag may contain a subjective word, thus we consider that there are two types of hashtags: sentiment and non-sentiment bearing. Additionally, we consider all hashtags to have a polarity which can be determined by examining the tweets that contain them. Therefore, in this study, we compare the effectiveness of sentiment and non-sentiment hashtags for classifying subjective tweets.

The main contributions of our paper are summarized as follows:

1. It demonstrates the effectiveness of combining different lexical resources to identify sentiment from non-sentiment bearing hashtags.
2. It presents different scoring algorithms for determining the sentiment polarity of hashtags.
3. It demonstrates the effectiveness of using the derived properties of hashtags, including their sentiment polarity, for the sentiment classification of tweets.
4. It shows that non-sentiment hashtags are more effective at classifying tweets as positive and negative, than sentiment hashtags.

The remainder of this paper is organized as follows. Section 2 summarizes previous studies on the sentiment analysis of tweets. Section 3 describes the development of the our approach. Section 4 discusses our experimental results, and compares these results with that of another study. Finally, Sect. 5 presents our conclusions and plans for future work.

2   Related Work

Sentiment analysis of tweets has garnered much research interest in recent years. A study by [2] demonstrated that a scoring algorithm can be used to accurately classify positive and negative tweets [2]. They applied the function to two separate datasets, Stanford [7] and Mejaj [5], which used emoticons, and sentiment suggestive words as sentiment labels, respectively. The scoring function calculated an overall score for each tweet by aggregating the difference in the positive and negative probabilities of unigrams, and assigning predefined weights to emoticons and punctuations. After applying stop word removal, stemming, spelling correction and noun identification, the function applied a Popularity Score in order to boost the scores of domain specific words. Tweets were determined to be positive (negative) if the sum of their sentiment scores was greater (less) than zero. Experimental results revealed that the Stanford and Mejaj datasets achieve accuracy scores of 87.2 % and 88.1 %, respectively. Also, these accuracy scores were comparable to that obtained using a SVM classifier.

In terms of the contribution of hashtags to the sentiment analysis of tweets, very few studies have focused on this task. Kouloumpis et al. [9] investigated Twitter hashtags for identifying positive, negative and neutral tweets such that the polarity of the tweet is determined by the hashtag. Using linguistic, lexical and microblogging features extracted from tweets, an AdaBoost.MH classifier achieved accuracy scores of 74 % and 75 % on hashtagged, and emoticon datasets, respectively. However, their study only focused on tweets containing a single hashtag. By contrast, Mohammad [10] analyzed self-labeled hashtagged emotional words in tweets, and concluded that they are good indicators of the sentiment of the entire tweet. In a later study, Mohammad et al. [11] developed a hashtag sentiment lexicon using a dataset of about 775,000 tweets and 78 hashtagged seed words. A tweet was assigned the same sentiment polarity if it contained any of the positive (negative) hashtagged seed words. By applying the hashtag lexicon to classify sentiment in tweets, the performance of the classifier increased by 3.8 %. Therefore, both studies demonstrate that hashtags can be useful in the sentiment analysis of tweets.

Wang et al. [17] applied a graph-based approach for classifying sentiment in hashtags as either positive or negative by incorporating hashtag co-occurrence information, their literal meaning, and the sentiment polarity distribution of tweets. By doing so, they showed that the polarity distribution of tweets can be combined with hashtag information for sentiment classification.

Rodrigues Barbosa et al. [14] performed a preliminary investigation into determining the effectiveness of hashtags in the sentiment analysis of tweets. The study focused specifically on using hashtags to detect and track online population sentiment. In order to do this, the authors studied hashtag propagation patterns, and the use of hashtags to express sentiment in tweets. Using a dataset of tweets on elections in Brazil, the authors manually identified frequent positive and negative hashtags. After performing analysis on the hashtags, the results revealed that in some cases hashtags were required for defining the sentiment of the tweet. Overall, this study concluded that hashtags may be useful for the sentiment analysis of tweets. By contrast, our study seeks to demonstrate conclusively that hashtags are accurate predictors of the sentiment of tweets.

3   Method

In order to investigate the effectiveness of hashtags as predictors of the overall sentiment of tweets, we divide the project into two main phases. In the first phase, we develop a modified lexicon-based approach to automatically classify hashtags as either sentiment or non-sentiment bearing. In the second phase, we apply supervised machine learning to classify tweets containing these hashtags as either positive or negative.

3.1   Phase 1: Classification of Hashtags

In the modified lexicon-based approach, the subjective words from different sentiment resources are used to detect subjectivity in the hashtags extracted from the tweets. Hashtags are stripped of their hash symbol, and their stems are found using a Regexp stemmer from Natural Language Processing Toolkit (NLTK)[6].

Sentiment resources refer to both opinion lexicons and word lists of sentiment terms. In our study, we use seven opinion lexicons (listed from smallest to largest): AFINN [12], SentiStrength [16], Bing Liu Lexicon [8], Subjectivity Lexicon [18], General Inquirer [15], NRC Hashtag Sentiment Lexicon [11], and SentiWordNet [1]. For each lexicon, we extract all positive and negative words. However, there are a few lexicons, in which we extract only the strongly subjective words. For SentiStrength Lexicon, we extract positive and negative words with semantic orientations greater than 2.0, and less than -2.0, respectively. For NRC Hashtag Sentiment Lexicon, we extract the top 500 adjectives for each sentiment class (positive and negative). For SentiWordNet, we consider only the adjectives (POS tags provided in the lexicon) that have scores for positivity or negativity, which are greater than or equal to 0.5.

We also use three lists of sentiment words: Steven Hein feeling words[7] which contains 4232 words, The Compass DeRose Guide to Emotion Words[8] which consists of 682 words, and sentiment bearing Twitter slangs and acronyms collected from various online sources[9,10]. Most of these words are not found in the other lexicons. Examples include "fab" for "fabulous", and "HAND" for "Have a Nice Day".

Using these 10 sentiment resources, a total of five aggregated lists of words are created after a series of experiments is performed on the training set to determine the selected combinations. These are described below.

1. FOW (Frequently Occurring Words) list contains the most subjective words. These 915 words have occurred in at least five resources. The threshold of five represents half of the total number of resources under consideration.
2. Stems of FOW contains the stems of all the opinion words in the FOW list. There are 893 words in this list.
3. MDW (More Discriminating Words) list contains strongly subjective words. These 7366 words occur in the smaller opinion lexicons and word lists: AFINN, SentiStrength, Bing Liu and Compass DeRose Guide as well as those which occur in 4 out of the 5 larger lexicons and word lists: NRC Hashtag Sentiment, SentiWordNet, General Inquirer, Subjectivity Lexicon and Steven Hein list of feeling words.
4. LDW (Less Discriminating Words) list consists of subjective words that occur in at least 2 but not exceeding 3 of the 5 larger lexicons and word lists. These 868 words are considered to be the least subjective.
5. Twitter slangs and acronyms which have been manually identified. This list also includes common interjections[11], giving a total of 308 words.

---

[6] www.nltk.org.

[7] http://eqi.org/fw.htm.

[8] http://www.derose.net/steve/resources/emotionwords/ewords.html.

[9] http://www.socialmediatoday.com/content/top-twitter-abbreviations-you-need-kn ow.

[10] http://www.webopedia.com/quick_ref/Twitter_Dictionary_Guide.asp.

[11] http://www.dailywritingtips.com/100-mostly-small-but-expressive-interjections/.

Model Development. The classification model uses a binary search algorithm to determine whether the hashtags in the training datasets meet <u>one</u> of the following criteria:

1. It is an opinion word or originates from an opinion word.
2. It contains an opinion word or feature.

Based on this criteria, the model is divided into two steps. Initially, each of the aggregated word lists are sorted alphabetically. In the first step, each hashtag is compared with each opinion word in the different word lists. Comparisons are also made between the stem of the hashtag and each opinion word. If a match is found, the search terminates. Otherwise, the search must continue into the second step.

The second step focuses on the hashtags that have not been matched after the first step. Our aim is to ascertain if the hashtag contains an opinion word (including a word originating from an opinion word) or feature. In order to do this, two recursive algorithms are employed to create substrings of the hashtag. Both algorithms return a list of substrings sorted in descending order of length. The resulting substrings are compared to the opinion words in the FOW, stems of FOW, and MDW lists because the substrings are smaller representations of the hashtag, and thus, we consider only matches to the most subjective words are considered. Additionally, we only consider substrings of the hashtag that contain 3 or more characters. Our two recursive algorithms are described below.

1. **reduce_hashtag** eliminates the rightmost character from the hashtag after each iteration. The remaining characters form the left substring, whereas the removed character(s) form the right substring. For example, the hashtag, "lovestory" has 10 substrings: "lovestor", "lovesto", "lovest", "estory", "loves", "story", "love", "tory", "lov", and "ory".
2. **remove_left** removes the leftmost character from the hashtag after each iteration. Using this algorithm, six relevant substrings of the pre-processed hashtag, "lovestory", are found: "ovestory", "vestory", "estory", "story", "tory", and "ory".

Initially, the reduce_hashtag algorithm is applied to produce a list of substrings. Starting with the largest substring, each one is compared to each opinion word in the FOW, stems of FOW and MDW lists, until a match is found. If the search is unsuccessful, then the remove_left algorithm is applied.

We then ascertain if the hashtag contains an opinion feature. In this study, an opinion feature is any non-word attribute in the hashtag that suggests the expression of a sentiment. Therefore, we consider only the presence of extra repeated letters (at least 3), exclamation or question marks.

Table 1 outlines the eight rules for determining whether a hashtag is sentiment bearing. If **any** of these rules is found to be true, then the hashtag is determined to be sentiment bearing. Otherwise, the hashtag is non-sentiment bearing.

Table 1. Rules for identifying sentiment hashtags

No.	Rules
1	Hashtag = opinion word
2	Hashtag = stem of an opinion word
3	Stem of the hashtag = an opinion word
4	Stem of the hashtag = stem of a FOW
5	Max(substring of the hashtag) = an opinion word
6	Stem of the max(substring of the hashtag) = stem of a FOW
7	Max(substring of the hashtag) = stem of an opinion word
8	Hashtag contains a sentiment feature

3.2 Phase 2: Classification of Tweets

In this phase, we develop different scoring algorithms that can be used in conjunction with various classifiers, in determining the sentiment polarity of tweets. We only consider tweets with hashtags.

Model 1. In this model, the total number of occurrences of each unique hashtag, is determined for each sentiment class. Each unique hashtag is assigned to the sentiment class, for which it has the highest frequency. This is the simplest model.

Model 2. This model uses a bag-of-words approach. Tweets in the training set are tokenized into unigrams. Usernames and URL links are replaced with generic tags [7]. Hashtags are extracted, and stored separately. Emoticons are identified, and replaced with tags to indicate their sentiment polarity. Similarly, negating words, repeated questions and exclamation marks are also extracted, and substituted with special tags. All other punctuation marks and stop words are removed from the dataset. Then, each unique word, $word_f$, in the tweet is used as a feature. The frequency of each word in the different sentiment classes is calculated. Then the positive and negative ratios are found using Eqs. 1, and 2. The positive ratio shown in Eq. 1 is defined as the difference between the number of positive tweets and the number of non-positive tweets that the word occurs in, divided by the number of positive tweets that contains the word.

$$positive\ ratio(word) = \frac{pos(word) - (neg(word))}{pos(word)} \tag{1}$$

The negative ratio shown in Eq. 2 refers to the difference between the number of negative tweets and the number of non-negative tweets that the word occurs in, divided by the number of negative tweets that contains the word.

$$negative\ ratio(word) = \frac{neg(word) - (pos(word))}{neg(word)} \tag{2}$$

The sentiment polarity of the word, $sp(word)$, is the maximum of the positive and negative ratios. The sentiment weight of each word, $word_{sw}$, is determined as the product of $sp(word)$ and $word_f$.

Additionally, emoticons, punctuation marks, and negating words are also incorporated as features into the model. Positive emoticons and exclamation marks are assigned a polarity of 1, whereas negative emoticons, question marks and negations are assigned a polarity of -1. The feature weight $feature_{fw}$ of each feature is described in Eq. 3

$$feature_{fw} = \frac{count(fw)}{frequency_{fw}} \times sp(feature) \tag{3}$$

where $count(fw)$, is the frequency of the word in the tweet, $frequency_{fw}$, is the total frequency in the dataset and $sp(feature)$, is the sentiment polarity of the word. Then, the sentiment score of each hashtag in the tweet is the weighted average of all the features (including the words) which is determined by Eq. 4 as

$$hastag\_score = \frac{\sum_{i=1}^{n} feature_{fw_i} \times sp(feature_i)}{\sum_{i=1}^{n} feature_{fw_i}} \tag{4}$$

where n, is the number of features. The sentiment score for each hashtag in the tweet ranges from -1 to 1. If the score is greater than 0, the hashtag is considered to be positive. Otherwise, the hashtag is considered to be negative.

In order to determine the overall sentiment of the hashtag in the training set, we count the number of times the hashtag is scored as positive or negative, and assign the sentiment class with the highest frequency.

At the end of the training phase, we have two derived properties for each hashtag: its frequency in the training set, and its sentiment polarity.

4   Experimental Results

4.1   Dataset

Our dataset consists of 71,836 unique tweets with hashtags, which are extracted using the Twitter API[12]. The tweets were collected during the FIFA World Cup 2014 using search terms (excluding hashtags) related to the football matches, in order to capture the opinions of the Twitter users during each game. We use Sentiment140 API[13] to automatically assign sentiment labels to the tweets in our dataset. Sentiment140 uses a Maximum Entropy classifier with an accuracy of 83 percent on a combination of unigrams and bigrams [7]. Positive, negative, and neutral tweets are assigned numerical values of 4, 2, and 0, respectively.

---

[12] http://www.dev.twitter.com/.
[13] http://help.sentiment140.com/api.

4.2   Evaluation Measures

Our evaluation measures are accuracy, precision, recall and f-measure. Accuracy measures the number of tweets (hashtags) for each class that are classified correctly. Precision determines the ratio of actual relevant tweets (hashtags) among predicted tweets (hashtags) for the sentiment category. Recall refers to the fraction of relevant tweets (hashtags) actually classified by the model. F-measure is the average of precision and recall.

4.3   Classification of Hashtags

Hashtags are extracted from the dataset and manually classified. Hashtags belonging to the same type are grouped. For each hashtag type, we selected all the tweets containing at least one hashtag of the respective type. Then, we divided this group of tweets equally into training and test sets. Table 2 shows the total number of hashtags extracted from the training and test sets.

Table 2. Training and test sets for each type of hashtag

Type	Train	Test	Total
Sentiment	1,368	1,376	2,744
Non-sentiment	3,070	3,142	6,212

In order to evaluate our model, we compare the hashtags extracted in the training and test sets. The hashtags in the test set is compared with the list of determined hashtags in the training set. If the hashtag is found in this list, the same class label is assigned. If it is not found, then similarity testing is performed where we compare their stems and length (threshold of 95 %) of the hashtags to determine a suitable match. Then, we compare the predicted class label assigned by the model to that of actual label of the hashtag assigned during manual annotation in order to evaluate our model.

Table 3 shows examples of the sentiment and non-sentiment hashtags that are identified by our model. Table 4 shows the precision, recall, f-measure, and accuracy metrics (in percent) obtained by our classification model.

Discussion. It can be observed from Table 4 that our model achieved higher percentages for accuracy, precision, recall and f-measure in identifying non-sentiment hashtags than sentiment hashtags. Therefore, our results suggest that it is easier to identify non-sentiment hashtags than sentiment hashtags.

In order to compare the performance of our model, we created models which used a single lexical resource in order to identify sentiment hashtags from non-sentiment hashtags. Figure 1 shows the accuracy scores for the top five models. It can be observed in Fig. 1 that our model (last column), which used combined

Table 3. Examples of sentiment and non-sentiment hashtags classified by our model

Type	Examples
Sentiment	#stressful, #helpmeunderstand,
	#shedoesntlookveryhappy, #strong
	#celebration #mindblowing
Non-sentiment	#budweiser, #dance
	#2014fifaworldcup, #children
	#teambrazil, #waiting

Table 4. Results for classification of hashtags

Hashtag type	Accuracy	Precision	Recall	F-measure
Sentiment	83.58	86.27	80.96	83.53
Non-sentiment	**83.83**	**94.25**	**84.93**	**89.35**

resources is the most accurate in identifying sentiment hashtags when compared with models which used a single lexical resource. For the identification of non-sentiment hashtags, our model performs comparably to one of the models which used a single lexical resource. Therefore, the experimental results show that using subjective words from different lexical resources is effective in boosting the identification of sentiment hashtags.

4.4 Classification of Tweets

We use the sentiment and non-sentiment hashtags that are classified by our model to select tweets that contain these hashtags. These tweets form our training and test set for each sentiment class. Table 5 show the number of tweets in our training and test sets, for each sentiment class.

Table 5. Tweets for sentiment classification

Dataset	Train	Test	Total
Positive	2,886	2,888	5,774
Negative	16,477	16,478	32,995

In order to classify positive and negative tweets in the test sets, we use the derived properties of the hashtags in the training sets. For each tweet in the test set, we determine if it contains at least one hashtag from the corresponding training set. Then, we assign the tweet the same sentiment polarity as the hashtag. If the tweet contains multiple hashtags from the training set, we apply two derived properties of the hashtags: the type of hashtag and, its frequency in

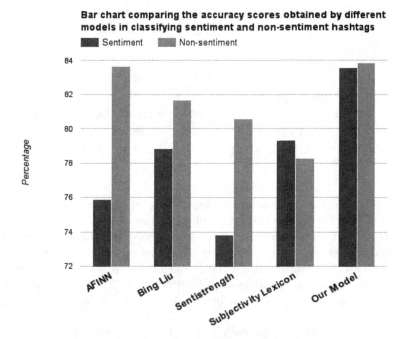

Fig. 1. Comparing the accuracy of our model to models using a single resource

the training set. For classifying tweets using sentiment hashtags, we determine the most subjective hashtag in the group by comparing the hashtags to opinion words in the FOW list. If one of the hashtags is found in this list, the tweet is assigned the same sentiment polarity as this hashtag. Otherwise, the tweet is assigned the sentiment polarity of the hashtag with the highest frequency. For classifying tweets using non-sentiment hashtags, we determine the most descriptive hashtag in the group by selecting the hashtag that is not determined to be a noun. We use a POS tagger in NLTK for python. Then the tweet is assigned the same sentiment polarity as this hashtag. Otherwise, the tweet is assigned the sentiment polarity of the hashtag with the highest frequency.

Table 6 shows the precision, recall, f-measure, and accuracy metrics (in percent) for our models on the test set, for each type of hashtag.

Discussion. It can be observed from Table 6 that both Model 1 and 2 for non-sentiment hashtags achieve higher accuracy, recall, precision, and f-measure in classifying tweets as positive and negative than Models 1 and 2 for sentiment hashtags. Therefore, non-sentiment hashtags are more effective in classifying tweets as positive and negative than sentiment hashtags.

Furthermore, our experimental results show that Model 1 outperforms Model 2 in classifying tweets using non-sentiment hashtags. For classifying tweets with sentiment hashtags, Model 1 achieves higher precision and f-measure than

Table 6. Results for classification of tweets

Hashtag type	Model	Accuracy	Precision	Recall	F-measure
Sentiment	Model 1	81.14	**76.94**	81.23	**77.77**
	Model 2	**81.72**	71.34	**81.84**	73.91
	Bakliwal et al. [2]	71.24	76.72	71.22	73.38
Non-sentiment	Model 1	**86.07**	**81.96**	**86.03**	**81.50**
	Model 2	85.97	73.89	85.92	79.46
	Bakliwal et al. [2]	71.70	81.13	71.73	75.22

Model 2. However, Model 2 achieves slightly higher accuracy and recall than Model 1. Overall, Model 1 is the better classification model.

In order to further evaluate our models, we apply the scoring algorithm created by Bakliwal et al. [2] to our dataset. Experimental results show that all of our models achieve higher accuracy, recall and f-measure scores than the model which applied the scoring algorithm by Bakliwal et al. [2]. However, the models created using the scoring algorithm by Bakliwal et al. [2], achieve the highest precision.

We then compare Model 1 for each hashtag type to four established classifiers, Naive Bayes, SVM, Maximum Entropy and C4.5. We use the WEKA implementation of these classifiers. We modify the training and test sets previously used for Model 1, using 1 and 0 to indicate the presence and absence of each hashtag in each tweet. Tables 7 and 8 shows the precision, recall, f-measure, and accuracy values (in percent) for the five classifiers for the test set on sentiment and non-sentiment hashtags, respectively.

Table 7. Results for classification using sentiment hashtags

Classifier	Accuracy	Precision	Recall	F-measure
Naive Bayes	80.81	77.20	80.80	78.20
SVM	**82.85**	**79.70**	**82.90**	**79.60**
Maximum Entropy	73.52	75.40	73.50	74.40
C4.5	82.78	80.10	82.80	76.90
Model 1	**81.14**	**76.94**	**81.23**	**77.77**

It can be observed from Tables 7 and 8 that our models performed quite comparably to the established classifiers. Additionally, all five models which applied non-sentiment hashtags achieve higher accuracy, precision, recall and f-measure scores than the models which applied sentiment hashtags. Therefore, this suggests that non-sentiment hashtags are more effective than sentiment hashtags in classifying tweets as positive or negative.

Table 8. Results for classification using non-sentiment hashtags

Classifier	Accuracy	Precision	Recall	F-measure
Naive Bayes	85.90	80.50	85.90	79.90
SVM	86.12	82.30	86.10	82.00
Maximum Entropy	85.74	81.70	85.70	**82.10**
C4.5	**86.41**	**83.30**	**86.40**	81.80
Model 1	**86.07**	**81.96**	**86.03**	81.50

Overall, all our experimental results show that non-sentiment hashtags are more effective in classifying tweets as positive and negative than sentiment hashtags. Additionally, Model 1 is determined to be the better model. Model 1 significantly outperforms the model created using the scoring algorithm by Bakliwal et al. [2], and performs comparably to that of the established classifiers, which demonstrates that our method is effective.

5   Conclusions and Future Work

In this paper, we evaluated the effectiveness of hashtags as accurate predictors of the sentiment of tweets. First, we applied a modified lexicon-based approach, which incorporated subjective words from different lexical resources, in order to accurately distinguish sentiment-bearing hashtags from non-sentiment hashtags. Using this model, we are able to achieve accuracy of 83.58 % and 83.83 % in identifying sentiment hashtags and non-sentiment hashtags, respectively. Furthermore, our accuracy surpassed those scores obtained using models that applied a single lexical resource.

Then, we applied the derived properties of hashtags to classify tweets as positive and negative. We developed and evaluated two separate classification models using training and test datasets of tweets. Our best models achieved accuracy scores of 81.14 % and 86.07 % in classifying tweets using sentiment hashtags and non-sentiment hashtags, respectively. Additionally, the performance of our models outperforms a previously developed algorithm [2] but is comparable to established classifiers. Finally, all our experimental results clearly indicate that non-sentiment hashtags are more effective than sentiment hashtags for the sentiment analysis of tweets.

In terms of future work, we will extend our work to include neutral tweets, and we plan to use hashtags for topic-based sentiment analysis.

References

1. Baccianella, S., Esuli, A., Sebastiani, F.: Sentiwordnet 3.0: an enhanced lexical resource for sentiment analysis and opinion mining. In: Proceedings of the Seventh International Conference on Language Resources and Evaluation, LREC 2010, Valletta, Malta (2010)
2. Bakliwal, A., Arora, P., Madhappan, S., Kapre, N., Singh, M., Varma, V.: Mining sentiments from tweets. In: Proceedings of the 3rd Workshop in Computational Approaches to Subjectivity and Sentiment Analysis, WASSA 2012, Portland, Oregon, pp. 11–18 (2012)
3. Bakliwal, A., Foster, J., van der Puil, J., O'Brien, R., Tounsi, L., Hughes, M.: Sentiment analysis of political tweets: towards an accurate classifier. In: Proceedings of the Workshop on Language Analysis in Social Media, Atlanta, Georgia, pp. 49–58 (2013)
4. Bifet, A., Frank, E.: Sentiment knowledge discovery in twitter streaming data. In: Pfahringer, B., Holmes, G., Hoffmann, A. (eds.) DS 2010. LNCS, vol. 6332, pp. 1–15. Springer, Heidelberg (2010)
5. Bora, N.N.: Summarizing public opinions in tweets. Int. J. Comput. Linguist. Appl. **3**(1), 41–55 (2012)
6. Davidov, D., Tsur, O., Rappoport, A.: Enhanced sentiment learning using twitter hashtags and smileys. In: Proceedings of the 23rd International Conference on Computational Linguistics: Posters, COLING 2010, Beijing, China, pp. 241–249 (2010)
7. Go, A., Bhayani, R., Huang, L.: Twitter sentiment classification using distant supervision. Technical report, Stanford University (2009)
8. Minqing, H., Liu, B.: Mining and summarizing customer reviews. In: Proceedings of the Tenth ACM SIGKDD International Conference on Knowledge Discovery and Data Mining, KDD 2004, Seattle, WA, USA, pp. 168–177 (2004)
9. Kouloumpis, E., Wilson, T., Moore, J.D.: Twitter sentiment analysis: the good the bad and the omg! In: Proceedings of the Fifth International Conference on Weblogs and Social Media, Barcelona, Catalonia, Spain (2011)
10. Mohammad, S.: #emotional tweets. In: *SEM 2012: The First Joint Conference on Lexical and Computational Semantics – vol. 1: Proceedings of the main conference andthe shared task, and vol. 2: Proceedings of the Sixth International Workshop on Semantic Evaluation (SemEval 2012), Montréal, Canada, pp. 246–255, 7–8 June 2012
11. Mohammad, S.M., Kiritchenko, S., Zhu, X.: Nrc-canada: building the state-of-the-art in sentiment analysis of tweets. CoRR, abs/1308.6242 (2013)
12. Nielsen, F.Å.: A new ANEW: evaluation of a word list for sentiment analysis in microblogs. CoRR, abs/1103.2903 (2011)
13. Pang, B., Lee, L., Vaithyanathan, S.: Thumbs up?: sentiment classification using machine learning techniques. In: Proceedings of the ACL-02 Conference on Empirical Methods in Natural Language Processing - vol. 10, EMNLP 2002, Stroudsburg, PA, USA, pp. 79–86 (2002)
14. Rodrigues Barbosa, G.A., Silva, I.S., Zaki, M., Meira Jr., W., Prates, R.O., Veloso, A.: Characterizing the effectiveness of twitter hashtags to detect and track online population sentiment. In: CHI 2012 Extended Abstracts on Human Factors in Computing Systems, CHI EA 2012, Austin, Texas, USA, pp. 2621–2626 (2012). ISBN 978-1-4503-1016-1

15. Stone, P.J., Dunphy, D.C., Smith, M.S., Ogilvie, D.M.: The General Inquirer: A Computer Approach to Content Analysis. MIT Press, Cambridge (1966)
16. Thelwall, M., Buckley, K., Paltoglou, G., Cai, D., Kappas, A.: Sentiment in short strength detection informal text. J. Am. Soc. Inf. Sci. Technol. **61**(12), 2544–2558 (2010)
17. Wang, X., Wei, F., Liu, X., Zhou, M., Zhang, M.: Topic sentiment analysis in twitter: a graph-based hashtag sentiment classification approach. In: Proceedings of the 20th ACM International Conference on Information and Knowledge Management, CIKM 2011, Glasgow, Scotland, UK, pp. 1031–1040 (2011)
18. Wilson, T., Wiebe, J., Hoffmann, P.: Recognizing contextual polarity in phrase-level sentiment analysis. In: Proceedings of the Conference on Human Language Technology and Empirical Methods in Natural Language Processing, HLT 2005, Vancouver, British Columbia, Canada, pp. 347–354 (2005)
19. Zhang, L., Ghosh, R., Dekhil, M., Hsu, M., Liu, B.: Combining lexicon-based and learning-based methods for twitter sentiment analysis. Technical report, HP Laboratories (2011)

On the Feasibility of Discovering Meta-Patterns from a Data Ensemble

Einoshin Suzuki[✉]

Department of Informatics, ISEE, Kyushu University, Fukuoka, Japan
suzuki@inf.kyushu-u.ac.jp

Abstract. We introduce meta-pattern discovery from a data ensemble, a new paradigm of pattern discovery which goes beyond the KDD process model. A data ensemble, which represents a set of data sets, seems to be more natural as a model of the big data (We focus on the volume and velocity aspects of the big data.). We propose two kinds of meta-patterns, each of which specifies patterns such as clusters for a set of data sets, for an unsupervised setting and a supervised one. Our solutions for these settings were shown to be feasible with one synthetic and two real data ensembles by experiments.

Keywords: Meta-pattern · Data ensemble · Data squashing · Sparse coding

1 Introduction

Typical huge data, including the big data, can be rather regarded as a set of data sets. For instance, some specific location data of cellular phone users can be regarded as an ensemble of 1 million data sets, i.e., a collection of data on 10000 users, each being monitored for 100 days. We call a set of data sets a data ensemble, which is beyond just a collection of data sets because they together describe the target. In this paper, we propose a new paradigm of pattern discovery which goes beyond the KDD (Knowledge Discovery in Databases) process model [4]. The paradigm skips the effort of selecting the data sets for pattern extraction and instead extract patterns from a data ensemble and merge them to discover new kinds of patterns. Relationships among the data sets would be described explicitly in the output as "meta patterns". Our motivation here is to explore the possibility of this new pattern mining rather than improving the state-of-the-art methods for individual, traditional problems. In this respect, our paradigm differs substantially from such seemingly-relevant data mining problems, e.g., tensor decomposition [6], multi-task learning [8], data approximation [5].

E. Suzuki—A part of this research was supported by Grant-in-Aid for Scientific Research 25280085 and 15K12100 from the Japanese Ministry of Education, Culture, Sports, Science and Technology.

© Springer International Publishing Switzerland 2015
N. Japkowicz and S. Matwin (Eds.): DS 2015, LNAI 9356, pp. 266–274, 2015.
DOI: 10.1007/978-3-319-24282-8_22

2   Meta-Pattern Discoveries from a Data Ensemble

2.1   Data Ensemble

A data ensemble Ω is a set of data sets $\Omega = \{D_1, D_2, \ldots, D_\nu\}$, where D_i and ν represent the ith data set and the number of the data sets, respectively. For instance, D_i can represent the location data of a specific cellular phone user on a specific day.

D_i consists of $n(i)$ instances, i.e., $D_i = \{\mathbf{d}_{i,1}, \mathbf{d}_{i,2}, \ldots, \mathbf{d}_{i,n(i)}\}$, where $\mathbf{d}_{i,j}$ represents the jth instance of D_i. We assume that it is a point $(v_{i,j,1}, v_{i,j,2}, \ldots, v_{i,j,m})$ in the Euclidean space described with m features $a_1, a_2, \ldots, a_m$ due to its generality. Here $v_{i,j,k}$ represents the value of a feature a_k of $\mathbf{d}_{i,j}$. An instance $\mathbf{d}_{i,j}$ is either unsupervised, i.e., $\mathbf{d}_{i,j} = (v_{i,j,1}, v_{i,j,2}, \ldots, v_{i,j,m})$, or supervised, i.e., $\mathbf{d}_{i,j} = (v_{i,j,1}, v_{i,j,2}, \ldots, v_{i,j,m}, y_{i,j})$, where $y_{i,j}$ represents the class label of $\mathbf{d}_{i,j}$.

2.2   Meta-Patterns in an Unsupervised Setting

In the unsupervised setting, we adopt clustering due to its generality and utility. Note that in our meta-patterns, a cluster is discovered from the data ensemble Ω and not given in the input. We extract a set of clusters from the whole data set $D_1 \cup D_2 \cup \cdots \cup D_\nu$ in Ω and define meta-patterns over the probability distribution on the clusters. Let the extracted clusters be $c_1, c_2, \ldots, c_\mu$, which is a partition of $D_1 \cup D_2 \cup \cdots \cup D_\nu$, then the probability distribution of D_i on the clusters is given by $(p_{i,1}, p_{i,2}, \ldots, p_{i,\mu})$, where $p_{i,j} = |c_j| / \sum_{i=1}^{\nu} n(i)$. We set a meta-pattern as specifying data sets with similar distributions. We call this kind of pattern

$$\cup_i D_i \sim (p_{i,1}, p_{i,2}, \ldots, p_{i,\mu}) \tag{1}$$

a cluster distribution meta-pattern, where $A \sim B$ represents that a set A of data sets follows a probability distribution B. It represents that a set $\cup_i D_i$ of data sets contain clusters $c_1, c_2, \ldots, c_\mu$ with a probability distribution $(p_{i,1}, p_{i,2}, \ldots, p_{i,\mu})$.

2.3   Meta-Patterns in a Supervised Setting

In the supervised setting, we opt for classification, which would be the most popular problem in this kind of learning. We restrict our attention to dictionary learning based on sparse coding [8] due to its affinity to pattern extraction which will be explained later. Sparse coding is an information representation for multi-dimensional observed data [8]. Sparse coding is defined as the operation of decomposing an observed vector $\mathbf{d}$ to a multiplication of $\mathbf{L}$ and a sparse weight vector (coefficient vector) $\mathbf{s}$, i.e., $\mathbf{d} \sim \mathbf{L}\mathbf{s}$.

By following [8] we induce a classifier $f(\mathbf{d}; \boldsymbol{\theta}_i)$ from D_i, where $\boldsymbol{\theta}_i$ represents the parameter vector. The value of $\boldsymbol{\theta}_i$ is approximated with latent components $\mathbf{L} \in \mathbb{R}^{\nu \times k}$ shared among classification tasks of inducing $f(\mathbf{d}; \boldsymbol{\theta}_i)$ from D_i ($i = 1, 2, \ldots, \nu$), where k is the dimensionality of the latent space. That is, the value of $\boldsymbol{\theta}_i$ is learnt so that it becomes close to the value of $\mathbf{L}\mathbf{s}_i$, i.e., $\boldsymbol{\theta}_i \simeq \mathbf{L}\mathbf{s}_i$, in terms of a sparse weight vector $\mathbf{s}_i \in \mathbb{R}^k$. Here each column vector of $\mathbf{L}$ corresponds to an atom, which corresponds to a principal component in PCA, and $\mathbf{L}$ is considered to be the dictionary. Note that the latent space here is learned from Ω to transfer knowledge among the classification tasks.

By following [8], we adopt the logistic regression classifier

$$f(\mathbf{d}; \boldsymbol{\theta}_i) = \mathrm{sgn}\left[\frac{1}{1 + \exp\left(-\theta_{i,0} - \theta_{i,1}a_1 - \cdots - \theta_{i,m}a_m\right)} - 0.5\right] \qquad (2)$$

to induce from D_i, where $\theta_{i,j}$ represents the jth coefficient and $\boldsymbol{\theta}_i = (\theta_{i,0}, \ldots, \theta_{i,m})$. Note that the logistic regression classifier has been popular for decades in various fields.

In our meta-pattern we adopt weight $s_{i,j}$ of dictionary learning based on sparse coding, where $\mathbf{s}_i = (s_{i,0}, \ldots, s_{i,k})$. Our choice corresponds to describing relationships among data sets in the latent space defined by the atoms in the dictionary $\mathbf{L}$. The philosophy of sparse coding dictates that the weight vector $\mathbf{s}_i$ is sparse, i.e., most of its elements are zero. Thus we adopt a non-zero weight $s_{i,j}$ $(\neq 0)$ as an element of the meta-pattern in this setting. Note that $s_{i,j} > 0$ $(j = 1, \ldots, m)$ represents that a_j has a positive influence for predicting the positive class while $s_{i,j} < 0$ $(j = 1, \ldots, m)$ represents that a_j has a positive influence for predicting the negative class. Thus we propose directional non-zero weight meta-pattern

$$\cup_i D_i \rightarrow \bigwedge_j [(s_{i,j} > 0) \text{ or } (s_{i,j} < 0)], \qquad (3)$$

where $\rightarrow$ represents the logical implication. Note that the data sets that follow a directional non-zero weight meta-pattern share the same characteristics in the latent space spanned by the elements with non-zero weights.

3   Meta-Pattern Discovery

3.1   Unsupervised Setting

To efficiently discover clusters $c_1, c_2, \ldots, c_\mu$ from $D_1 \cup D_2 \cup \cdots \cup D_\nu$, we adopt data squashing [2]. In our problem, the approach squashes the ν data sets into a set $\mathcal{C}$ of micro clusters, each of which consists of information on similar instances. Then the approach agglomerates similar micro clusters to obtain the output clusters $c_1, c_2, \ldots, c_\mu$.

BIRCH [11] is an early work of data squashing for clustering, which is simpler and hence requires less tuning than its successors. It employs the CF (clustering feature) vector, which consists of the number and the first and second moments of examples, as the condensed representation of data. The CF vectors are managed by the CF tree, which is a height-balanced tree with three parameters, i.e., branching factor β_{internal} for an internal node, branching factor β_{leaf} for a leaf[1],

---

[1] In this paper we use the same values and denote them β.

and the diameter θ of the CF vectors in a leaf. Note that we could adopt a more recent clustering algorithm, e.g., CluStream [1], as well as options of BIRCH such as outlier filtering. These possibilities would be necessary when we handle a time-evolving data ensemble or a data ensemble with a considerable number of outliers. We leave these possibilities to our future work.

We show the pseudo-code below. We first initialize the CF tree T and a linked list L to which we add the leaves of T. Procedure addInstanceTree($\mathbf{d}_{i,j}, T, L, \theta$) adds instance $\mathbf{d}_{i,j}$ to T, updates L, and returns the updated tree. Procedure mergeSimilarClusters($merged, L, \theta$) merges the cells in L with similar micro clusters based on the value of θ. Here a pair of CF vectors are judged similar if the average inter-cluster distance between them is less than θ.

Procedure transformListDSVec(L, H) transforms L into a matrix H, of which (i, j) element represents the occurrence probability of the learnt jth cluster in terms of the examples in D_i. For instance, if L contains three clusters and D_1 contains 20 % of c_0, 0 % of c_1, and 80 % of c_2, then the second row of H is $(0.2, 0, 0.8)$. Procedure transformDSVecMetaPattern(H, ϕ) transforms H into a set of cluster distribution meta-patterns. The procedure regards that D_i and D_j ($i < j$) belong to the same meta-pattern if the KL divergence[2] of the ith row of H to the jth row of H is less than a user-specified threshold ϕ, which we call data set agglomeration parameter. Note that the information theoretic approach is known to be robust to various kinds of probabilistic distributions and often provides a sound interpretation.

Procedure addInstanceTree($\mathbf{d}_{i,j}, T, L, \theta$) first assigns the instance $\mathbf{d}_{i,j}$ to the root node of the CF tree T and updates the CF vector of the root node so that it includes $\mathbf{d}_{i,j}$. Then it iteratively assigns $\mathbf{d}_{i,j}$ to the closest child node in terms of the average inter-cluster distance and repeats the update until it reaches a parent of a leaf node. If the average inter-cluster distance between $\mathbf{d}_{i,j}$ and the closest child leaf node is below θ, $\mathbf{d}_{i,j}$ is absorbed in the leaf node[3], which undergoes the same update procedure. Otherwise, a new leaf node to which only $\mathbf{d}_{i,j}$ is assigned is created.

```
algorithm: Cluster distribution meta-pattern discovery
Input: data ensemble Ω, thresholds θ, φ
Output: set of meta-patterns
```
T = (empty tree)
L = (empty list)
For $i=1$ To ν do
    For $j = 1$ To $n(i)$
       T = addInstanceTree($\mathbf{d}_{i,j}, T, L, \theta$)
Do
   $merged$ = FALSE
   mergeSimilarClusters($merged, L, \theta$)
While($merged$)

---

[2] We adopt the standard procedure of using the Laplace correction.
[3] This condition is our modification to the original BIRCH.

transformListDSVec(L, H)
Output transformDSVecMetaPattern(H, ϕ).

3.2   Supervised Setting

For discovering directional non-zero weight meta-patterns[4], we adopt ELLA (Efficient Lifelong Learning Algorithm) [8], which is a lifelong learning algorithm based on dictionary learning with sparse coding. Note that ELLA due to its approximation has shown to be much faster than strict methods and to achieve comparable performance [8]. Lifelong Learning is an extension of online multitask learning and its objective is to leverage the overall prediction accuracy by transferring knowledge among domains each time given a new task [8]. Lifelong learning targets at a series of supervised learning tasks $\mathcal{Z}_1, \mathcal{Z}_2, \ldots, \mathcal{Z}_\nu$. In our case, the ith classification task takes D_i as the input.

Given the training data of each task, ELLA updates $\mathbf{L}$ and $\mathbf{s}_i$ to minimize the following objective function $e_\nu(\mathbf{L})$ for the ν tasks it has received.

$$
e_\nu(\mathbf{L}) = \frac{1}{\nu} \sum_{i=1}^{\nu} \arg\min_{\mathbf{s}_i} \left\{ \frac{1}{n(i)} \sum_{j=1}^{n(i)} \mathcal{L}\left(f\left(\mathbf{d}_{i,j}; \mathbf{L}\mathbf{s}_i\right), y_{i,j}\right) + \mu \|\mathbf{s}_i\|_1 \right\} + \lambda \|\mathbf{L}\|_F^2
\tag{4}
$$

Here $\mathcal{L}$ is the given loss function and the L^1-norm of $\mathbf{s}_i$ is used as a convex approximation of the sparseness of the vector. $\|\mathbf{A}\|_F = \sqrt{\sum_{i=1}^{m} \sum_{j=1}^{n} |a_{ij}|^2}$ represents the Flobenius norm of matrix $\mathbf{A} = (a_{ij})$ and μ and λ are parameters.

After obtaining the weight vectors $\mathbf{s}_i$ ($i = 1, 2, \ldots, \nu$), the task of discovering the directional non-zero weight meta-patterns is straightforward. Note that the pattern adopts logical implication and not probabilistic implication, making the use of the support threshold unnecessary. Note also that most of the weights are 0 in sparse coding, which would result in a manageable number of meta-patterns.

4   Experiments

4.1   Unsupervised Synthetic Data Ensemble

We used a PC with Intel Core i7-3960X CPU of clock speed 3.30–3.90 GHz and 16 GB RAM. The OS was Ubuntu 12.04.3 and the compiler gcc 4.6 with -O3 option. We tested[5] the performance of our method for the unsupervised setting firstly with synthetic data, as our problem requires the ground truth as the answer. The synthetic data were generated in the 3D space, i.e., $m = 3$, which is common for evaluating a distance-based method such as [9,11] to avoid the curse of dimensionality. Note that a variety of effective dimension reduction techniques

---

[4] Note that this solution is independent from the one in the previous section. A semi-supervised, hybrid solution is beyond the scope of this paper.

[5] We use the past tense for our past actions.

exist, ranging from unsupervised ones [7] to semi-supervised ones [10]. Our goal was to evaluate methods from the viewpoint of estimating the true distribution of the examples in a data ensemble with the discovered cluster distribution meta-patterns. We, in this case, employed the recall and the precision of the cluster distribution meta-patterns as evaluation criteria.

The synthetic data concern 27 clusters $\mathcal{N}((x_0, y_0, z_0), 1)$, where each of x_0, y_0 and z_0 takes a value either $\gamma, 0$, or $-\gamma$. We generated 18 data sets, i.e., $\nu = 18$, each of which is an equi-probable mixture of 3 to 6 of the 27 clusters. Note also that when γ becomes small, the clusters become close to each other as γ also represents the distance from a cluster center to the centers of its closest clusters. For instance, $\gamma = 3$ means that two closest clusters overlap each other by nearly 50 %, as about 99.7 % of examples belong to the range of $\mu \pm 3\sigma$. One concern was the effective range of ϕ, which turned out to be wide in our preliminary experiments. It turned out that ϕ has only a small influence on the results so we set ϕ equal to θ throughout this paper.

We compared our method with the k-means algorithm, as it is one of the most frequently used algorithms in clustering. Our clusters are spherical and equi-probable, which justifies its usage. As common with the algorithm, we ran the k-means algorithm with randomly set k seeds multiple times and report the clustering result with the smallest $\sum_{i-1}^{k} \sum_{x \in C_i} D(x, c_i)^2$, where c_i is the mean of the cluster to which x belongs. The number of the random restart was fixed to 10 throughout the experiments[6]. For a fair comparison, we agglomerated the resulting clusters with the list agglomeration procedure of our method in Sect. 3.1 using the same parameter θ. This agglomeration typically leverages the performance of the k-means algorithm when k is larger than the number of true clusters.

Figure 1 up shows the precision and recall for $\gamma = 6, 5, 4, 3$ with varying noise levels. The parameters were set $n(i) = 2000$, $\theta = 3$, $\beta = 15$, $\phi = 0.3$, $k = 27$ and noise deviations $\mathcal{N}((0, 0, 0), 3)$ were added to the 18 data sets. We see that the k-means algorithm outperforms our method in both recall and precision in many cases, which is not surprising as the former treats the examples as they are while our method squashes them. The run times of the k-means algorithm was 40 times longer than our method, which is not a serious problem under these conditions as each run is completed in less than 10 s. However, the k-means algorithm has a fatal flaw in its scalability. The compiler failed with $n(i) = 4000$ and $\nu = 18$ because the (half) distance matrix contains about $2.59 * 10^9$ distances. On the other hand, even with $n(i) = 4000000$, the run time of our method was about 546 s, which we believe acceptable and even relatively short as a pattern mining problem. Figure 1 down shows the results of the experiments under extremely hard conditions, as the noise deviation is increased to 5σ. Basically we see the

---

[6] Preliminary experiments showed that the number of the random restart has a minor influence to the performance as long as it is not extremely small.

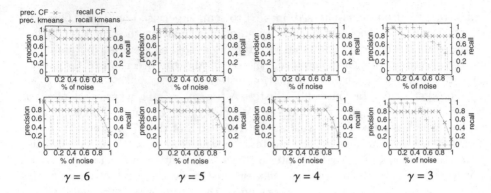

Fig. 1. Results with the 3D domain ($n(i) = 2000$, $\theta = 3$, $\beta = 15$, $\phi = 0.3$, $k = 27$) with noise deviations $\mathcal{N}((0,0,0),3)$ (up) and $\mathcal{N}((0,0,0),5)$ (down). Precisions are plotted with lines - points while recalls are displayed with impulses.

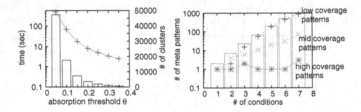

Fig. 2. Results of discovering cluster distribution meta-patterns (Left) and directional non-zero weight meta-patterns (Right) from 100 facial expression data

same tendencies with worse performance for both methods. We believe that these conditions seldom occur in reality[7].

Note that most of these experiments suffer from extremely bad conditions, e.g., noise level above 40 %, some even 100 %, many clusters overlap, examples can be shifted much farther than the next cluster even with a noise deviation of 3σ, which is considered as "modest" in our experiments. These results are surprisingly well given these bad conditions, partly because of the large number of examples, i.e., $n(i) = 2000 \times \nu = 18$, and the fact that we evaluated meta-patterns in our problem instead of examples in clustering. A few number of the recalls and precisions might appear to be inconsistent in terms of the difficulties of the clustering problems but we think that they are due to the fact that the resulting CF tree depends on the order of the examples.

We also investigated other parameters. For our method, we have found that β is not influential under these conditions. Precisions and recalls are usually the same. Run time can increase several dozens of percent, which is not a problem because they are less than 0.3 s. For the k-means algorithm, we provided a correct

---

[7] Due to the good performance under these conditions, we believe that our method outperforms sampling-based k-means algorithms as well as state-of-the-art methods.

value, $k = 27$, throughout the experiments in Fig. 1. We tried $k = 40$ under some conditions and the results were similar, as similar clusters were merged with our list agglomeration procedure with ϕ as stated above. However, setting k below the correct number of clusters cause fatal problems, resulting very low precision and recall.

4.2 Unsupervised Facial Data Ensemble

We tested the feasibility of our method for the unsupervised setting with a real data set[8]. We constructed facial expression benchmark data of 100 persons using Kinect face tracking application [3]. We devised multi-lingual instruction sheets on 25 expressions, collected data from 115 persons, and carefully inspected and labeled the outcome to construct the data. The benchmark data consist of 263,106 instances, each of which is described with 17 features. Among the 17 features, we used 6 animation units (AUs), each of which represents a deviation from the neutral face and is highly useful in such an analysis. The images of a person is considered as the data set for the person, resulting in a data ensemble of $\nu = 100$ data sets.

Branching factor $\beta_{internal} = \beta_{external}$ were set to 30 throughout the experiments in this and the next section. We varied the absorption threshold θ and measured the run time and the number of discovered meta-patterns. Figure 2 left shows the results of the experiments. The results show that our solution for discovering cluster distribution meta-patterns is at least feasible: the run time is at most 8.5 min and the numbers of meta-patters range from 684 to about 47300. The interpretation of the meta-patterns are not as easy as these evaluation criteria because it involves the interpretation of the learned clusters and our intuition on human faces sometimes hinders it.

4.3 Supervised Facial Data Ensemble

We tested the feasibility of our method for the supervised setting with a real data set[9]. For the 100 person facial expression data set, out of the 263,106 instances, we labeled 62,500 of them as 1 of the 25 expressions. We used only the labeled images to form a data ensemble of $\nu = 100$ data sets with again the 6 AU features.

We classified the discovered meta-patterns in terms of the number of the conditions in the premise and for each of them we show the distribution of the number of data sets the meta-pattern covers. Figure 2 right shows the results, where a box represents the total number of meta-patterns and the dagger, cross, and star respectively represents the number of low coverage (1–32 data sets), mid coverage (33–56 data sets), and high coverage (57–100 data sets) meta-patterns. Again, our solution turned out to be feasible in this case, i.e., the number of

---

[8] As feasibility study, we did not compare our method with other methods.
[9] We used the default setting of ELLA http://www.seas.upenn.edu/~eeaton/publications.html.

meta-patterns are relatively small in number (below 1000), especially if we go for finding meta-patterns which cover many data sets.

References

1. Aggarwal, C.C., Han, J., Wang, J., Yu, P.S.: A framework for clustering evolving data streams. In: Proceedings of VLDB 2003, pp. 81–92 (2003)
2. DuMouchel, W., Volinsky, C., Johnson, T., Cortes, C., Pregibon, D.: Squashing flat files flatter. In: Proceedings of KDD 1999, pp. 6–15 (1999)
3. Erna, A., Yu, L., Zhao, K., Chen, W., Suzuki, E.: Facial expression data constructed with Kinect and their clustering stability. In: Ślęzak, D., Schaefer, G., Vuong, S.T., Kim, Y.-S. (eds.) AMT 2014. LNCS, vol. 8610, pp. 421–431. Springer, Heidelberg (2014)
4. Fayyad, U.M., Piatetsky-Shapiro, G., Smyth, P.: From data mining to knowledge discovery: an overview. In: Advances in Knowledge Discovery and Data Mining, pp. 1–34. AAAI/MIT Press, Menlo Park (1996)
5. Feldman, D., Schmidt, M., Sohler, C.: Turning big data into tiny data: constant-size coresets for k-means, PCA and projective clustering. In: Proceedings of SODA 2013, pp. 1434–1453 (2013)
6. Kolda, T.G., Bader, B.W.: Tensor decompositions and applications. SIAM Rev. **51**(3), 455–500 (2009)
7. Roweis, S.T., Saul, L.K.: Nonlinear dimensionality reduction by locally linear embedding. Science **290**(5500), 2323–2326 (2000)
8. Ruvolo, P., Eaton, E.: ELLA: an efficient lifelong learning algorithm. In: Proceedings of ICML, vol. 1, pp. 507–515 (2013)
9. Seidl, T., Assent, I., Kranen, P., Krieger, R., Herrmann, J.: Indexing density models for incremental learning and anytime classification on data streams. In: Proceedings of EDBT 2009, pp. 311–322 (2009)
10. Zhang, D., Zhou, Z.-H., Chen, S.: Semi-supervised dimensionality reduction. In: Proceedings of SDM 2007, pp. 629–634 (2007)
11. Zhang, T., Ramakrishnan, R., Livny, M.: BIRCH: a new data clustering algorithm and its applications. Data Min. Knowl. Discov. **1**(2), 141–182 (1997)

An Algorithm for Influence Maximization in a Two-Terminal Series Parallel Graph and its Application to a Real Network

Koji Tabata[✉], Atsuyoshi Nakamura, and Mineichi Kudo

Graduate School of Information Science and Technology,
Hokkaido University, Sapporo, Japan
ktabata@main.ist.hokudai.ac.jp

Abstract. We developed an algorithm to exactly solve an influence maximization problem (MAXINF) for a two-terminal series parallel graph (TTSPG) in the independent cascade model. The class of TTSPGs can be considered as a class wider than that of trees, only for which an efficient exact solver of this problem has been developed so far. Our algorithm calculates candidate node sets in the divide-and-conquer manner keeping the number of them as small as possible by efficiently eliminating unnecessary ones in merge of subproblems' solutions. Furthermore, we propose a way of converting an arbitrary network to a TTSPG with edges important for propagation to apply our method to real networks. According to our empirical results, our method is significantly faster than the greedy approximation algorithm for MAXINF of a TTSPG. We also demonstrate improvement of solutions by converting to TTSPGs instead of trees using real networks made from DBLP datasets.

Keywords: Influence maximization · Two-terminal series parallel graph

1 Introduction

The information or influence propagation on networks has gathered attention in this decade. The research on this area is useful for word-of-mouth marketing [1,3], epidemics analysis [4], innovation diffusion [8] and so on.

In this paper, we study the influence maximization problem (MAXINF) on Independent Cascade (IC) model formulated by [5], whose objective is to find the most influential k nodes from a given network for a given number $k \in \mathbb{N}$. Here IC model is one of the most popular models of the influence diffusion dynamics. MAXINF is one of the most important problems in the area of network analysis because of its rich applications. A problem of finding the best testers for new products in word-of-mouth marketing is an instance of this problem.

[5] has proposed a greedy approximation algorithm for MAXINF. Their algorithm finds a set of k initially activated nodes by repeatedly adding the node that increases the effect most. The approximation ratio of the solution is guaranteed by the analysis using a sub-modular function. Their algorithm, however,

© Springer International Publishing Switzerland 2015
N. Japkowicz and S. Matwin (Eds.): DS 2015, LNAI 9356, pp. 275–283, 2015.
DOI: 10.1007/978-3-319-24282-8_23

must calculate the effect for a given initial active nodes many times, and they estimated it by a large number of simulations in their experiments because its fast exact calculation has not been known yet. On the other hand, [6] proposed an exact algorithm for a tree that runs in polynomial time.

We propose an exact method for a *two-terminal series parallel graph* (TTSPG). Since any weighted directed tree can be converted to a weighted TTSPG with the same propagation structure by adding a new node v and 0-weighted edges from all the leaves to v, the class of TTSPGs can be considered to be larger than the class of trees. In our algorithm, the effect on a TTSPG for a set of initial active nodes is represented as a linear function of the activation probability of its source node. This enables the algorithm to calculate the effect on each subTTSPG without knowing the activation probability of its source node that depends on what upstream nodes are selected as initial active nodes. Thus, effect calculation in the divide-and-conquer manner becomes possible. The number of candidate sets of initial active nodes are exponential to k but we have to check only the sets whose local effect functions are maximal, and the number of such *maximal* sets is not large according to the result of our experiments using randomly generated TTSPGs. Our algorithm calculates candidate node sets in the divide-and-conquer manner efficiently by eliminating non-maximal ones in merge of subproblems' solutions. We also propose a way of obtaining a TTSPG from a general directed graph to apply our algorithm to a real network.

According to the result using synthetically generated random TTSPGs, our exact algorithm is significantly faster than the greedy approximation algorithm developed by [7]. In the experiments on networks constructed from DBLP records, more influential authors are extracted by our method compared with the method using a spanning tree developed by [6] in terms of the h- and g-index sum.

2   Preliminary

2.1   Influence Maximization in Independent Cascade Model

Independent Cascade (IC) Model is one of the most popular information propagation models. An edge-weighted graph $G = (V, E, p)$ is used in this model, where V and E are the set of nodes and directed edges, respectively, and p is a weight function of edges that satisfies $p(e) \in [0, 1]$ for all edges e. Nodes correspond to individuals and the edges correspond to relationship between individuals. Information propagates from a node to its neighbor nodes via edges only. The weight $p(e)$ of an edge e is the success probability of propagation via the edge e. We assume that the success of the propagation via one edge is independent from that via any other edge. Active state propagates step by step, that is, if a node v become active at time round t, a neighbor inactive node u of v turns into active state at time round $t + 1$ via an edge $v \to u$ by probability $p(v \to u)$. If the node v fails to activate u at time round $t + 1$, v doesn't try to activate u anymore. Once a node becomes active state, it never turns into inactive in the future. The propagation process starts from given initial active nodes at time round 0 and ends by time round $|V|$. For an initial active set of nodes $S \subset V$, its *effect*, which is denoted by $\sigma(S)$, is defined as the expected number of active nodes at the end of the state propagation process.

An influence maximization problem (MAXINF) is formalized as follows.

Problem 1 (MAXINF [5]). *Given a weighted directed graph* $G = (V, E, p)$ *and* $k \in \mathbb{N}$, *find a* k-*element subset* S *of* V *that maximizes* $\sigma(S)$.

2.2   Two-Terminal Series Parallel Graph

A two-terminal series parallel graph (TTSPG) is defined as follows.

Definition 1 (TTSPG [9]). *A two-terminal series parallel graph (TTSPG) is defined recursively as follows.*

1. *A directed graph* G *consisting of two vertices* v_1 *and* v_2 *joined by an edge* $v_1 \rightarrow v_2$ *is a TTSPG. The node* v_1 *(v_2) is called the source (sink) of* G.
2. *If* G_1 *and* G_2 *are TTSPGs, so the directed graph obtained by either of the following operations.*
   - *(Parallel Composition): identify the source of* G_1 *with the source of* G_2 *and the sink of* G_1 *with the sink of* G_2.
   - *(Series Composition): identify the sink of* G_1 *with the source of* G_2.

By Definition 1, any TTSPG with more than two nodes can be decomposed in parallel or series, and we call such decomposition as *SP decomposition*. By applying SP decomposition repeatedly, TTSPG can be decomposed into its component non-decomposable subgraphs with one edge in polynomial time [9].

3   Algorithm for a TTSPG

Let $G = (V, E, p)$ be a TTSPG. As the number k of initially activated nodes, we assume $k < |V|$ because $S = V$ is the solution of MAXINF otherwise, where $|V|$ denotes the number of elements in V. Since any node subset S which contains the sink node cannot be a unique solution when $k < |V|$, we only have to search the solution S among k-element subsets of V that exclude the sink node.

3.1   Local Effect Function of Source Activation Probability

By SP decomposition, a TTSPG $G = (V, E, p)$ can be decomposed into $G_1 = (V_1, E_1, p)$ and $G_2 = (V_2, E_2, p)$ in parallel or series. Let us try to solve MAXINF for G and k in the divide-and-conquer manner using this fact. Consider the case that the decomposition is series. In this case, the activation probability of the source s_1 of G_1 is 0 or 1 depending on whether s_1 is chosen as an initial active node or not. The activation probability of the source s_2 of G_2, however, can take a variety of values depending on what subset of V_1 is selected as the set of initial active nodes in G_1 when s_2 is not chosen as an initial active node: the number of values taken by it can be $\Omega(((|V_1| - 1)/k)^k)$. This prevents the problem from being efficiently solved by a divide-and conquer approach. Note that the number of possible values for the activation probability of any node is at most its depth plus two in the case with a tree, which enables an efficient algorithm of MAXINF for a tree.

To overcome the above problem, we propose a method in which the effect of the set $S_2 \subseteq V_2$ of initial active nodes in G_2 is calculated as a function of the activation probability p_0 of s_2 without knowing the specific values of p_0. We introduce two functions σ_G^S and q_G^S of p_0 defined for each node subset S of G.

Definition 2. *Let $G = (V, E, p)$ be a TTSPG with a sink node t. For a node subset $S \subseteq V \backslash \{t\}$, $\sigma_G^S(p_0)$ and $q_G^S(p_0)$ are defined as the expected number of active non-sink nodes and the probability to have an active sink node, respectively, at the end of the state propagation process in which all the nodes in S are initially activated and the source node is also activated (initially) with probability of p_0.*

The effect $\sigma(S)$ of S on G can be expressed as $\sigma(S) = \sigma_G^S(0) + q_G^S(0)$ using notations introduced above. It is trivial that σ_G^S and q_G^S are monotonically increasing functions. Note that $\sigma_G^S(p_0)$ and $q_G^S(p_0)$ are constantly equal to $\sigma_G^S(1)$ and $q_G^S(1)$, respectively, when the source node of G is included in S. By the following theorem, we know that σ_G^S and q_G^S are representable by the form of $ap_0 + b$ using some constant reals a and b.

Theorem 1. *For a TTSPG $G = (V, E, p)$ and $k < |V|$, let S be a k-element subset of V that excludes the sink node of G. Then, σ_G^S and q_G^S are linear.*

Corollary 1. *Assume that a TTSPG $G = (V, E, p)$ can be decomposed into two TTSPGs $G_1 = (V_1, E_1, p)$ and $G_2 = (V_2, E_2, p)$ by an SP decomposition. Let $S_1 = S \cap V_1$ and $S_2 = S \cap V_2$ for $S \subseteq V$. If all the parameters of linear functions of $\sigma_{G_1}^{S_1}(x) = a_1 x + b_1$, $q_{G_1}^{S_1}(x) = c_1 x + d_1$, $\sigma_{G_2}^{S_2}(x) = a_2 x + b_2$ and $q_{G_2}^{S_2}(x) = c_2 x + d_2$ are known, then parameters a, b, c and d of linear functions of $\sigma_G^S(x) = ax + b$ and $q_G^S(x) = cx + d$ can be calculated in $O(1)$ time.*

For a TTSPG $G = (V, E, p)$ and two subsets $S_1, S_2 \subseteq V$, let us write $(S_1, \sigma_G^{S_1}, q_G^{S_1}) \leq (S_2, \sigma_G^{S_2}, q_G^{S_2})$ if $\sigma_G^{S_1}(p_0) \leq \sigma_G^{S_2}(p_0)$ and $q_G^{S_1}(p_0) \leq q_G^{S_2}(p_0)$ hold for all $p_0 \in [0, 1]$. We say that a triplet (S, σ_G^S, q_G^S) is *maximal* in $\mathcal{S}$ if $(S, \sigma_G^S.q_G^S) \in \mathcal{S}$ and $(S, \sigma_G^S, q_G^S) \nleq (S', \sigma_G^{S'}, q_G^{S'})$ for any $S' \in \mathcal{S} \backslash \{S\}$.

3.2   Algorithm

For a given TTSPG $G = (V, E, p)$ and a natural number $k < |V|$, let $\mathcal{S}_{all}(G, k) = \{(S, \sigma_G^S, q_G^S) | S \subseteq V \backslash \{t\}, |S| = k\}$, where t is the sink node of G. Then, the set S^* of the most influential k nodes is expressed as $S^* = \underset{S:(S,\sigma_G^S,q_G^S)\in\mathcal{S}_{all}(G,k)}{\arg\max} (\sigma_G^S(0) + q_G^S(0))$.

Unfortunately, $\mathcal{S}_{all}(G, k)$ cannot be enumerated efficiently because $\mathcal{S}_{all}(G, k)$ contains $\Omega(((n-1)/k)^k)$ subsets. Considering the fact that S for non-maximal triplets (S, σ_G^S, q_G^S) cannot be S^*, the algorithm only have to check the effect of S for (S, σ_G^S, q_G^S) in

$$\mathcal{S}(G,k) = \{(S,\sigma_G^S,q_G^S) \in \mathcal{S}_{\text{all}}(G,k) \mid (S,\sigma_G^S,q_G^S) \text{ is maximal in } \mathcal{S}_{\text{all}}(G,k)\}.$$

There has been yet no theoretical analysis on how much reduced the set $\mathcal{S}(G,k)$ is from $\mathcal{S}_{\text{all}}(G,k)$, but the reduction seems significantly large according to the following our empirical result. we randomly generated 100 TTSPGs (V,E,p) with $|V| = 800$ and $p \equiv 0.1$ in the way explained in Sect. 4.1 and calculate $|\mathcal{S}(G,k)|$. Their largest size is 104 and their average size is 43.2, which is overwhelmingly smaller than $|\mathcal{S}_{\text{all}}(G,5)| = (799 \cdot 798 \cdot 797 \cdot 796 \cdot 795)/(5 \cdot 4 \cdot 3 \cdot 2 \cdot 1) \approx 2.7$ trillion.

So, in the rest of this subsection, we explain how efficiently our algorithm calculates $\mathcal{S}(G,k)$. Let $\mathcal{S}_{G,1}^k$ denote the set of all the elements in $\mathcal{S}(G,k)$ that contains the source of G and let $\mathcal{S}_{G,0}^k$ denote the set of all the other elements in $\mathcal{S}(G,k)$. Consider the problem of calculating $\{(\mathcal{S}_{G,0}^\ell,\mathcal{S}_{G,1}^\ell)\}_{\ell=0,1,\dots,k}$ instead of $\mathcal{S}(G,k)$ only. Since $\mathcal{S}(G,k) = \mathcal{S}_{G,0}^k \cup \mathcal{S}_{G,1}^k$, this is a finer problem. This finer problem can be solved recursively by making use of SP decomposition of G. Let $G_1 = (V_1,E_1,p)$ and $G_2 = (V_2,E_2,p)$ be TTSPGs that is made by some SP decomposition of G. Define k_1 and k_2 as $\min\{|V_1|-1,k\}$ and $\min\{|V_2|-1,k\}$, respectively. Assume that $\{(\mathcal{S}_{G_1,0}^\ell,\mathcal{S}_{G_1,1}^\ell)\}_{\ell=0,1,\dots,k_1}$ and $\{(\mathcal{S}_{G_2,0}^\ell,\mathcal{S}_{G_2,1}^\ell)\}_{\ell=0,1,\dots,k_2}$ have been already obtained. We show that $\{(\mathcal{S}_{G,0}^\ell,\mathcal{S}_{G,1}^\ell)\}_{\ell=0,1,\dots,k}$ can be obtained from those.

Let us consider the case with series decomposition. We can assume that G_1 is positioned upstream of G_2 without loss of generality in this case. Then, for each $\ell \in \{0,1,\dots,k\}$ and $x \in \{0,1\}$, $\mathcal{S}_{G,x}^\ell$ can be calculated from $\mathcal{S}_{G_1,x}^{\ell-k'}$ and $\mathcal{S}_{G_2,0}^{k'} \cup \mathcal{S}_{G_2,1}^{k'}$ for $k' \in [0,\ell] \cap [\ell-k_1,k_2]$ as

$$\mathcal{S}_{G,x}^\ell = \{(S,\sigma_G^S,q_G^S) \in \overline{\mathcal{S}}_{G,x}^\ell \mid (S,\sigma_G^S,q_G^S) \text{ is maximal in } \overline{\mathcal{S}}_{G,x}^\ell\}, \text{ where}$$

$$\overline{\mathcal{S}}_{G,x}^\ell = \bigcup_{k'=\max\{0,\ell-k_1\}}^{\min\{\ell,k_2\}} \{(S_1 \cup S_2, \sigma_G^{S_1 \cup S_2}, q_G^{S_1 \cup S_2}) \mid (S_1, \sigma_{G_1}^{S_1}, q_{G_1}^{S_1}) \in \mathcal{S}_{G_1,x}^{\ell-k'},$$

$$(S_2, \sigma_{G_2}^{S_2}, q_{G_2}^{S_2}) \in \mathcal{S}_{G_2,0}^{k'} \cup \mathcal{S}_{G_2,1}^{k'}\}. \tag{1}$$

Consider the case that G can be made from G_1 and G_2 by parallel composition. In this case, for each $\ell \in \{0,1,\dots,k\}$, $\mathcal{S}_{G,1}^\ell$ can be calculated from $\mathcal{S}_{G_1,1}^{\ell-k'+1}$ and $\mathcal{S}_{G_2,1}^{k'}$ for $k' \in [1,\ell] \cap [\ell-k_1+1,k_2]$ as

$$\mathcal{S}_{G,1}^\ell = \{(S,\sigma_G^S,q_G^S) \in \overline{\mathcal{S}}_{G,1}^\ell \mid (S,\sigma_G^S,q_G^S) \text{ is maximal in } \overline{\mathcal{S}}_{G,1}^\ell\}, \text{ where}$$

$$\overline{\mathcal{S}}_{G,1}^\ell = \bigcup_{k'=\max\{1,\ell-k_1+1\}}^{\min\{\ell,k_2\}} \{(S_1 \cup S_2, \sigma_G^{S_1 \cup S_2}, q_G^{S_1 \cup S_2}) \mid (S_1, \sigma_{G_1}^{S_1}, q_{G_1}^{S_1}) \in \mathcal{S}_{G_1,1}^{\ell-k'+1},$$

$$(S_2, \sigma_{G_2}^{S_2}, q_{G_2}^{S_2}) \in \mathcal{S}_{G_2,1}^{k'}\}, \tag{2}$$

and $\mathcal{S}_{G,0}^{\ell}$ can be calculated from $\mathcal{S}_{G_1,0}^{\ell-k'}$ and $\mathcal{S}_{G_2,0}^{k'}$ for $k' \in [\ell - k_1, k_2] \cap [0, \ell]$ as

$$\mathcal{S}_{G,0}^{\ell} = \{(S, \sigma_G^S, q_G^S) \in \overline{\mathcal{S}}_{G,0}^{\ell} \mid (S, \sigma_G^S, q_G^S) \text{ is maximal in } \overline{\mathcal{S}}_{G,0}^{\ell}\}, \text{ where}$$

$$\overline{\mathcal{S}}_{G,0}^{\ell} = \bigcup_{k'=\max\{0,\ell-k_1\}}^{\min\{\ell,k_2\}} \{(S_1 \cup S_2, \sigma_G^{S_1 \cup S_2}, q_G^{S_1 \cup S_2}) \mid (S_1, \sigma_{G_1}^{S_1}, q_{G_1}^{S_1}) \in \mathcal{S}_{G_1,0}^{\ell-k'},$$

$$(S_2, \sigma_{G_2}^{S_2}, q_{G_2}^{S_2}) \in \mathcal{S}_{G_2,0}^{k'}\}. \quad (3)$$

Note that at the formulas (1), (2) and (3), $\sigma_G^{S_1 \cup S_2}$ and $q_G^{S_1 \cup S_2}$ can be calculated in $O(1)$ from $\sigma_{G_1}^{S_1}$, $\sigma_{G_2}^{S_2}$, $q_{G_1}^{S_1}$ and $q_{G_2}^{S_2}$ as guaranteed by Corollary 1.

For a non-decomposable TTSPG $G = (\{s,t\}, \{s \to t\}, p)$,

$$\mathcal{S}_{G,0}^0 = \{(\emptyset, p_0, p(s \to t)p_0)\}, \ \mathcal{S}_{G,1}^0 = \emptyset, \ \mathcal{S}_{G,0}^1 = \emptyset \text{ and } \mathcal{S}_{G,1}^1 = \{(\{s\}, 1, p(s \to t))\}$$

hold. The correctness of our algorithm can be proved by mathematical induction on the number of edges of G using this fact and the formulas (1), (2) and (3).

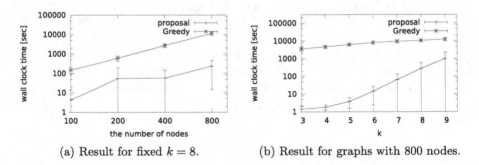

(a) Result for fixed $k = 8$.   (b) Result for graphs with 800 nodes.

Fig. 1. Running times and their 95 % confidence intervals of our algorithm and Greedy for MAXINF with $k = 8$ (a) and in the graphs with 800 nodes (b).

4   Experiments

4.1   Experiments for TTSPGs

In this subsection, we describe empirical evaluation on calculation time and accuracy of our method comparing to those of the greedy approximation method developed by [7].

Dataset. We generated n-node TTSPGs by combining two subTTSPGs using "series" or "parallel" composition with probability 0.5. Each of the subTTSPGs is also generated by the above generating method so that the number of nodes after combining them becomes n, that is, this generating method is a recursive procedure. The number of nodes of each subTTSPG is selected from a uniform distribution. For each number of nodes $n \in \{100, 200, 400, 800\}$, we generated 5 TTSPGs. The weights of the edges are set to a constant value $p \equiv 0.1$.

Comparative Algorithm. Since no algorithm to exactly solve MAXINF for a TTSPG has been developed so far, we compare the performance of our algorithm to that of a hill-climbing approximate algorithm for a general network developed by [5] and improved by [7], which we call Greedy for short. To estimate the influence spread we run 10,000 Monte Carlo simulations in all experiments.

Results. The running time dependency on the number of nodes in a given graph is shown in Fig. 1(a). In this experiment, the number of the initial active nodes are fixed to 8. Our algorithm is significantly faster than Greedy: at least about 50 times faster for all the numbers of nodes in our setting ($n = 100, 200, 400, 800$). In Fig. 1(b), we show the running time dependency on the number of initial active nodes. The number of nodes in a given graph is fixed to 800 in the experiment. Our algorithm is faster than Greedy for all the numbers of initial active nodes in our setting ($k = 3, 4, \ldots, 9$). The growth rate of the running time of our algorithm, however, is larger than that of Greedy.

4.2   Experiments for Real Networks

In order to demonstrate the effectiveness of our approach to application for the general networks, we conducted experiments using real networks. Real networks used in our experiments are the collaborative networks made from the snapshot of the DBLP database on April. 4th, 2014. We extracted the sets of influential researchers from the collaborative networks using our approach and compared them with those extracted by the conventional approach using a spanning tree developed by [6]. We made a weighted directed graphs (V, E, p) of conference A from the set D of papers recorded in the DBLP database as follows. Let $W (\subseteq D)$ denote the set of papers presented in conference A in some past year and let V denote the set of authors of the papers in W. Define $m(w)$ as the set of authors of a paper w for each paper $w \in W$. Then, an edge set E and a weight function p is defined as follows.

$$E = \{u \to v \mid \{u, v\} \subseteq m(w) \text{ for some } w \in W\}$$
$$p(u \to v) = \frac{|\{w \in W \mid \{u, v\} \subseteq m(w)\}|}{|\{w \in W \mid v \in m(w)\}|}$$

In order to obtain a good approximate solution for the original graph $G = (V, E, p)$, the set E' of edges in the spanning tree $T = (V, E', p)$ are selected so as to maximize its *likelihood* $\prod_{e \in E'} p(e)$ in their method. Such a spanning tree can be found by solving the problem of finding the minimum spanning tree for the graph $G' = (V, E, -\ln p)$, which is known to be solved in $O(|E| + |V| \ln |V|)$ time [2]. Different from trees, there may be no spanning TTSPGs even for the graphs in which there is a node that is reachable to any node. Such graphs, however, have a spanning pseudo-TTSPG, which is defined as follows.

Definition 3. *Let $G = (V, E)$ be a directed graph and let $V_t(\subseteq V)$ be the set of nodes with no outgoing edge. A directed graph G is said to be a pseudo-TTSPG if the directed graph $(V \cup \{w\}, E \cup \{v \to w \mid v \in V_t\})$ is a TTSPG for $w \notin V$.*

If a spanning tree exists for G, a spanning pseudo-TTSPG also exists because a directed tree is a pseudo-TTSPG. For a given weighted pseudo-TTSPG (V, E', p), a corresponding weighted TTSPG $(V \cup \{w\}, E' \cup \{v \to w \mid v \in V_t\}, p')$, which can be seen as a graph with the same propagation structure, is obtained easily by adding 0-weight edges $v \to w$ to E' for all the nodes v in V_t, where V_t and w are notations used in Definition 3, and weight function p' is defined as $p'(e) = p(e)$ for $e \in E'$ and $p'(e) = 0$ otherwise.

Therefore, in order to use our algorithm to approximately solve MAXINF for a general weighted directed graph G, we only have to find a spanning pseudo-TTSPG with many edges and high likelihood. Since any efficient algorithm to find a maximum-likelihood spanning pseudo-TTSPG with a maximal set of edges has not been known yet, we took the following greedy approach for a given weighted directed graph (V, E, p). (Step1) Calculate the spanning tree (V, E', p) with the maximal likelihood using the algorithm of [2]. (Step2) For each $e \in E \setminus E'$, do the following in decreasing order of $p(e)$: if $(V, E' \cup \{e\}, p)$ is still a pseudo-TTSPG, add e to E'.

We made graphs for several major conferences on machine learning and data mining and conducted experiments using them, but we report the results only for ACML due to space limitation. The numbers of nodes and edges are 341 and 1,146 in ACML graph.

The results of our experiments are shown in Table 1. The number k of initial active nodes was set to 10. There are three authors unique to the set of the influential authors extracted by each the approach, and more influential authors seem to be extracted by using spanning pseudo-TTSPGs in terms of the h-index sum and also g-index sum. These indices are known as the most popular measures to quantify researcher's performance.

Moreover, we estimated the effects on original graphs of these solutions over 1,000,000 times of Monte Carlo simulation. The estimated effects of the solutions by using a tree and a spanning pseudo-TTSPG are 83.565798 and 83.599365, respectively.

Table 1. Authors extracted as the solution of MaxInf: The solutions for a spanning pseudo-TTSPG and a spanning tree are shown. We used the h- and g-indices provided by ArnetMiner (http://arnetminer.org/) on Feb. 4, 2015. (CI: confidence interval)

Extracted authors	h-&g-idx		Tree	TTSPG
Qiang Yang	50	85		✓
T. G. Dietterich	47	122		✓
Zhi-Hua Zhou	45	82	✓	✓
Aapo Hyvärinen	28	81	✓	✓
Kristian Kersting	26	47	✓	✓
M. Sugiyama	26	46	✓	✓
Alan Fern	21	39	✓	
Steven C. H. Hoi	19	33	✓	
Jan Ramon	14	27	✓	✓
M. Ghavamzadeh	14	26		✓
Yang Zhang	12	15	✓	✓
Kouzou Ohara	9	13	✓	✓
Qian Zhou	9	11	✓	
h-idx sum			209	271
g-idx sum			394	544
Estimated effect (99 % CI)			83.56	83.60
			±0.02	±0.02

The effects of solutions by pseudo-TTSPG are higher and there is a 99 % chance of a statistically significant difference between them are detected by Welch's method. We also obtained the similar result for other datasets in our experiments.

Acknowledgement. This work was partially supported by JSPS KAKENHI Grant Number 25280079.

References

1. Domingos, P., Richardson, M.: Mining the network value of customers. In: Proceedings of the KDD, pp. 57–66. ACM (2001)
2. Gabow, H.N., Galil, Z., Spencer, T., Tarjan, R.E.: Efficient algorithms for finding minimum spanning trees in undirected and directed graphs. Combinatorica **6**, 109–122 (1986)
3. Goldenberg, J., Libai, B., Muller, E.: Talk of the network: a complex systems look at the underlying process of word-of-mouth. Mark. lett. **12**(3), 211–223 (2001)
4. Hethcote, H.W.: The mathematics of infectious diseases. SIAM Rev. **42**(4), 599–653 (2000)
5. Kempe, D., Kleinberg, J., Tardos, É.: Maximizing the spread of influence through a social network. In: Proceedings of the ninth ACM SIGKDD International Conference on Knowledge Discovery and Data Mining, pp. 137–146. ACM (2003)
6. Lappas, T., Terzi, E., Gunopulos, D., Mannila, H.: Finding effectors in social networks. In: Proceedings of the 16th ACM SIGKDD International Conference on Knowledge Discovery and Data Mining, pp. 1059–1068. ACM (2010)
7. Leskovec, J., Krause, A., Guestrin, C., Faloutsos, C., VanBriesen, J., Glance, N.: Cost-effective outbreak detection in networks. In: Proceedings of the 13th ACM SIGKDD International Conference on Knowledge Discovery and Data Mining, pp. 420–429. ACM (2007)
8. Rogers, E.M.: Diffusion of Innovations. Simon and Schuster, New York (2010)
9. Valdes, J., Tarjan, R.E., Lawler, E.L.: The recognition of series parallel digraphs. SIAM J. Comput. **11**(2), 298–313 (1982)

Benchmarking Stream Clustering for Churn Detection in Dynamic Networks

Serdar Baran Tatar$^{(\boxtimes)}$, Andrew McIntyre, Nur Zincir-Heywood,
and Malcolm Heywood

Dalhousie University, Halifax, NS B3H 4R2, Canada
{tatar,armcnty,zincir,mheywood}@cs.dal.ca

Abstract. Retaining users and customers is one of the most important
challenges for the service industry from mobile communications to online
gaming. As the users of these services form dynamic networks that grow
in size, predicting 'churners' becomes harder and harder. In this work, we
explore the use of anomaly detection for churn prediction. To this end,
we evaluate bio-inspired and deterministic online clustering algorithms
on both cell phone and online gaming data sets. We discuss the results of
each technique from the perspective of: feature identification, sensitivity
analysis of the parameters as well as their capacity to detect churn.

Keywords: Churn detection · Performance measures · Clustering

1 Introduction

Customers are continuously offered alternative products and services from a
range of service providers in addition to their own, i.e. customers are potentially
only retained for short periods of time. Hence, the threat of low brand loyalty
may decrease the use of services or even end up with a loss of customers. Those
customers, who terminate their subscription of a service, are called churners
and the rate of churn in a certain period of time is referred to a churn rate.
Companies endeavour to keep churn rate as low as possible.

Churn detection represents the capacity to predict when a customer might
'churn' and therefore provide the provider with the capacity to take preventative
action, e.g. make service discounts. Churn detection is therefore of significance
to a broad range of industries, such as: telecommunication, insurance, finance,
Internet service providers, online services, and TV providers.

From the detection perspective, data sets are used to characterize the churn
behavior of customers taking a particular service. A cross section of attributes
are collected in an attempt to characterize the perceptions and demographics
of customers, as well as their interactions with the company [3]. One of the
problems of churn data sets is their unbalanced nature; generally, only small
portion of the data consists of churning customers [23]. Nevertheless, predicting
that small amount of churn can be a valuable source for companies to retain
their customers who are about to quit the service [20].

© Springer International Publishing Switzerland 2015
N. Japkowicz and S. Matwin (Eds.): DS 2015, LNAI 9356, pp. 284–298, 2015.
DOI: 10.1007/978-3-319-24282-8_24

In addition, the underlying causes that motivate users to switch between service providers are not necessarily static. Hence, merely constructing predictive models from historical data does not represent a feasible approach. The non-stationary aspect of the task implies that some form of online or streaming algorithm should to be assumed.

In this study, we benchmark a recently proposed bio-inspired approach to constructing clusters from streaming data, Flockstream [8] algorithm, and compare against three deterministic clustering algorithms available from the Massive online Analysis repository [4]: ClusTree, CluStream, and DenStream. The aim of this research is to give a comprehensive analysis of these algorithms in terms of clustering performance for churn detection on continuously streaming and evolving data. Algorithms are evaluated by experiments on three data sets; KDD 2009 churn data set which is publicly available, and two commercial data sets. In the evaluation phase, we discuss the results of all algorithms based on well-known performance metrics.

The paper is organized as follows. Section 2 summarizes the literature review. Section 3 presents the algorithms employed. Section 4 details the experimental evaluations on our real life data sets. Finally, conclusions are drawn and the future work is discussed in Sect. 5.

2   Literature Survey

A wide variety of classification techniques have been applied to churn detection (prediction). According to the results of the churn-modeling tournament presented by Neslin et al. in [17], logistic regression and decision trees are the two most popular methods with a usage ratio of 68 % in total. Lee et al. argue that k-NN based classification using time series performs better than the other classification techniques [14]. They used the data gathered from one of the largest telecommunication companies in Taiwan over a four months period. As a neural network approach, Mozer et al. [16] studied a wireless telecommunication data set containing approximately 47, 000 subscribers. The experimental tests are performed on two randomly selected groups of subscribers. The results show that the churn rate in treatment group in which potential subscribers were contacted by the company is 40 % less than the control group where no action was taken by the company for the subscribers. Huang et al. [11] compared decision tree, neural network and SVM methods for churn modelling. They used the telecommunication data set where roughly 23 % of the data consisted of churners and results indicate that neural networks and multi-class SVMs performed better. Another comparison is made by Zhao and Dang [23] using artificial neural network (ANN), decision tree (C4.5), logistic regression and naive Bayesian classifiers on financial data set containing information about one of the commercial banks' VIP customers. In the evaluation of the techniques, they consider the accuracy, hit, and covering rate, as well as the lift coefficient which is calculated as a ratio between accuracy rate and customer churn rate. According to the results, SVM shows the best performance compared to other techniques. GA is another technique used by Pendharkar [18] with a combination of neural networks. The data set involves nearly 200,000 records from a wireless company and evaluations are made in terms of ROC performance metric. A comparative study between GA-based ANN algorithm developed by researcher and statistical z-score classification model indicates that the GA-based model performs better than the statistical one. Instead of classification techniques, the effect of fuzzy c-means clustering for separating churning customers is studied by Karahoca and Karahoca in [12]. As in many studies, they ran the algorithm on telecommunication data which includes randomly selected loyal and churner subscribers. In the comparison with the data mining approaches (Ridor, ANFIS and decision trees) it is observed that the proposed clustering technique is better than the other algorithms with respect to sensitivity, specificity, precision, and correctness performance metrics.

Working on churn prediction itself is impractical for long term analysis when we consider the changing behavior of the subscribers. For that reason, we have decided to consider churn detection problem with the challenges of streaming data. In this context, three well known stream clustering algorithms (CluStream [1], DenStream [5], ClusTree [13], and Flockstream [8]) are discussed in detail in the next section.

3   Methodology

This study is an attempt to quantify to what degree state-of-the-art stream clustering techniques are applicable for addressing the challenging real-world task of churn detection. In this context, we conducted our research on three data sets gathered from both commercial and public sources. The first is the KDD Cup 2009 churn data set, which is publicly available and used in many studies related to classification and clustering tasks [21]. On the other hand, we gathered two more data sets from a leading online gaming company. The need for commercial confidentiality implies that throughout the study we will refer to this data as 'commercial' small/big. The data sets are characterized in more detail in Sect. 3.1.

Benchmarking will be performed across four stream clustering algorithms: CluStream [1], DenStream [5], ClusTree [13], and Flockstream [8]. The first three algorithms are sourced via the Massive Online Analysis (MOA) toolbox [4], which is a free open source software to perform clustering and classification techniques for data stream mining. Conversely, the Flockstream algorithm was implemented using libraries in Java. Sections 3.2 to 3.5 summarize the properties of algorithms for constructing clusters from streaming data.

3.1   Data Sets

KDD Cup 2009, characterized in detail by Guyon et al. [10], was presented by the French Telecommunication company, Orange, for the KDD Cup 2009 challenge. The aim of the challenge was to predict propensity of the customers to change their provider. Two data sets were presented as churn data, one large and one small. In this work, we picked the small data set which consists of 50, 000 instances and 230 attributes. The data includes a large amount of missing values ($\sim$60 %). Additionally, the first 190 variables are numerical, whereas the last 40 variables are categorical. Due to the privacy concerns, this data set is not fully documented. Moreover, categorical variables were replaced with meaningless codes. For that reason, the data is preprocessed before being used in our experiments such that the categorical attributes and features containing no value for all instances are deleted. Additionally, the mean of each feature is calculated column-wise and missing values are filled with the mean of the corresponding feature. Before running experiments on the data sets, they are normalized such that the range of the values is limited between 0 and 1, or $x(t) = \frac{x(t) - x_{min}}{x_{max} - x_{min}}$, where $x(t)$ denotes the independent variable at location t in the stream. Naturally, the process of normalization assumes that prior task specific attribute range information is available (x_{min}, x_{max}) to facilitate this process.

In the case of the online gaming data two data sets are used and referred to as 'big' and 'small'. They are composed of a set of activities by customers during the game. The main purpose is to detect churners within the upcoming $24 - 48$ h time window. The data sets are unbalanced, with only $\approx$ 9 % and $\approx$ 8 % churn rate for small and big data sets, respectively. The small data set consists of 1, 669, 593 data instances, whereas the large data set has 9, 618, 868 instances. There are 16 numerical features for each data instance. At the request of the company, we can not share the meaning of these features but basically, each instance represents one play with a collection of player behavior during the game. Naturally, more than one play might be played by each player. An identification number for each player helps us to group players and preprocess them to extract their behaviours throughout the game. Similar to the KDD 2009 data set, values are normalized before being processed. Information about data sets is summarized in Table 1.

Table 1. Information of data sets.

Data sets	# Instances	# Features	# Classes	Size (MB)	Churn rate (%)
Commercial-Small	1,669,593	16	2	118.8	9
Commercial-Big	9,618,868	16	2	809.1	8
KDD Cup 2009	50,000	175	2	122.1	7

3.2 CluStream

CluStream divides the clustering process into two separate phases. Phase 1, online micro-clustering, identifies summary statistics from the data. Subsequently, the statistics can be used to analyze clusters in accordance with the user demand, which is known as the offline phase. Naturally, if the stream is divided into non-overlapping windows, or 'chunks', then it is possible to construct clusters on an 'incremental' basis.

In order to minimize memory requirements, Zhang et al. suggested that features of data points can be represented as a characteristic vector [22]. This characteristic vector, or *Clustering Feature* (CF), has three components and is defined as $CF = (N, LS, SS)$. Here, N, LS, and SS represent the total number, linear sum $(\sum_{i=1}^{N} \boldsymbol{X}_i)$, and squared sum $(\sum_{i=1}^{N} \boldsymbol{X}_i^2)$ of the data points in the cluster, respectively.

When a new data point $\boldsymbol{x}_i$ arrives to the cluster, total number of points N is incremented by 1, $\boldsymbol{x}_i$ and $\boldsymbol{x}_i^2$ are added to $\boldsymbol{LS}$ and SS, respectively. If two CFs are need to be merged, then a new CF vector is comprised of sum of their components respectively such that, $CF_1 + CF_2 = (N_1 + N_2, LS_1 + LS_2, SS_1 + SS_1)$.

In CluStream, data points are processed in a time interval and each of them has a timestamp. Based on the above CF structure, CluStream introduces *micro-clusters*. A micro-cluster can be represented with the $(2 * d + 3)$ tuple for a set of points $X_{i_1}, \ldots, X_{i_N}$ and time stamps $T_{i_1}, \ldots, T_{i_N}$ as $(\overline{CF2^x}, \overline{CF1^x}, CF1^t, CF2^t, N)$. The benefit of this extension is to facilitate the access of saved micro-clusters at different time intervals. In this respect, users are able to extract old clusters from the history and examine them on demand.

The generation and maintenance of micro-clusters is addressed handled in two phases. The online phase begins with a sufficient amount of micro-clusters generated by using standard k-means algorithm. After that the system starts to accept new data points. When a new point arrives to the system, there are two cases in order to achieve maintenance: (1) Either the new point joins a current micro-cluster (2) or a new micro-cluster is initiated by including the new point.

In the offline phase, the aim is to provide in-depth analysis of micro-clusters to the users in a given time horizon. Hence, the summary statistics are prepared in the online phase, it is easy to extract relevant micro-clusters. Pyramidal time frame ensures that there is always a snapshot between the current time t_c and the time horizon h', which is an admissible horizon within the interval (t_c, h). A more detailed description of CluStream can be found in [1].

3.3   DenStream

Cao et al. [5] proposed a density-based clustering algorithm for streaming data. It extends the *core point* concept introduced with *DBSCAN* [7], by using the micro-clustering technique from CluStream (Sect. 3.2). However, unlike CluStream, the DenStream algorithm does not partition the stream into distinct blocks of data. Instead, a *damped window model* is used to determine the importance of historical data; thus, the importance of a data point gradually decreases according to the fading function $f(t) = 2^{-\lambda \cdot t}$, where $\lambda > 0$. In addition to the weight of data points, the data stream has its own weight, which is specified as a constant $W = v/1 - 2^{-\lambda}$, where v is the speed of the stream, i.e. the number of points arrive to the system in a one time unit. In order to separate (streaming) data points into different regions of density, the authors employ three concepts: *core-micro-cluster*, *potential core-micro-cluster*, and *outlier-micro-cluster*.

A core-micro-cluster (c-micro-cluster) is defined as $CMC(w, c, r)$ at time t for a group of close points $p_{i_1}, \ldots, p_{i_n}$ with time stamps $T_{i_1}, \ldots, T_{i_n}$. The components in $CMC(w, c, r)$ denote the weight, center, and radius of the c-micro-cluster, respectively. The authors propose that the set of c-micro-cluster can be used to represent the clusters with an arbitrary shape. However, the clusters need to be distinguished from outliers. Therefore, they introduced 'potential' core-micro-cluster, and outlier-micro-cluster, similar to those in [2] in terms of incremental computation.

Analogously to CluStream, there are two phases; the online phase where the micro-clusters are updated and maintained, and the offline phase where the clusters are extracted on user demand. In the first phase, assume that the data point p is one of the streaming data coming into the system. There are three scenarios respectively:

1. Merging p into the nearest p-micro cluster c_p such that, if it satisfies the condition that the new radius r_p of c_p is less than or equal to ε.
2. Merging p into the nearest o-micro cluster c_o if it satisfies the condition that the new radius r_o of c_o is less than or equal to ε, i.e. $r_o \leq \varepsilon$, then merging occurs. If the weight of c_o is greater than $\beta\mu$, i.e. $w \geq \beta\mu$, then it is promoted to p-micro-clustering.
3. If one of first two conditions is not satisfied, then a new o-micro-cluster is generated by p.

In the offline phase, the DenStream algorithm makes use of the variant of DBSCAN to find the final clusters. A more detailed information on DenStream can be found in [5].

3.4   ClusTree

Kranen et al. [13] have introduced a self-adaptive clustering algorithm, ClusTree, for mining data streams. They proposed a parameter-free solution which is able to adapt for different stream speeds (lengths of non-overlapping window). Micro-clustering and the R-Tree data structure [9] form the basis of the ClusTree algorithm. As per Sect. 3.2, the clustering feature (CF) tuple is used to store a summary of information related to the data stream. The tree structure makes it possible to maintain micro-clusters at different levels of a hierarchy. The main idea is to place an arriving object to the optimal micro-cluster, searching the tree from the root to leaf node.

Insertion of an object is a continuous process, every data point travels from root through the leaf nodes by choosing the subtree with the closest mean. In the case that an object can not reach the leaf nodes, the process is interrupted and the current CF is saved to the buffer of the subtree. Whenever that subtree is accessed by another object, then the saved entry is taken as a hitchhiker. Unless their paths differ from each other, they descend together. If their paths need to be separated, then the hitchhiker is saved again to the buffer and the current insertion continues on its path. When we assume that an object reaches the leaf node, then it causes a split if there is still time. Otherwise, the closest entries are merged and their ids are saved to a list as a pair.

In the updating process, all entries in the node are updated considering their CF, buffer and last update time. If a leaf node needs to be split, then the least significant entry in the system can be discarded. In that situation, the related entry is subtracted from the path through the root. This ensures that no entry or CF is discarded if an object is added to it before the last snapshot. Moreover, it guarantees that each entry is stored in at least one snapshot. A more detailed description of the ClusTree algorithm can be found in [13].

3.5   Flockstream

Forestiero et al., introduced a bio-inspired, agent-based approach for single pass stream clustering [8]. They adapt the micro-clustering approach from the Den-Stream algorithm but unlike the proceeding algorithms, a single-pass paradigm is adopted instead of a two phase approach. Analogously to DenStream, a damped window model is chosen to fade the importance of historical data points throughout the process. The *flocking model*, first proposed by [19] and developed further by [6] defines a set of *agents* that interact with each other in an environment. Agents can only interact with their neighbouring agents in their 'visibility range'. Additionally, they keep some distance between them to avoid collisions. Their movements in the environment are coordinated based on three rules: *alignment*, *separation*, and *cohesion*.

The algorithm defines two spaces: a d-dimensional feature space represented as $\mathcal{R}^d$, and a 2-dimensional Cartesian space or the virtual space, $\mathcal{R}_v^2$. The *feature space* is the data space in which each data point is stored. On the other hand, the *virtual space* is a discrete toroidal grid system consisting of a finite number of cells in which agents move and interact with each other. Each cell contains only one agent at a time. Every agent deployed to the virtual space represents a data point from the feature space. An Agent A is defined as $A = (P, v)$, such that P is the position of the agent A in $\mathcal{R}_v^2$, i.e. $P = (x, y)$ and v is the velocity vector of the agent A, i.e. $v = (m, \theta)$.

The basic relationships between agents and data is summarized as follows. Let p_c be a data point in feature space which is represented by the agent A_c in the virtual space. Each agent can interact with only the neighboring agents in a range with radius R_1 and may defend itself from collision with other agents within the range of R_2. Assume that the neighboring agents in A_c's visibility range are denoted as $F_1, \ldots, F_n$. The distance between two agents, i.e. $d_v(A_i, F_i)$, is the Euclidean distance between positions of the agents in the virtual space, i.e. $P_{F_i} = (x_{F_i}, y_{F_i})$, $P_{A_i} = (x_{A_i}, y_{A_i})$. On the other hand, $dist(p_c, p_i)$ specifies the Euclidean distance between the data points p_c and p_i, where p_i is the data point of the neighboring agent F_i. Thus, the similarity between two agents is defined such that, if $dist(p_c, p_i) \leq \varepsilon$, then two points are assumed to be similar. The maximum threshold value ε is used later to specify the radius of a micro-cluster.

The flocking behavior itself is the combination of velocity vectors: v_{ar}, v_{sr}, v_{cr}. Before the each movement of the agent A_c, these velocities are combined together to find the *target* velocity, i.e. $v_{ar} + v_{sr} + v_{cr}$, and normalized to obtain a unit vector. The final velocity vector determines the movement of the agent A_c. See [8] for the details of the MSF flocking rules. •

Flockstream employes a combination of three types of agent to perform the clustering process:

1. *basic* agents, which represent the new data points arriving to the system;
2. *p-representative* agents, which represent the p-micro-clusters; and,
3. *o-representative* agents, which represent the o-micro-clusters.

Clustering itself is performed in two steps: (1) initialization where the initial set of basic agents is established; and, (2) micro-cluster maintenance and clustering takes place for the all three types of agents present in the virtual space. Naturally, initialization is only performed once relative to some initial stream content, whereas step 2 is performed thereafter as the remainder of the stream passes. We conclude our discussion of the Flockstream algorithm by detailing these processes as per the following subsections.

Initialization. At the initialization phase, a predetermined number of basic agents are created and deployed to random positions in the virtual space. Initially, the velocity vectors of basic agents $v = (m, \theta)$ are assigned such that the magnitude m equals to 1 and the angle θ is a random value within the range $[0, 360]$. Agent identifiers are incrementally assigned from 1 to n, where n is the total number of data points.

For a predefined number of iterations, the agents move and interact with each other on the virtual space simultaneously according to the MSF flocking rules [8]. Each basic agent is influenced by the neighboring agents in its visibility range. At every iteration, the velocity vector of each basic agent is calculated and assigned for the next move. During this process, similar agents (based on a Euclidean distance between the data points from the feature space) are more likely to move together as a flock, whereas dissimilar agents move away from the flocks. Although an agent can be involved in a flock, it may disjoin due to a change in flocking behavior or effect of the MSF rules. At the end of this phase, we have two kinds of basic agents; those that belong to one of the flocks and those that do not.

At the end of the initialization phase, any flocks identified are characterized by representative agents, i.e. the basic agents corresponding to each flock will be represented by a single representative agent. Summary statistics of the basic agents in the flocks are computed and stored by the corresponding representative agent. Finally, these basic agents are discarded from the system. Given that representative agents are now expressing the cluster properties of the data stream, representative agents are divided into two types, p-representative and o-representative agents in accordance with the p and o micro-cluster definitions in DenStream (Sect. 3.3).

Maintenance and Clustering. When initialization has finished, there are three types of agents remaining in the system; p-representative, o-representative, and basic agents that were not involved in a flock at the initialization process. The purpose of this phase is both to maintain the p and o representative agents associated with the p and o micro-clusters in the feature space, and to perform online clustering. At the next iteration, a new allocation of basic agents is accepted to the system. Note that, the stream speed and maximum number of iterations specified by the user at the beginning of the process – in effect defining a non-overlapping window from which the next data 'chunk' is sampled. Similar to the initialization phase, agents move around the environment with respect to the MSF rules. However, because the types of the agents are not the same, agent similarity is defined as follows:

- Basic $\rightarrow$ Basic : In the case that basic agent A, associated with a data point $p_A \in \mathcal{R}^d$ meets basic agent B, associated with a data point $p_B \in \mathcal{R}^d$, then the two are considered similar when $dist\,(p_A, p_B) \leq \varepsilon$.
- Basic $\rightarrow$ Representative : In the case that a basic agent A meets either a p-representative B (i.e. p-micro-cluster c_p^B) or an o-representative B (i.e. o-micro-cluster c_o^B) then we add p_A to the copy of micro-cluster c_p^B or c_o^B to obtain a new radius. If the new radius r_p or $r_o \leq \varepsilon$, then they are considered similar.
- Representative $\rightarrow$ Basic : In the case that a p- or o-representative agent A, meets a basic agent B associated with data point p_B, then they are considered similar if the Euclidean distance between the center of the micro-cluster and p_B is less than ε.
- Representative $\rightarrow$ Representative : In the case that a p-representative A (i.e., p-micro-cluster c_p^A) or an o-representative A (i.e., o-micro-cluster c_o^A) meets another representative agent, then they are considered to be similar if the Euclidean distance between centers of the micro-clusters less than ε.

The above similarity functions enable agents to move around the virtual space. On completing a fixed number of iterations relative to the current block of data from the stream Flockstream 'labels' the next batch of stream content using the current content. Algorithm 1 summarizes the complete process assumed for Flockstream maintenance/clustering.

4   Evaluation

In this section, we present the evaluation metrics, which are used to compare the algorithms. The criteria of best can be stated as generating the purest possible clusters with a reasonable precision in lowest possible execution time.

4.1   Parameterization

The DenStream parameters are a subset of those used for Flockstream; hence, they assume a common process for parameterization. In the case of CluStream and ClusTree, the recommendations from the MOA distribution are assumed. Table 2 summarizes the resulting parameterization for the three MOA sourced

Algorithm 1. Flockstream cluster maintenance and identification.

```
1: for each flock F in the Virtual Space do
2:      check the type of each agent in F
3:      if the type of all agents is basic then
4:          if number of agents ≥ μ then
5:              create a new p-representative agent
6:          else if number of agents ≤ μ then
7:              create a new o-representative agent
8:          end if
9:      end if
10:     if there is only one representative agent A_r in F then
11:         insert all other basic agents to A_r
12:         if A_r is p-representative ∧ its weight ω ≤ βμ then
13:             diminish A_r to o-representative
14:         else if A_r is o-representative ∧ its weight ω ≥ βμ then
15:             promote A_r to p-representative
16:         end if
17:     end if
18:     if there is more than one representative agent in F then
19:         merge representative agents and insert basic agents into it
20:         label new representative agent as a swarm
21:     end if
22: end for
```

algorithms. Flockstream assumes the parameters of DenStream, plus: (1) *Max-Iterations* defining the maximum number of iterations; and, (2) d defining the size of the virtual space. The authors of Flockstream suggest that if the stream speed is v (aka size of the non-overlapping window interface to the data stream), then the size of the virtual space parameterized such that $d \times d \geq 4 \times v$. The maximum stream speed used in our experiments is 1000. Therefore, the *minimum* size of the virtual space would be 64×64. A virtual space value of 100×100 was adopted in order to reduce congestion resulting from agent immobility.

Flockstream includes two additional parameters which are *MaxIterations* for the maximum number of iterations and d for the dimensions of the virtual space. They are fixed to 800 and 100, respectively. The authors of Flockstream suggest that if the stream speed is v, then dimensions of virtual space can be specified such that $d \times d \geq 4 \times v$. The maximum stream speed used in our experiments is 1000. Therefore, dimensions of virtual space should be roughly more than 64×64. However, we picked a 100×100 virtual space to reduce congestion resulting from immobility. Another difference between parameters is the offline multiplier used in DenStream to calculate the final value of epsilon. In Flockstream, we only use epsilon. The following sections cover the topics of evaluation metrics.

Table 2. Descriptions of parameters of each algorithm from MOA

Algorithms	Parameters	Descriptions
CluStream	-h (d: 1000)	Range of the window
	-k (d: 100)	Maximum number of micro kernels to use
	-t (d: 2)	Multiplier for the kernel radius
	-M	Evaluate the underlying micro-clustering instead of the macro-clustering
DenStream	-h (d: 1000)	Range of the window
	-e (d: 0.02)	Defines the epsilon neighbourhood
	-b (d: 0.2)	Beta (β) constant
	-m (d: 1)	Mu (μ) constant
	-i (d: 1000)	Number of points to use for initialization
	-o (d: 2)	Offline multiplier for epsilon
	-l (d: 0.25)	Lambda (λ) constant
	-s (d: 100)	Number of incoming points per time unit
	-M	Evaluate the underlying micro-clustering instead of the macro-clustering
ClusTree	-h (d: 1000)	Range of the window
	-H (d: 8)	The maximal height of the tree
	-M	Evaluate the underlying micro-clustering instead of the macro-clustering
FlockStream	-d (d: 100)	Dimensions of Virtual Space
	-e (d: 0.1)	Defines the epsilon neighbourhood
	-b (d: 0.2)	Beta (β) constant
	-m (d: 10)	Mu (μ) constant
	-i (d: 300)	Number of points to use for initialization
	-x (d: 800)	Maximum number of iterations
	-l (d: 0.25)	Lambda (λ) constant
	-s (d: 300)	Number of incoming points per time unit

4.2   Evaluation Metrics

All algorithms used in this study are evaluated according to three performance metrics: *macro-purity*, *micro-precision*, and *recall*. MOA refers to these metrics as *Purity*, *F1-P*, and *F1-R* respectively [15]. Micro and macro measures are calculated in the same manner except with respect to the degree of support for clusters in the confusion matrix. Specifically, the macro confusion matrix groups all the clusters that share the same majority label in the same micro confusion matrix row. In the following, CR_i represents clusters identified by the algorithm, whereas CS_j represents the (true) class value; thus $|CS_j|$ are the number of classes. In every identified cluster, v_{i_j} represents the number of data points identified each class. The precision, recall and $F_1 - Score$ of clusters identified by the algorithms are defined as follows.

$$precision_{CR_i} = \frac{\max(v_{i1}, \ldots, v_{im})}{\sum_{j=1}^{m} v_{ij}} \tag{1}$$

$$recall_{CR_i} = \frac{\max(v_{i1}, \ldots, v_{im})}{\sum_{i=1}^{n} v_{ij}} \tag{2}$$

$$F_1 - Score_{CR_i} = 2 \cdot \frac{precision_{CR_i} \cdot recall_{CR_i}}{precision_{CR_i} + recall_{CR_i}} \tag{3}$$

Purity and F1-P metrics can now be defined in terms of the total number clusters identified by each algorithm, n, or:

$$Purity = \frac{\sum_{i}^{n} precision_{CR_i}}{n} \tag{4}$$

$$F1 - P = \frac{\sum_{i=1}^{n} F_1 - Score_{CR_i}}{n} \tag{5}$$

The final performance measure of the MOA used in the experiments is F1-R. This is estimated from the maximum $F_1 - Score$ for each value v_{ij} in the class CS_j as follows:

$$F1 - Score_{v_{ij}} = 2 \cdot \frac{precision_{v_{ij}} \cdot recall_{v_{ij}}}{precision_{v_{ij}} + recall_{v_{ij}}} \tag{6}$$

$$Max\ F1 - Score_{CS_j} = \max\left(F1 - Score_{v_{1j}}, \ldots, F1 - Score_{v_{nj}}\right) \tag{7}$$

$$F1 - R = \frac{\sum_{j=1}^{m} Max\ F1 - Score_{CS_j}}{m}. \tag{8}$$

4.3   Evaluation Results

In this section, all related experiments are done with the MOA Release 2014.04. For DenStream and Flockstream algorithms, the epsilon values are chosen as 0.1 and 0.8 for the experiments on the commercial and public data sets, respectively. Other parameters are set as discussed at the beginning of Sect. 4.

Tables 3, 4 and 5 summarize the performances of algorithms on Commercial-Big, Commercial-Small, and KDD 2009 churn data sets respectively. The evaluation metrics assume those used in MOA, or F1-P, F1-R, and Purity, where larger values are better.

Flockstream generally provides lower precision and higher recall than the MOA algorithms. From the purity point of view, it returns 25 % better purity than DenStream. The CluStream and ClusTree results are essentially equivalent.

Table 4 outlines the results of the bigger commercial data set. Flockstream again achieves the highest purity and recall, albeit at the expense of precision.

Table 3. Results of algorithms on commercial-small data set

Algorithms	F1-P	F1-R	Purity
Flockstream	0.01	0.40	0.96
DenStream	0.32	0.25	0.83
CluStream	0.04	0.03	0.54
ClusTree	0.04	0.03	0.53

Table 4. Results of algorithms on commercial-big churn data set

Algorithms	F1-P	F1-R	Purity
Flockstream	0.01	0.44	0.98
DenStream	0.14	0.11	0.85
CluStream	0.08	0.06	0.72
ClusTree	0.09	0.07	0.73

Both Clustree and CluStream provide improvements relative to their respective behavior under the 'small' commercial data set, but fail to approach the performance of either DenStream or Flockstream.

The results of public data set underlines the performance difference between Flockstream and the remaining algorithms. The purity of the Flockstream on public data remains the highest. Indeed, the performance differential in purity was typically 70%. The F1-R (recall) for Flockstream also remains strong. Performance by the other algorithms remains weak throughout, with all three of the returning algorithms returning poor values for F1-P and F1-R. As a general remark, we note that there is always a significant disparity between the Purity and the F1 metrics. This is a reflection of the macro nature of the Purity metric. Hence, for strong macro performance all that is necessary is that the clusters be 'pure' (not mix multiple classes). Strong micro performance in addition requires that the number of clusters matches the number of classes. Hence a micro metric incrementally penalizes the use of more clusters than classes.

Table 5. Results of algorithms on KDD 2009 churn data set

Algorithms	F1-P	F1-R	Purity
Flockstream	0.0328	0.475	0.93
DenStream	4.67×10^{-4}	2.33×10^{-4}	0.21
CluStream	2.15×10^{-4}	1.07×10^{-4}	0.20
ClusTree	6.82×10^{-4}	3.41×10^{-4}	0.19

5   Conclusions and the Future Work

The purpose of the study was to evaluate the performance of the Flockstream algorithm against state-of-the-art stream clustering algorithms available from MOA (CluStream, DenStream, and ClusTree) in task of the churn detection. Benchmarking on three data sets quantifies performance differences in terms of the micro precision and recall and macro purity of clusters identified over the course of the stream. In our experiments, we observed that the epsilon value plays a vital role in the performance of Flockstream. Larger epsilon values cause irrelevant data points to aggregate in the same cluster. In contrast, only very similar data points became a swarm, whereas a large amount of data points remain in the system at a lower epsilon value. Therefore, the choice of epsilon value is vitally important for Flockstream functionality.

Based on the results of all the algorithms on both the commercial and the public data sets, it is clear that Flockstream presents remarkable results, especially on purity. While DenStream is the closest on the commercial data sets, there is a still a considerable difference between them. In the experiments on the public data set, the superiority of Flockstream on the purity is most significant. Additionally, it is observed that the unbalanced data sets reveal better differentiation on data points in found clusters. When we examine results from the precision point of view, both DenStream and Flockstream produce the best results. As precision values are evaluated using micro performance metric, all the algorithms analyzed are characterized by lower values. That said, Flockstream is still more effective under F1-R (recall) metric implying that although many more clusters are created than classes, the clusters must be particularly pure.

In conclusion, the effect of clustering techniques on churn detection research is notable. This approach can be applied to various service industries in addition to the gaming and the telecommunication sectors. In the future, we plan to investigate how to speed up the Flockstream algorithm as well as how to improve the precision measurements.

Acknowledgement. This research is supported by the Mitacs Accelerate Internship grant, and is conducted as part of the Dalhousie NIMS Lab at: https://projects.cs.dal.ca/projectx.

References

1. Aggarwal, C.C., Han, J., Wang, J., Yu, P.S.: A framework for clustering evolving data streams. In: Proceedings of the 29th International Conference on Very Large Data Bases, vol. 29, pp. 81–92. VLDB Endowment (2003)
2. Aggarwal, C.C., Han, J., Wang, J., Yu, P.S.: A framework for projected clustering of high dimensional data streams. In: Proceedings of the Thirtieth International Conference on Very Large Data Bases, vol. 30, pp. 852–863. VLDB Endowment (2004)
3. Ali, Ö.G., Arıtürk, U.: Dynamic churn prediction framework with more effective use of rare event data: The case of private banking. Expert Syst. Appl. **41**(17), 7889–7903 (2014)

4. Bifet, A., Holmes, G., Pfahringer, B., Kranen, P., Kremer, H., Jansen, T., Seidl, T.: Moa: Massive online analysis, a framework for stream classification and clustering (2010)
5. Cao, F., Ester, M., Qian, W., Zhou, A.: Density-based clustering over an evolving data stream with noise. In: SDM, vol. 6, pp. 328–339. SIAM (2006)
6. Eberhart, R.C., Shi, Y., Kennedy, J.: Swarm Intelligence. Elsevier, London (2001)
7. Ester, M., Kriegel, H.P., Sander, J., Xu, X.: A density-based algorithm for discovering clusters in large spatial databases with noise. Kdd. **96**, 226–231 (1996)
8. Forestiero, A., Pizzuti, C., Spezzano, G.: Flockstream: a bio-inspired algorithm for clustering evolving data streams. In: 21st International Conference on Tools with Artificial Intelligence, ICTAI 2009, pp. 1–8. IEEE (2009)
9. Guttman, A.: R-trees: a dynamic index structure for spatial searching, vol. 14. ACM (1984)
10. Guyon, I., Lemaire, V., Boullé, M., Dror, G., Vogel, D.: Analysis of the kdd cup 2009: Fast scoring on a large orange customer database (2009)
11. Huang, B.Q., Kechadi, T.M., Buckley, B., Kiernan, G., Keogh, E., Rashid, T.: A new feature set with new window techniques for customer churn prediction in land-line telecommunications. Expert Syst. Appl. **37**(5), 3657–3665 (2010)
12. Karahoca, A., Karahoca, D.: Gsm churn management by using fuzzy c-means clustering and adaptive neuro fuzzy inference system. Expert Syst. Appl. **38**(3), 1814–1822 (2011)
13. Kranen, P., Assent, I., Baldauf, C., Seidl, T.: The clustree: indexing micro-clusters for anytime stream mining. Knowl. Inf. Syst. **29**(2), 249–272 (2011)
14. Lee, Y.H., Wei, C.P., Cheng, T.H., Yang, C.T.: Nearest-neighbor-based approach to time-series classification. Decis. Support Syst. **53**(1), 207–217 (2012)
15. Moise, G., Sander, J., Ester, M.: P3c: A robust projected clustering algorithm. In: Sixth International Conference on Data Mining, 2006, ICDM 2006, pp. 414–425. IEEE (2006)
16. Mozer, M.C., Wolniewicz, R., Grimes, D.B., Johnson, E., Kaushansky, H.: Predicting subscriber dissatisfaction and improving retention in the wireless telecommunications industry. IEEE Trans. Neural Netw. **11**(3), 690–696 (2000)
17. Neslin, S.A., Gupta, S., Kamakura, W., Lu, J., Mason, C.H.: Defection detection: Measuring and understanding the predictive accuracy of customer churn models. J. Mark. Res. **43**(2), 204–211 (2006)
18. Pendharkar, P.C.: Genetic algorithm based neural network approaches for predicting churn in cellular wireless network services. Expert Syst. Appl. **36**(3), 6714–6720 (2009)
19. Reynolds, C.W.: Flocks, herds and schools: A distributed behavioral model. ACM Siggraph Comput. Graph. **21**(4), 25–34 (1987)
20. Verbeke, W., Dejaeger, K., Martens, D., Hur, J., Baesens, B.: New insights into churn prediction in the telecommunication sector: A profit driven data mining approach. Eur. J. Oper. Res. **218**(1), 211–229 (2012)
21. Vogel, D., Guyon, I.: Kdd cup 2009: Customer relationship prediction
22. Zhang, T., Ramakrishnan, R., Livny, M.: Birch: an efficient data clustering method for very large databases. In: ACM SIGMOD Record, vol. 25, pp. 103–114. ACM (1996)
23. Zhao, J., Dang, X.H.: Bank customer churn prediction based on support vector machine: taking a commercial bank's vip customer churn as the example. In: 4th International Conference on Wireless Communications, Networking and Mobile Computing, WiCOM 2008, pp. 1–4. IEEE (2008)

Canonical Correlation Methods for Exploring Microbe-Environment Interactions in Deep Subsurface

Viivi Uurtio[1,2], Malin Bomberg[3], Kristian Nybo[1,2], Merja Itävaara[3], and Juho Rousu[1,2(✉)]

[1] Helsinki Institute for Information Technology HIIT Department of Computer Science, Aalto University, P.O.Box 15400, FI-00076 Aalto, Finland
{viivi.uurtio,kristian.nybo,juho.rousu}@aalto.fi
[2] Helsinki Institute for Information Technology HIIT Department of Computer Science, Aalto University, Konemiehentie 2, 02150 Espoo, Finland
[3] VTT Technical Research Centre of Finland, Espoo, Finland
{malin.bomberg,merja.itavaara}@vtt.fi

Abstract. In this study, we apply non-linear kernelized canonical correlation analysis (KCCA) as well as primal-dual sparse canonical correlation analysis (SCCA) to the discovery of correlations between sulphate reducing bacterial taxa and their geochemical environment in the deep biosphere. For visualization of canonical patterns, we demonstrate the applicability of the correlation plot technique on kernelized data. Finally, we provide an extension to the visual analysis by clustergrams. The presented framework and visualization tools enabled extraction of latent canonical correlation patterns between the salinity of the groundwater and the bacterial taxonomic orders *Desulfobacterales*, *Desulfovibrionales* and *Clostridiales*.

Keywords: Canonical correlation · Kernel methods · Sparsity · Deep biosphere

1 Introduction

Multivariate analysis methods are becoming increasingly popular in uncovering the complex network of microbe-environment interactions. Various settings have been studied concerning the human microbiome [14], soil microbes related to agricultural practice [15] and microbiota in sediments associated with europhication [16]. Canonical correlation analysis (CCA) [14–16] and combinations of univariate and multivariate regression [14] including principal component analysis (PCA) [15] have been among the popular methods. Despite their popularity, these methods are limited by the assumption of linear dependencies among the variables and the fact that the resulting models are often overly complicated for human interpretation.

© Springer International Publishing Switzerland 2015
N. Japkowicz and S. Matwin (Eds.): DS 2015, LNAI 9356, pp. 299–307, 2015.
DOI: 10.1007/978-3-319-24282-8_25

In this paper, we examine sulphate reducing bacteria (SRB)-environment data arising from the deep biosphere research. Our data originates from deep bedrock drill holes of the Fennoscandian shield. There, SRB are observed up to several kilometer's deep [7]. SRB affect their anoxic living habitats for example by producing corrosive hydrogen sulfides. In deep geological storage of nuclear waste they may impact the long-term safety of the spent nuclear fuel storage canisters and other metallic radioactive waste [11]. In order to efficiently abate or estimate the effects of SRB the factors driving the SRB communities residing deep in the bedrock environment the physicochemical parameters driving the SRB communities must be identified. Better understanding of the deep biosphere has potential ramifications to application fields such as climate research [13] and biotechnology [8].

In microbe-environment interaction studies, the sample size is generally relatively modest and the number of variables is large. Thus, we choose to analyse microbe-environment interactions by kernel CCA (KCCA) [4] and sparse CCA (SCCA) [5], recent extensions of CCA, designed to tackle high-dimensional data through regularization, to extract non-linear dependencies and to find sparse solutions facilitating interpretation. In order to ensure statistical validity of the results, we apply in addition cross-validation to optimize model hyperparameters and randomization through permutation tests to determine the statistical significance of the discovered patterns.

Visualization of the results of multivariate analysis is challenging due to the typical high dimensionality. Frequent visualization approaches of projection-based methods, such as PCA or CCA, include score plots [15] biplots [14,16], and, more recently, correlation plots [2,3,10], all designed to project the data on a two-dimensional scatter plot, where similarity of variables or data points can be visually deduced. In this study, we first show that the correlation plot technology naturally extends to kernelized CCA variants, and go on to introduce a new clustergram visualization to represent the results of CCA-based methods, including kernelized ones. A clustergram provides an alternative dimension to the analysis of the results since it does not suffer from the problem of visual clutter that occurs in correlation plots when multiple variables have similar correlation coefficients with the projections.

2   Canonical Correlation Analysis Methods

We first present the Canonical Correlation Analysis (CCA) methods used in this paper. Let the data matrices X_a and X_b, of sizes $n \times p$ and $n \times q$, denote the views a and b respectively. The row vectors $\mathbf{x}_a^k \in \mathbb{R}^p$ and $\mathbf{x}_b^k \in \mathbb{R}^q$ for $k = 1, 2, \ldots, n$ denote the sets of empirical observations, or samples, of X_a and X_b respectively and the column vectors $\mathbf{a}_i \in \mathbb{R}^n$ for $i = 1, 2, \ldots, p$ and $\mathbf{b}_j \in \mathbb{R}^n$ for $j = 1, 2, \ldots, q$ denote centered variable vectors of the n samples respectively.

In canonical correlation analysis [6], two projection directions $\mathbf{w}_a \in \mathbb{R}^p$ and $\mathbf{w}_b \in \mathbb{R}^q$ that maximize the correlation

$$\rho = \max_{\mathbf{w}_a, \mathbf{w}_b} \frac{\mathbf{w}_a^T X_a^T X_b \mathbf{w}_b}{\|\mathbf{w}_a^T X_a\| \|\mathbf{w}_b^T X_b\|} \tag{1}$$

between the two datasets are sought for. Extending the CCA framework, recent years have put forward regularized, sparse and kernelized variants of CCA, widening the applicability of the method and to overcome limitations of CCA on high-dimensional problems [4,5].

Kernel Canonical Correlation Analysis (KCCA). [4] performs CCA by first mapping the original observations through a feature map $\phi_a : \mathbb{R}^p \mapsto \mathcal{H}_a$ to a Hilbert Space $\mathcal{H}_a$. The similarity of the objects is captured by a symmetric positive semi-definite kernel function, corresponding to the inner product in $\mathcal{H}_a$

$$K_a(\mathbf{x}_a^i, \mathbf{x}_a^j) = \langle \phi_a(\mathbf{x}_a^i), \phi_a(\mathbf{x}_a^j) \rangle_{\mathcal{H}_a}.$$

Using kernels K_a and K_b to map the objects in view a and b, respectively, one can express the KCCA objective by [4]

$$\rho = \max_{\alpha, \beta} \frac{\alpha^T K_a K_b \beta}{\sqrt{\alpha^T K_a^2 \alpha \cdot \beta^T K_b^2 \beta}}, \tag{2}$$

where $\alpha, \beta \in \mathbb{R}^n$ denote the dual variables that assign weights to the training examples. However, this optimisation problem results in a trivial correlation coefficient of value 1 when either K_a or K_b is invertible. Following [4], we solve this problem through partial Gram-Schmidt orthogonalisation (PGSO) to reduce the dimensionality of the kernels and by penalising the norms of the weight vectors by a convex combination of constraints based on Partial Least Squares to enforce non-trivial learning of the projection directions:

$$\rho = \max_{\alpha, \beta} \frac{\alpha^T \tilde{K}_a \tilde{K}_b \beta}{\sqrt{(\alpha^T \tilde{K}_a^2 \alpha + \kappa \alpha^T \tilde{K}_a \alpha) \cdot (\beta^T \tilde{K}_b^2 \beta + \kappa \beta^T \tilde{K}_b \beta)}}$$

Above, the kernel matrices are substituted by product of lower-triangular matrices R_a (resp. R_b) arising from PGSO approximation: $\tilde{K}_a = R_a R_a^T \cong K_a$ and $\tilde{K}_b = R_b R_b^T \cong K_b$, respectively.

Primal-Dual Sparse Canonical Correlation Analysis (SCCA). [5] seeks to maximise the correlation among subsets of features by discarding the features that do not contribute to the correlation sufficiently in comparison to others. In primal-dual SCCA, one of the views is given in the primal representation (using features) while the other is given in dual (using kernels). We denote the non-kernelized view by X_a and the kernelized view by K_b. The primal weights for X_a are denoted by $\mathbf{w}_a$ and the dual weights for K_b by β.

$$\rho = \max_{\mathbf{w}_a, \beta} \frac{\mathbf{w}_a^T X_a^T K_b \beta}{\sqrt{\mathbf{w}_a^T X^2 \mathbf{w}_a \cdot \beta^T K_b^2 \beta}}$$

The correlation is maximised between the vectors $X_a \mathbf{w}_a$ and $K_b \boldsymbol{\beta}$. This is equivalent to minimising the 2-norm between the vectors subject to $||K_b \boldsymbol{\beta}||^2 = 1$. Since this would not result in a convex optimisation problem, the constraint is replaced by $||\boldsymbol{\beta}||_\infty = 1$. In order to force a non-trivial solution, the dual weight of one selected example that has an index k is fixed to $\beta_k = 1$ and the a constraint of 1-norm is put on the remaining entries of dual weight vector, denoted by $\tilde{\boldsymbol{\beta}} = (\beta_\ell)_{\ell \neq k}$. The 1-norm of $\mathbf{w}_a$ is also constrained to favour a sparse solution in view a. The final optimisation problem is then

$$\min_{\mathbf{w}_a, \boldsymbol{\beta}} ||X_a \mathbf{w}_a - K_b \boldsymbol{\beta}||^2 + \mu ||\mathbf{w}_a||_1 + \gamma ||\tilde{\boldsymbol{\beta}}||_1 \qquad (3)$$

subject to $||\boldsymbol{\beta}||_\infty = 1$ where μ and γ are regularisation parameters controlling the trade-off between the function objective and the level of sparsity. The scalar μ represents the level of sparsity that controls how many of the features in X_a are discarded. The parameter γ is determined directly from the data in K_b.

3   Experiments

3.1   Data

Data consists of 43 deep bedrock groundwater samples obtained at different time points from three different sites around Finland: 15 and 11 samples from Outokumpu, Finland in 2007 and 2009, respectively, 13 samples from Olkiluoto in years 2009–2013, one from Onkalo, Finland, and three samples from Palmottu, Finland. Bacterial species were identified by dissimilatory sulphate reduction dsrB marker gene targeting which is used to identify specifically sulphate reducing microbial species. Denaturing Gradient Gel Electrophoresis (DGGE) was used to separate taxonomically variable genes which were used to construct operational taxonomic units (OTUs). The 58 DGGE bands corresponded to the binary bacterial variables that were paired with 15 geochemical variables.

3.2   Training Settings

In the SCCA algorithm [5] the constraint $||\boldsymbol{\beta}||_\infty = 1$ is fulfilled by selecting a seed example $\boldsymbol{x}_b^k$, and fixing its dual variable at $\beta_k = 1$. Following [12], we used spectral clustering to compute three medoids from the second view b to be used as candidate seeds. We compared two settings of kernel combinations. First, we performed linear analysis by applying linear kernel $K(\mathbf{x}, \mathbf{x}') = \mathbf{x}^T \mathbf{x}'$ to both views in KCCA and the kernelized view of SCCA. These settings will be referred to as L-KCCA and L-SCCA respectively. In the second setting, we used Gaussian kernel $K(\mathbf{x}_b, \mathbf{x}_b') = exp(-\frac{||\mathbf{x}_b - \mathbf{x}_b'||^2}{2\sigma^2})$ on the second (geochemical data) view b and let the first view (microbial communities) view to remain linear for both SCCA and KCCA. The motivation for this setup was to keep the data representation comparable for the two methods, whilst allowing us to impose primal sparsity on the microbial community view. These settings will be referred to as G-KCCA and G-SCCA respectively.

Table 1. Results obtained at the optimal parameter values that yielded a maximal predictive canonical correlation coefficient.

	Projections		
	1	2	3
L-SCCA	0.863	0.835	0.826
G-SCCA	0.910	0.908	0.751
L-KCCA	0.838	0.802	0.798
G-KCCA	0.965	0.962	0.948

3.3 Parameter Estimation and Statistical Significance Testing

We optimized hyperparameter μ that controls sparsity of view a in L-SCCA and G-SCCA, as well as the width of the Gaussian kernel σ in G-SCCA by 3-fold cross-validation. The same kernel width σ was also applied to G-KCCA. The model selection criterion was predictive canonical correlation, that is, the canonical correlation of test fold, using the canonical weights computed from the training fold.

The canonical correlation coefficients given the optimized hyperparameters of the first three leading projection directions are shown in Table 1. We observe that correlation coefficient of the leading projection ($k = 1$) is the greatest, as expected. The use of Gaussian kernels improve the correlations for both SCCA and KCCA. The statistical significance of the canonical correlation coefficients, given the optimized hyperparameters, were estimated using permutation tests [12]. A background data distribution consistent with the null hypothesis, H_0 : "There is no correlation between the two views of the data", was generated by permuting the rows of one view 500 times and computing the correlation coefficients for such randomized data. According to the permutation tests, the leading three canonical correlations obtained from the dataset were statistically significant in all settings at 99 % significance level. We note that the present setup corrects for multiple testing with respect to the optimal projection directions ($\mathbf{w}_a, \mathbf{w}_b$). However, it omits multiple testing correction of the hyperparameters (μ, σ) due to large computational resource requirement of permutation tests.

3.4 Visualization of the Correlations

We visualized the canonical projections arising from SCCA and KCCA with correlation plots (c.f. [3,9,10]) based on Pearson's correlation coefficients of single variables and the projections. In particular, here we show that the technique is immediately applicable to kernelized projections, despite the fact that we do not have access to the projection weights of the variables.

Within view a, correlation coefficient between values of a single variable and the primal ($\mathbf{s}_a^k = X_a \mathbf{w}_a^k$) and dual ($\mathbf{s}_a^k = K_a \boldsymbol{\alpha}_k$) representation of the k'th canonical projection scores are computed by

$$\rho(\mathbf{a}_i, \mathbf{s}_a^k) = \frac{\langle \mathbf{a}_i, X_a \mathbf{w}_a^k \rangle}{||\mathbf{a}_i|| ||X_a \mathbf{w}_a^k||} = \frac{\langle \mathbf{a}_i, K_a \boldsymbol{\alpha}_k \rangle}{||\mathbf{a}_i|| ||K_a \boldsymbol{\alpha}_k||} \tag{4}$$

Correlations of the two leading canonical projections in view a are used as coordinates for plotting the single variables $\mathbf{a}_i$ in view a by $(\rho(\mathbf{a}_i, \mathbf{s}_a^1)), \rho(\mathbf{a}_i, \mathbf{s}_a^2))$. The correlations and coordinates regarding view b are computed analogously.

A correlation plot showing the relations between the variables in the data, obtained by L-SCCA, is shown in Fig. 1. For example, a high correlation is observed between the DGGE band number 57, that represents the *Peptococcaceae* bacterial family, and Ca^{2+} measurements. On the other hand, there is a high negative correlation between the DGGE band number 68, that represents the *Desulfobacteraceae* family, and Ca^{2+}.

The similarities and dissimilarities between the samples can be analysed by score plots. In a score plot, the axes are the first two leading projections, $\mathbf{s}_a^k = X_a \mathbf{w}_a^k$ and $\mathbf{s}_a^k = K_a \boldsymbol{\alpha}_k$ for $k = 1, 2$ for SCCA and KCCA respectively [2]. In this case, the clusters of the samples can be interpreted by their positions in relation to the variables on the correlation plot.

A score plot on the results of L-SCCA on the data is shown in Fig. 2. In general, samples obtained from the same site at the same time point cluster together. The positions of the samples on the score plot can be explained by analysing which of the variables on the correlation plot are found in the same position. The samples obtained from very deep, OK-2-23, OK-2-21 and OK-2-19, are located in similar positions with respect to the projection axes as the

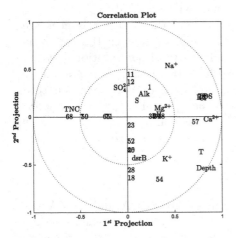

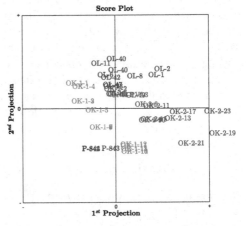

Fig. 1. Correlation plot showing L-SCCA results on dataset. The numbers represent the DGGE bands of the bacterial species and the geochemical measurements are given by their names.

Fig. 2. Score plot showing L-SCCA results on dataset. The names of the different sampling sites are abbreviated by OK-1, OK-2, OL, P and ONK for Outokumpu 1, Outokumpu 2, Olkiluoto, Palmottu and the ONKALO respectively. The drill hole number is given after the name.

depth variable on the correlation plot. In addition, since salinity and temperature increase with depth, also the temperature and Ca^{2+} variables explain the separation of OK-2-23, OK-2-21 and OK-2-19 from the other samples.

Clustergram Visualization. In a correlation plot, visual clutter may occur since the coordinates are defined by the correlation coefficients of the variables with the two projections. In particular, this is a problem in KCCA, where no sparsity is enforced on the number of variables. Also, it is not easy to obtain an overall picture of correlations picked up by a set of projections by examining correlation plots. Here, we propose using clustergrams, frequently used in gene expression data analysis [1], in a novel way: to visualize the overall correlation of two sets of variables in a set of canonical projections. Clustergrams combine heatmaps and hierarchical clustering for visualization. To compute an entry $cg(i,j)$ in the clustergram heatmap for two variables $\mathbf{a}_i$ and $\mathbf{b}_j$ and k leading canonical projections, we compute

$$cg(i,j) = \frac{\langle \rho(i,\ell), \rho(j,\ell) \rangle}{||\rho(i,\ell)|| \, ||\rho(i,\ell)||}, \tag{5}$$

where $\rho(i,\ell)$ denotes the correlation of $\mathbf{a}_i$ with the ℓ'th canonical projection.

Clustergrams representing results of L-SCCA and G-SCCA are shown in Figs. 3 and 4. Both methods find similar correlation patterns but the Gaussian kernel induces more sparsity on the results than the linear kernel. When comparing the two clustergrams, the DGGE band 54 that represents the *Peptococcaceae* taxonomic family correlates positively with Ca^{2+}, depth and temperature (T). In addition, the DGGE bands 7, 59 and 68 representing the *Desulfobacteraceae* and *Desulfovibrionaceae* families, correlate negatively with depth, Cl^-, Ca^{2+}, total dissolved solids (TDS), electrical conductivity (EC) and temperature, in both clustergrams.

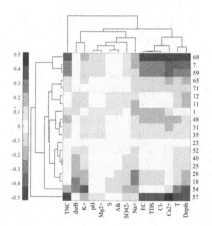

Fig. 3. Clustergram showing L-SCCA results.

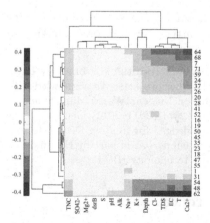

Fig. 4. Clustergram showing G-SCCA results.

4   Discussion

In this paper, we have studied primal-dual sparse (SCCA) and kernel canonical correlation (KCCA) analysis of the deep groundwater SRB communities and their geochemical environment. We have presented a data analysis framework including model selection, parameter tuning, statistical testing through permutation tests that allowed us to distill statistically significant patterns, despite a relatively modest-sized dataset. For visualization, we showed that correlation plots [3,9,10] are also applicable for kernelized setting in primal-dual SCCA and KCCA models. Finally, we introduced an alternative way to summarize the correlations in two or more projections through the use of clustergrams which provide an accessible overview to the correlations induced by the CCA projections. Indeed, the sparsifying effect of the Gaussian kernel in SCCA was first spotted by the authors from the clustergram.

Analyzing the models, higher canonical correlation coefficients were observed for the Gaussian kernel than the for linear kernel, which indicates that the data contains significant non-linear dependencies. We also observed that predictive canonical correlation coefficient assessed through cross-validation provided a good model selection criterion for SCCA.

The results in this paper help in characterizing the sulphate-reducing bacterial communities and their biochemical processes in their habitat. The discovered canonical patterns related the salinity of the groundwater, defined by the geochemical measurements Ca^{2+}, Cl^-, total dissolved solids and electrical conductivity, to the bacterial taxonomic orders *Desulfobacterales*, *Desulfovibrionales* and *Clostridiales*. Salinity seemed to be a unique characteristic of each of the drill hole sites based on the sample clusters of the score plots. In general, depth and temperature measurements co-occurred close together on the correlation plot which was expected, since temperature is known to increase with increasing depth below ground surface.

The software used to produce the results in this paper are available for download at https://github.com/aalto-ics-kepaco/DeepBiosphere.

Acknowledgements. Microbiology data used in this article has been generated in several earlier projects. We acknowledge financial support from Finnish Research Programme on Nuclear Waste Management KYT2010 (2006–2010) (GEOMOL project) and KYT2014 (2011–2014) Geobioinfo and GEOMICRO projects. Finnish Academy is acknowledged for funding Deep Life project (2009–2014). The work by Viivi Uurtio has been supported in part by Helsinki Doctoral Network in Information and Communication Technology HICT.

References

1. Eisen, M.B., Spellman, P.T., Brown, P.O., Botstein, D.: Cluster analysis and display of genome-wide expression patterns. PNAS **95**(25), 14863–14868 (1998)

2. González, I., Déjean, S., Martin, P.G., Gonçalves, O., Besse, P., Baccini, A.: Highlighting relationships between heterogeneous biological data through graphical displays based on regularized canonical correlation analysis. J. Biol. Syst. **17**(02), 173–199 (2009)
3. González, I., Lê Cao, K.A., Davis, M.J., Déjean, S.: Visualising associations between paired omic data sets. BioData Min. **5**(1), 1–23 (2012)
4. Hardoon, D., Szedmak, S., Shawe-Taylor, J.: Canonical correlation analysis: An overview with application to learning methods. Neural Comput. **16**(12), 2639–2664 (2004)
5. Hardoon, D.R., Shawe-Taylor, J.: Sparse canonical correlation analysis. Mach. Learn. **83**(3), 331–353 (2011)
6. Hotelling, H.: Relations between two sets of variates. Biometrika **28**(3–4), 321–377 (1936)
7. Itävaara, M., Nyyssönen, M., Kapanen, A., Nousiainen, A., Ahonen, L., Kukkonen, I.: Characterization of bacterial diversity to a depth of 1500 m in the outokumpu deep borehole, fennoscandian shield. FEMS Microbiol. Ecol. **77**(2), 295–309 (2011)
8. Kalogerakis, N., Arff, J., Banat, I.M., et al.: The role of environmental biotechnology in exploring, exploiting, monitoring, preserving, protecting and decontaminating the marine environment. New Biotechnol. **32**(1), 157–167 (2015)
9. Lê Cao, K.A., Martin, P.G., Robert-Granié, C., Besse, P.: Sparse canonical methods for biological data integration: application to a cross-platform study. BMC Bioinformatics **10**(1), 34 (2009)
10. Mevik, B.H., Wehrens, R.: The pls package: principal component and partial least squares regression in r. J. Stat. Softw. **18**(2), 1–24 (2007)
11. Rajala, P., Carpén, L., Vepsäläinen, M., Raulio, M., Sohlberg, E., Bomberg, M.: Microbially induced corrosion of carbon steel in deep groundwater environment. Front. Microbiol. **6**, 647 (2015)
12. Rousu, J., Agranoff, D.D., Sodeinde, O., Shawe-Taylor, J., Fernandez-Reyes, D.: Biomarker discovery by sparse canonical correlation analysis of complex clinical phenotypes of tuberculosis and malaria. PLoS Comput. Biol. **9**(4), e1003018 (2013)
13. Waldron, P.J., Petsch, S.T., Martini, A.M., Nüsslein, K.: Salinity constraints on subsurface archaeal diversity and methanogenesis in sedimentary rock rich in organic matter. Appl. Environ. Microbiol. **73**(13), 4171–4179 (2007)
14. Wang, X., Eijkemans, M.J., Wallinga, J., Biesbroek, G., Trzciński, K., Sanders, E.A., Bogaert, D.: Multivariate approach for studying interactions between environmental variables and microbial communities. PloS One **7**(11), e50267 (2012)
15. Ye, R., Wright, A.L.: Multivariate analysis of chemical and microbial properties in histosols as influenced by land-use types. Soil and Tillage Res. **110**(1), 94–100 (2010)
16. Zeng, J., Yang, L., Li, J., Liang, Y., Xiao, L., Jiang, L., Zhao, D.: Vertical distribution of bacterial community structure in the sediments of two eutrophic lakes revealed by denaturing gradient gel electrophoresis (dgge) and multivariate analysis techniques. World J. Microbiol. Biotechnol. **25**(2), 225–233 (2009)

KeCo: Kernel-Based Online Co-agreement Algorithm

Laurens Wiel[1]($\boxtimes$), Tom Heskes[1], and Evgeni Levin[2]

[1] Institute for Computing and Information Sciences, Radboud University,
Nijmegen, The Netherlands
laurensvandewiel@student.ru.nl, t.heskes@science.ru.nl
[2] The Netherlands Organization for Applied Scientific Research,
Zeist, The Netherlands
evgeni.levin@tno.nl

Abstract. We propose a kernel-based online semi-supervised algorithm
that is applicable for large scale learning tasks. In particular, we use a
multi-view learning framework and a co-agreement strategy to take into
account unlabelled data and to improve classification performance of the
algorithm. Unlike the standard online methods our algorithm is naturally applicable to many real-world situations where data is available in
multiple representations. In addition our online algorithm allows learning
non-linear relations in the data via kernel functions, that are efficiently
embedded into the formulation of the algorithm. We test performance
of the algorithm on several large-scale LIBSVM and UCI benchmark
datasets and demonstrate improved performance in comparison to standard online learning methods. Last but not least, we make a Python
implementation of our algorithm available for download (Available at
https://github.com/laurensvdwiel/KeCo).

Keywords: Kernel · Non-linear · Online · Large-scale · Semi-supervised ·
Co-agreement · Multi-view · Classification

1 Introduction

Semi-supervised learning algorithms have gained more and more attention in
recent years as they allow the use of large amounts of easily accessible unlabelled data. One of the most elegant approaches to take unlabelled data into
account is based on *multi-view* framework [1]. Multi-view learning algorithms
split the attributes into independent sets and an algorithm is learnt based on
these different 'views'. The goal of the learning process consists in finding a
prediction function for every view performing well on the labelled data and so
that all prediction functions agree on the unlabelled data. Closely related to
this approach is the *co-agreement* framework, where the same idea of agreement
maximization between the predictors is central. Briefly stated, algorithms based
upon this approach search for hypotheses from different views, such that the
training error of each hypothesis on the labelled data is small and, at the same

© Springer International Publishing Switzerland 2015
N. Japkowicz and S. Matwin (Eds.): DS 2015, LNAI 9356, pp. 308–315, 2015.
DOI: 10.1007/978-3-319-24282-8_26

time, the hypotheses give similar predictions for the unlabelled data. Within this framework, the disagreement among the predictors is taken into account via a co-regularization term. Empirical results show that the co-regularization approach works well for domain adaptation [2], classification [3,4], regression [5], and clustering [6] tasks. Moreover, theoretical investigations demonstrate that the co-regularization approach reduces the Rademacher complexity by an amount that depends on the 'distance' between the views [7,8].

Recently an online multi-view algorithm has been proposed in [9]. The algorithm can operate in the semi-supervised regime by using *co-regularization*. The reported performance of the algorithm is promising compared to other state-of-the-art supervised approaches that do not make use of co-regularization, such as PEGASOS [10] and SPD [11]. Our work presents substantial improvements to the online algorithm proposed in [9] by formulating a novel co-agreement learning strategy (explained in Sect. 3) and use of the non-linear kernel function, which enables the method to learn *non-linear relations* from the data.

2 Preliminaries

Prior to introducing the co-agreement framework and our algorithm we describe some standard notations and settings that are frequently used in online learning.

Consider a training set $D = (X, Y)$ of size N, originating from a set $\{(\mathbf{x}_i, y_i)\}_{i=1}^{N}$ of data points where $X = (\mathbf{x}_1, \ldots, \mathbf{x}_N)^{\mathbf{T}} \in \mathcal{X}^N$ and $Y = (y_1, \ldots, y_N)^{\mathbf{T}} \in \{-1, 1\}$. Thus, a single data point $(\mathbf{x}, y)$ where $\mathbf{x} = (x_1, \ldots, x_m)$ is a feature vector of size m, with features defined as $\mathbf{x}_i \in \mathcal{X}$ and a label $y \in \{-1, 1\}$. A *loss function* $\mathcal{L}(y, f(\mathbf{x}))$ measures the quality of the prediction $f(\mathbf{x})$ if the actual label is y. The learning task now becomes a minimization task, which constitutes a typical *binary-classification* learning setting.

In practice, we often would like not only to achieve the smallest possible loss (via error minimization) but also to ensure good generalization of the model. A popular mechanism to accomplish this is *regularization*, which is applied in state-of-the-art online learning algorithms such as GURLS [12], PEGASOS [10], SPD [11] and LPC [13].

Regularization enhances the minimization of a loss function by imposing additional restrictions on specific properties (e.g. smoothness, sparsity, etc.) of the prediction function $f(\mathbf{x})$. This is done by adding a regularization term with an appropriate parameter λ, which denotes how much regularization should be applied.

Online learning methods are frequently formulated in the primal setting for computational efficiency. However, the dual formulation (e.g. [10]) makes it possible to learn non-linear relations in the data. Predictions then depend on a linear combination of weights and kernel function evaluations:

$$\hat{y}_i = \frac{1}{\lambda t} \sum_{j=1, j \neq i}^{p} \alpha[j] y_j K(\mathbf{x}_i, \mathbf{x}_j), \tag{1}$$

where $\hat{y}_i$ is the predicted label for example $\mathbf{x}_i$, α represents the sample weight vector, y_j is the actual label for example $\mathbf{x}_j$, λ is the regularization parameter, t is the number of observed examples at the moment of prediction, and $1/(\lambda t)$ represents the step size. Furthermore, the α-vector is a sparse vector of size N with p non-zero elements, containing discrete weight values to indicate the importance of specific examples. The kernel functions make it possible to create rich feature spaces at low computational cost.

To achieve optimal prediction in (1), [10] proposes to minimize

$$F(\alpha) = \min_{\alpha} \frac{\lambda}{2} \sum_{i,j=1}^{p} \alpha[i]\alpha[j]K(\mathbf{x}_i, \mathbf{x}_j) + \frac{1}{p} \sum_{i=1}^{p} \max\left\{0, 1 - y_i \sum_{j=1}^{p} \alpha[j]K(\mathbf{x}_i, \mathbf{x}_j)\right\}, \quad (2)$$

where the last term corresponds to hinge loss. The setting above is frequently referred to as minimization in a dual setting.

2.1 Multiple Views

The multi-view paradigm [1] is particularly suited for learning from datasets having more than a single data representation. A classic example is a web document classification task [1], where documents are represented via two different views - one that is based on the links and another one based on the text in each document. As another example, complex, structured data with multiple representations are frequently encountered in the biomedical domain, making multi-view methods a natural application choice. Although in many circumstances the individual data representation can be sufficient for training a model, a combination of the multiple views can lead to more robust and accurate predictions compared to the ones obtained via the individual views [14].

Online learning in the multi-view setting was recently introduced in [9]. Informally, building a prediction model for each view can be considered as training several 'judges', whose advice can be combined for improved evaluation. Views can be built by randomly choosing various features from the dataset. However, more informative views are usually considered such as those that use different feature sources (e.g. in a medical diagnosis application, separating the blood sample data from the x-ray results).

We extend the description of an example $(\mathbf{x}, y)$ to cope with multiple views as with $\mathbf{x} = (\mathbf{x}^{(1)}, ...\mathbf{x}^{(V)})$, where V is the maximum number of views and each $\mathbf{x}^{(i)}$ represents the ith subset of the features in $\mathbf{x}$. The predicted label for a view i is denoted as $\hat{y}^{(i)}$. Our learning algorithm is presented in the following section.

3   Kernel-Based Online Co-agreement Algorithm

First we discuss the preliminaries on how to deal with predictions in the case of unlabelled examples.

(Semi-)supervised prediction for an example $\mathbf{x}_i$ in a single view n after t iterations of the KeCo algorithm is defined as:

$$\hat{y}_i^{(n)} = \frac{1}{\lambda(t-1)} \sum_{j=1,j\neq i}^{p} \alpha_{t-1}^{(n)}[j] z_j^{(n)} K(\mathbf{x}_i^{(n)}, \mathbf{x}_j^{(n)}), \tag{3}$$

where $z_i^{(n)}$, given the example to be labelled or unlabelled, represents either the label or the co-agreement for example i:

$$z_i^{(n)} = \begin{cases} y_i & \text{if } x_i \in S \\ c_i^{(n)} & \text{if } x_i \in \hat{S}. \end{cases} \tag{4}$$

Here S represents the set containing all labelled examples, $\hat{S}$ the set containing all unlabelled examples and $c_i^{(n)}$ is the co-agreement (as described in the next paragraph) for sample i in view n. In the case where $\hat{S} = \emptyset$, (3) turns into a supervised multi-view variant of the prediction function used in the PEGASOS algorithm [10].

Co-agreement is a strategy to estimate an '*agreement*' on the label of an example $\mathbf{x}_i$. This is done by making use of different views of the examples and training on each of those views separately in order to create multiple models. We consider these models as 'judges'. Co-agreement represents the agreement between the 'judges' in order to label an example. This is done for each view n by measuring agreement of all the views over the prediction of the example, while excluding the current view n:

$$c_i^{(n)} = \text{sign} \left(\sum_{v=1,v\neq n}^{V} \hat{y}_i^{(v)} \right). \tag{5}$$

We propose to extend the hinge loss with the use of the co-agreement (5) in the case of unlabelled examples as:

$$\max\{0, 1 - z_i^{(n)} \hat{y}_i^{(n)}\}, \tag{6}$$

with $z_i^{(n)}$ as described in (4). Again when $z_i^{(n)} \hat{y}_i^{(n)} \geq 1$, there is *zero* loss, allowing us to generalize the rule to $z_i^{(n)} \hat{y}_i^{(n)} < 1$ and retain the possibility to use it as part of the optimization goal (2).

3.1   KeCo Algorithm

We propose a kernel-based online co-agreement algorithm, which we name KeCo, that is applicable to large-scale semi-supervised binary classification tasks.

S = The set containing all labelled examples

$\hat{S}$ = The set containing all unlabelled examples

V = The number of views

λ = The regularization parameter

T = The maximum number of iterations

$\alpha_t^{(n)}$ = α -vector for view n at iteration t

$\hat{y}_i^{(n)}$ = Predicted label for example i in view n , see (3)

$\mathbf{x}_i^{(n)}$ = Feature vector for example i in view n

$z_i^{(n)}$ = Represents the label or co-agreement label for example i, see (4)

Algorithm 1. KeCo Algorithm

Input: $S, \hat{S}, V, \lambda, T$

1: **for** $t = 1, 2, \ldots, T$ **do**
2:     Choose $i \in \{1, \ldots, |S \cup \hat{S}|\}$ uniformly at random
3:     **for all** $j \in \{1, \ldots, |S \cup \hat{S}|\} \wedge j \neq i, n \in \{1, \ldots, V\}$ **do**
4:         $\alpha_{t+1}^{(n)}[j] \leftarrow \alpha_t^{(n)}[j]$
5:     **for all** $n \in \{1, \ldots, V\}$ **do**
6:         $\hat{y}_i^{(n)} \leftarrow \frac{1}{\lambda t} \sum_{j=1, j \neq i}^{p} \alpha_t^{(n)}[j] z_j^{(n)} K(\mathbf{x}_i^{(n)}, \mathbf{x}_j^{(n)})$
7:     **for all** $n \in \{1, \ldots, V\}$ **do**
8:         **if** $z_i^{(n)} \hat{y}_i^{(n)} < 1$ **then**
9:             $\alpha_{t+1}^{(n)}[i] \leftarrow \alpha_t^{(n)}[i] + 1$
10:        **else**
11:            $\alpha_{t+1}^{(n)}[i] \leftarrow \alpha_t^{(n)}[i]$

Output: $[\alpha_{T+1}^{(1)}, \ldots, \alpha_{T+1}^{(V)}]$

Initially, $\alpha_0^{(n)}[1] = \ldots = \alpha_0^{(n)}[\,|S \cup \hat{S}|\,] = 0$ for all $n \in \{1, \ldots, V\}$.

In the formulation above we could consider having a regularization parameter λ for each different view. This, however, would notably increase the number of hyper parameters and hence the time required for an optimal model calculation (e.g. grid search via cross-validation).

The computational complexity of a prediction (see (3) and Algorithm 1, line 6), may require as many as $\min(t, p)$ kernel evaluations.

4   Experimental Set-Up and Results

We evaluate the performance of KeCo and compare it with the PEGASOS [10] algorithm by measuring the AUC [15] achieved on the testing data.

We have evaluated the performance during experiments on the SVMGUIDE1, SVMGUIDE3 [16] and Ionosphere datasets. SVMGUIDE1 and SVMGUIDE3 are publicly available through LIBSVM [17], Ionosphere is available through the UCI [18] repository. KeCo is designed for partially-labelled, large-scale datasets with multiple views that contain non-linear relations. These datasets do not necessarily cohere to all of these claims, but for the sake of these experiments are considered as such.

Each experiment is repeated five times and the reported AUC is an average over the performance of these five experiments. In each experiment we first randomly select 20 % of the data to be considered as labelled examples and use a classic train-test split of the data, where we randomly select 70 % to be used for training and the remaining 30 % of the data is used for testing.

The training consists of a 10-fold cross-validation with a grid search over the parameter sets where $T = 1000$, $\lambda \in \{10^{-10}, 10^{-9}, 10^{-8}, 10^{-7}, 10^{-6}, 10^{-5}, 10^{-4}, 10^{-3}, 10^{-2}, 10^{-1}\}$ and for the Gaussian kernel, $\sigma \in \{2^{-4}, 2^{-3}, 2^{-2}, 2^{-1}, 2^0, 2^1, 2^2, 2^3\}$. The optimal grid found during this process is used to construct a model on the training set and that model is evaluated on the test set in order to generate the result for an experiment. We only evaluate a 2-view scheme for KeCo and we construct the views to represent a randomly selected 75 % of the original set of features.

Furthermore, we used the following strategy to evaluate the performance of the algorithm. Semi-supervised strategy (KeCo) first uses 50 % of the maximum iterations T to train on the labelled samples present in the partially labelled dataset, followed by a repeating sequence that considers 1 labelled followed by 1 unlabelled sample until the number of iterations reaches the value for T. The supervised strategy (Pegasos) uses the exact same sequence as KeCo, except that it skips the learning from the unlabelled sample.

The result of the experiments can be found in Table 1 where we see the Gaussian kernel outperforming the linear one and in both cases achieving a statistical significant improvement for semi-supervised learning with respect to the supervised setting. For the implementation of the experimental setup we have made use of the scikit-learn framework [19].

Table 1. Results of the semi-supervised KeCo algorithm, in a 2-view setting, with a comparison to the fully supervised PEGASOS algorithm on 20 % labelled versions of the SVMGUIDE1, SVMGUIDE3 and Ionosphere datasets. The comparisons have been made using a Linear and a Gaussian kernel.

Kernel	linear		Gaussian	
Method	Pegasos	KeCo	Pegasos	KeCo
SVMGUIDE1	0.8778	0.8959	0.9231	**0.9238**
SVMGUIDE3	0.6491	0.6183	0.7192	**0.7331**
Ionosphere	0.7627	0.8012	0.8953	**0.9107**

5   Conclusion

This work presents a kernel-based online co-agreement algorithm applicable to large scale semi-supervised classification tasks.

Our algorithm is related to online methods such as such as PEGASOS [10], LaSVM [20] and GURLS [12] and unlike many of these methods is naturally applicable for multi-view datasets. In the empirical evaluation we demonstrate that our method consistently performs well on publicly available datasets as well as notably outperforms supervised learning algorithms.

In particular we demonstrate that a kernel-based version of the method in combination with a co-agreement over a multi-view learning regime makes it possible to achieve more optimal results then the fully supervised state-of-the-art PEGASOS algorithm, given a partially labelled dataset. Last but not least, we make available an efficient implementation of our algorithm coded in Python[1].

Our algorithm can be extended to be applicable to various learning tasks. For instance, it can straightforwardly be adapted for the task of large scale kernel-based online regression analysis.

In the near future we also aim to pursue a theoretical analysis of kernel-based online co-agreement algorithms. For example, we aim to investigate the consistency of multi-view online learning, and provide results on the rate of convergence of the algorithm as the sample size increases.

Acknowledgments. This research was financially supported by Top Institute Food and Nutrition, Wageningen, The Netherlands (Project RE-002).

References

1. Blum, A., Mitchell, T.: Combining labeled and unlabeled data with co-training. In: Proceedings of the Eleventh Annual Conference on Computational Learning Theory, COLT 1998, pp. 92–100. ACM, New York (1998)
2. Kumar, A., Saha, A., Daume, H.: Co-regularization based semi-supervised domain adaptation. In: Lafferty, J., Williams, C., Shawe-Taylor, J., Zemel, R., Culotta, A. (eds.) Advances in Neural Information Processing Systems 23, pp. 478–486. Curran Associates Inc., UK (2010)
3. Sindhwani, V., Niyogi, P., Belkin, M.: A co-regularization approach to semi-supervised learning with multiple views. In: Proceedings of ICML Workshop on Learning with Multiple Views (2005)
4. Goldberg, A.B., Li, M., Zhu, X.: Online manifold regularization: a new learning setting and empirical study. In: Daelemans, W., Goethals, B., Morik, K. (eds.) ECML PKDD 2008, Part I. LNCS (LNAI), vol. 5211, pp. 393–407. Springer, Heidelberg (2008)
5. Brefeld, U., Gärtner, T., Scheffer, T., Wrobel, S.: Efficient co-regularised least squares regression. In: Proceedings of the 23rd International Conference on Machine Learning, ICML 2006, pp. 137–144. ACM, New York (2006)
6. Brefeld, U., Scheffer, T.: Co-EM support vector learning. In: Proceedings of the Twenty-first International Conference on Machine Learning, ICML 2004, p. 16. ACM, New York (2004)
7. Rosenberg, D., Bartlett, P.L.: The rademacher complexity of co-regularized kernel classes. In: Meila, M., Shen, X. (eds.) Proceedings of the Eleventh International Conference on Artificial Intelligence and Statistics, pp. 396–403 (2007)

---

[1] Available at https://github.com/laurensvdwiel/KeCo.

8. Sindhwani, V., Rosenberg, D.S.: An RKHS for multi-view learning and manifold co-regularization. In: Proceedings of the 25th International Conference on Machine Learning, ICML 2008, pp. 976–983. ACM, New York (2008)

9. de Ruijter, T., Tsivtsivadze, E., Heskes, T.: Online co-regularized algorithms. In: Ganascia, J.-G., Lenca, P., Petit, J.-M. (eds.) DS 2012. LNCS, vol. 7569, pp. 184–193. Springer, Heidelberg (2012)

10. Shalev-Shwartz, S., Singer, Y., Srebro, N.: Pegasos: primal estimated sub-GrAdient SOlver for SVM. In: Proceedings of the 24th International Conference on Machine Learning, ICML 2007, pp. 807–814. ACM, New York (2007)

11. Sculley, D.: Large scale learning to rank. In: NIPS Workshop on Advances in Ranking, pp. 1–6 (2009)

12. Tacchetti, A., Mallapragada, P., Santoro, M., Rosasco, L.: GURLS: a toolbox for large scale multiclass learning. In: NIPS 2011 workshop on parallel and large-scale machine learning. http://lcsl.mit.edu/#/downloads/gurls

13. Fürnkranz, J., Hüllermeier, E.: Preference learning and ranking by pairwise comparison. Prefer. Learn. 1, 65–82 (2010)

14. Dasgupta, S., Littman, M.L., McAllester, D.: PAC generalization bounds for co-training. Adv. Neural Inform. Proc. Syst. 1, 375–382 (2002)

15. Cortes, C., Mohri, M.: AUC optimization vs. error rate minimization. Adv. Neural Inform. Proc. Syst. 16(16), 313–320 (2004)

16. Hsu, C.W., Chang, C.C., Lin, C.J., et al.: A practical guide to support vector classification (2003)

17. Chang, C.C., Lin, C.J.: LIBSVM: a library for support vector machines. ACM Transactions on Intelligent Systems and Technology 2, 27:1–27:27 (2011). Software available at http://www.csie.ntu.edu.tw/~cjlin/libsvm

18. Lichman, M.: UCI machine learning repository (2013). http://archive.ics.uci.edu/ml

19. Pedregosa, F., Varoquaux, G., Gramfort, A., Michel, V., Thirion, B., Grisel, O., Blondel, M., Prettenhofer, P., Weiss, R., Dubourg, V., Vanderplas, J., Passos, A., Cournapeau, D., Brucher, M., Perrot, M., Duchesnay, E.: Scikit-learn: machine learning in Python. J. Mach. Learn. Res. 12, 2825–2830 (2011)

20. Bottou, L., Bordes, A., Ertekin, S.: LASVM (2009). http://mloss.org/software/view/23/

Tree PCA for Extracting Dominant Substructures from Labeled Rooted Trees

Tomoya Yamazaki[1]([✉]), Akihiro Yamamoto[1], and Tetsuji Kuboyama[2]

[1] Graduate School of Informatics, Kyoto University Yoshida-Honmachi,
Sakyo-ku, Kyoto 606-8501, Japan
t.yamazaki@iip.ist.i.kyoto-u.ac.jp, akihiro@i.kyoto-u.ac.jp
[2] Computer Centre, Gakushuin University, 1-5-1 Mejiro, Toshima-ku,
Tokyo 171-8588, Japan
ori-ds2015@tk.cc.gakushuin.ac.jp

Abstract. We propose novel principal component analysis (PCA) for rooted labeled trees to discover dominant substructures from a collection of trees. The principal components of trees are defined in analogy to the ordinal principal component analysis on numerical vectors. Our methods substantially extend earlier work, in which the input data are restricted to binary trees or rooted unlabeled trees with unique vertex indexing, and the principal components are also restricted to the form of paths. In contrast, our extension allows the input data to accept general rooted labeled trees, and the principal components to have more expressive forms of subtrees instead of paths. For this extension, we can employ the technique of flexible tree matching; various mappings used in tree edit distance algorithms. We design an efficient algorithm using top-down mappings based on our framework, and show the applicability of our algorithm by applying it to extract dominant patterns from a set of glycan structures.

1 Introduction

Capturing the characteristic features of a given data set is one of the fundamental problems in data mining. A popular method for high dimensional numerical vector data is *Principal Component Analysis* (*PCA*, for short) proposed by Pearson [8]. In PCA, the features are subspaces, and the projected subspaces are extracted so that the amount of information of the original data set is retained as possible. We want to apply PCA also to non-numerical data such as tree structure data for extracting dominant features in a set of data. Since PCA requires a feature space and a distance on the space, we have to tailor a suitable feature space and a distance to capture common patterns in tree structures.

PCA for tree structure data was first formulated by Wang and Marron [11]. They applied it to binary trees representing the brain artery structures obtained from MRA images. Ayding et al. [2] proposed an efficient algorithm to compute principal components for unlabeled rooted binary trees. They further extended the method to unlabeled rooted ordered trees with indexing [1], like a *k-way tree*

© Springer International Publishing Switzerland 2015
N. Japkowicz and S. Matwin (Eds.): DS 2015, LNAI 9356, pp. 316–323, 2015.
DOI: 10.1007/978-3-319-24282-8_27

Table 1. The comparison of PCAs for three types of input data.

	Ordinal PCA	Previous methods [1,2]	Proposed methods
Input data	Numerical vectors	Unlabeled rooted ordered trees with indexing	Labeled rooted ordered/unordered trees
Feature space	Euclidean space $\mathbb{R}^n$	Union of all given data called a support tree	A set of subtrees (a generalized support tree)
Projected space	$d(< n)$ dimensional hyperplanes	Paths called tree-lines	Subtrees in the feature space
Criterion	Maximum amount of variation	Minimum sum of indel distances	Minimum sum of indel distances based on a mapping
Origin point	The origin of coordinate	The root vertex of the support tree	The root vertex of the generalized support tree

indexing [3]. In their methods, the total space of the input data set is defined as the *support tree* which is the smallest supertree including all members of the data set as subtrees. The support tree is defined as the union of all trees in a given data set on the assumption that all tree structures share the same index schema uniquely. The projected space is defined as a *tree-line*, which is a sequence of subtrees $\{l_0, \ldots, l_k\}$ where l_0 is a given subtree and l_i is defined from l_{i-1} by the addition of a single vertex to the same direction. We can treat the tree space like the Euclidean space when we regard l_0 as the origin, the tree-space as a two dimensional total space, and the tree-line as a one dimensional axis.

In this paper, we extend the idea due to [1] and introduce PCA for labeled rooted unordered trees *without indexing*; i.e., our methods do not rely on the strong assumption above. Our idea is to use a *mapping*, a set of pairs of vertices with some restrictions, to express principal components. In the previous work [1,2], the expression of principal components is restricted to paths on trees, while our methods allow principal components to have more expressive forms of subtrees by taking advantage of mappings. The notion of mappings was originally introduced for defining the distance between trees [9]. The mapping is regarded as a common substructure between two trees, and many variants of tree edit distance are formulated by the classes of mappings [7]. In this paper, we introduce a general schema for defining PCA for labeled rooted trees and show an algorithm using top-down mappings as an instance of the schema. We apply it to a glycan structure data set, and compute principal components for extracting dominant patterns in the structures. We confirm its validity by classifying the glycan data, and evaluate the accuracy. The comparison of PCA properties among the conventional numerical vector, previous methods and our methods is shown in Table 1.

2   Preliminary

A *rooted tree* (*tree*, for short) is a connected directed acyclic graph in which every vertex is connected from a *root* vertex. A tree $T = (V, E, r, \alpha)$ is a labeled rooted unordered tree, where V is a set of vertices, E is a set of edges, r is a vertex in V called the *root*, and α is a *label function* defined as $\alpha : V \to \Sigma$, assuming an alphabet to be Σ. The label of $v \in V$ is denoted by $l(v)$. We write $v \in T$ instead of $v \in V$. A *forest* $F = \{T_1, \ldots, T_n\}$ is a set of trees. If $|F| = 1$, we identify $F = \{T_1\}$ with $F = T_1$. The ancestor-descendant relation is denoted by $<$, and for $v, w \in T$, $v < w$ means that w is an ancestor of v. The *depth* of a vertex v is defined as $\mathrm{dep}(v) = |\{w \mid v \le w\}|$. The sibling relation is denoted by $\prec$, and for $v, w \in T$, $v \prec w$ means that w is a right sibling vertex of v. The parent vertex of $v \in V \setminus \{r\}$ is denoted by $\mathrm{parent}(v)$.

For two trees T_1 and T_2, the following operations are called *edit operations* : *deletion* and *insertion* of a vertex $v \in T_1$, and *substitution* of the label of $v \in T_1$ for the label of $w \in T_2$. The costs of edit operations, deletion, insertion and substitution, are denoted by $\gamma(v \to \lambda), \gamma(\lambda \to w)$ and $\gamma(v \to w)$, respectively. An *edit distance* is the minimum sum of the costs for transforming T_1 to T_2 if all of the costs of edit operations are the same. When $\gamma(v \to w) \ge \gamma(v \to \lambda) + \gamma(\lambda \to w)$, that is, the edit distance without substitution operations, is called an *indel* (insertion-deletion) distance.

A *mapping* $M \subseteq V_1 \times V_2$ is a set of pairs of vertices for two trees $T_1 = (V_1, E_1)$ and $T_2 = (V_2, E_2)$. Various types of mappings have been proposed [7]. They are distinguished by their restrictions such as an ancestor-descendent relation. We show two instances of mappings; i.e., a Tai mapping and a top-down mapping. **Tai Mapping** [9]: Let $T_1 = (V_1, E_1)$ and $T_2 = (V_2, E_2)$ be rooted ordered trees, a set $M \subseteq V_1 \times V_2$ is called a *Tai mapping* on ordered trees if any pairs $(v_1, v_2), (w_1, w_2) \in M$ satisfy all of the following conditions.

$$v_1 = w_1 \Longleftrightarrow v_2 = w_2 \quad \text{(one-to-one relation)},$$
$$v_1 < w_1 \Longleftrightarrow v_2 < w_2 \quad \text{(ancestor-descendant preservation)}, \text{and}$$
$$v_1 \prec w_1 \Longleftrightarrow v_2 \prec w_2 \quad \text{(sibling order preservation)}.$$

If T_1 and T_2 are unordered trees, the third condition is not considered.

Top-Down Mapping [12]: A Tai mapping M between two trees T_1 and T_2 is a *top-down mapping* if for any pair of vertices $(v, w) \in M$ such that both of v and w are not root vertices, there exists a pair $(\mathrm{parent}(v), \mathrm{parent}(w)) \in M$.

The sets of vertices of V_1 and V_2 including in a mapping M are respectively denoted by $M|^{T_1}$ and $M|^{T_2}$ which are defined as $M|^{T_1} = \{v \in V_1 \mid \exists w \in V_2, (v, w) \in M\}$ and $M|^{T_2} = \{w \in V_2 \mid \exists v \in V_1, (v, w) \in M\}$. The total cost of edit operations for M is $\gamma(M) = \sum_{(v,w) \in M} \gamma(v \to w) + \sum_{v \in M|^{T_1}} \gamma(v \to \lambda) + \sum_{w \in M|^{T_2}} \gamma(\lambda \to w)$. Calculating the edit distance between T_1 and T_2 is equivalent to finding a mapping M minimizing the cost $\gamma(M)$ [9]. We call such mappings by *optimal mappings*. Below, we assume a rule R for selecting an optimal mapping from the set of all optimal mappings. Given a set

$FS = \{F_1, \ldots, F_n\}$ of forests, and a mapping M, we define a forest FS^M recursively as follows:

$$FS^M = \begin{cases} F_1 & \text{if } |FS| = 1, \\ \{F_n, \{F_1, \ldots, F_{n-1}\}^M\} = M(F_n, \{F_1, \ldots, F_{n-1}\}^M) & \text{otherwise.} \end{cases}$$

where $M(F_1, F_2)$ is the forest induced by an optimal mapping M between F_1 and F_2 following the rule R.

3   Tree PCA by Top-Down Mappings

3.1   New Schema for Tree PCA

In this section, we introduce new methods for extracting principal components from labeled rooted unordered trees without indexing and we give a new schema for formulating principal components based on the following three contents. The first is a distance metric $d_M(\cdot, \cdot)$ based on a mapping M, and the second is the total space of given data set $\mathcal{T}$, denoted by $TS(\mathcal{T})$. The total space $TS(\mathcal{T})$ is a set of trees. The third is the set of all components, denoted by $AC(\mathcal{E})$, for any elements $\mathcal{E}$ of $TS(\mathcal{T})$. For example, the set of all components of a path is a tree-line.

The projection of the tree T onto the union of $AC(t_1) \uplus \cdots \uplus AC(t_k)$ is defined by using the *inclusion-exclusion principle* as follows:

$$P_{\{t_1 \ldots t_k\}}(T) \equiv \underset{PS \in AC(t_1) \uplus \cdots \uplus AC(t_k)}{\arg \min} \sum_{n=1}^{|PS|} (-1)^{n+1} \sum_{ps \in \{U \subset PS \mid |U| = n\}} d_M(T, ps^M), \tag{1}$$

where $S \uplus T$ denotes the set $\{\{s, t\} \mid s \in S, t \in T\}$. The k-th principal component is defined as

$$PC_k \equiv \underset{\mathcal{E} \in TS(\mathcal{T})}{\arg \min} \sum_{T \in \mathcal{T}} \sum_{t \in P_{\{PC_1 \ldots PC_{k-1}, \mathcal{E}\}}(T)} d_M(T, t). \tag{2}$$

3.2   Path Features by Top-Down Mappings

In this section, we show how to apply the top-down mappings to the new schema. The sequence of labels of the path from the root r to v is defined as $\text{Path}(v) \equiv \langle \alpha(r)\alpha(\text{parent}^{\text{dep}(v)-2}(v)) \ldots \alpha(\text{parent}^1(v))\alpha(v)\rangle$, where $\text{parent}^n(v)$ is defined as $\underbrace{\text{parent}(\text{parent}(\ldots \text{parent}(v)))}_{n}$, and thus $\text{parent}^1(v) = \text{parent}(v)$.

The set $\text{Fiber}(T) \equiv \{\text{Path}(v) \mid v \in \text{Leaf}(T)\}$ of paths is called the *fiber* of T. We define the total space as the *support fiber* of an input data set $\mathcal{T}$, denoted by $SF(\mathcal{T})$ while the total space is defined as the support tree in the previous section. The support fiber representing the total space is defined as $SF(\mathcal{T}) = \bigcup_{T \in \mathcal{T}} \text{Fiber}(T)$.

Given a path P, the tree-line composed of P is $TL(P) = \bigcup_{v \in P} \mathrm{Path}(v)$, where v is a vertex of path P. Given two trees T_1 and T_2, an indel distance based on a top-down mapping between T_1 and T_2 is denoted by $d_{\mathrm{TD}}(T_1, T_2)$. The top-down mapping without substitution operations following the rule R is denoted by $M_{\mathrm{TD}}^{T_1, T_2}$. In other words, $M_{\mathrm{TD}}^{T_1, T_2}$ is a set of pairs of vertices corresponding with the both of labels of vertices from the root vertex completely.

Algorithm 1. Making a super tree of $SF(\mathcal{T})$

INPUT : the set of path $SF(\mathcal{T})$
Support tree ST initialized with a single dummy vertex v_d.
for Path P in $SF(\mathcal{T})$ **do**
    $L_{\mathrm{child}} \leftarrow \{l(v_c) \mid v_c \in \mathrm{children}(v_d)\}$ /*list of child vertices*/
    Add P to the single vertex with ϵ label until $\mathrm{dep}(P) = \mathrm{dep}(SF(\mathcal{T}))$.
    for Each vertex $v \in P$ from the root vertex to a leaf vertex **do**
        if L_{child} contains $l(v)$ **then**
            $L_{\mathrm{child}} \leftarrow \{l(v_c) \mid v_c \in \mathrm{children}(v_e)\}$ where $v_e \in L_{\mathrm{child}}$ satisfies $l(v_e) = l(v)$.
        else
            A subpath SP $\leftarrow$ a subtree of P whose root vertex is v.
            Add SP to the parent vertex of L_{child}.
            Break the inner loop.
        end if
    end for
end for
OUTPUT : ST.

Therefore, we can extract principal component paths by adapting M to M_{TD}, $d_M(\cdot, \cdot)$ to $d_{\mathrm{TD}}(\cdot, \cdot)$, $TS(\mathcal{T})$ to $SF(\mathcal{T})$, and $AC(\mathcal{E})$ to $TL(P)$ where P is a path. Our method extends the previous methods by Alfaro et al. [1] and Aydin et al. [2]. Our method based on the top-down mappings can apply to ordered trees if we just give the label of a vertex to a sibling information.

3.3 An Algorithm for PCA by Top-Down Mappings

In this subsection, we give an algorithm for extracting principal component based on the top-down mappings. First, we make a super tree from the each path of $SF(\mathcal{T})$ and we show the algorithm for making such super tree in Algorithm 1. The algorithm is similar to making a prefix tree representing an upper common subtree. In Algorithm 1, for a tree T, the maximum depth and maximum number of leaves of T is denoted by $\mathrm{dep}(T)$ and $\mathrm{Leaf}(T)$, respectively. The label meaning the empty is denoted by $\epsilon \notin \Sigma$. We show an example of the input data set in Fig. 1a and the super tree of the input data set in Fig. 1. We generalize $ST(\mathcal{T})$ in accordance with Algorithm 1 and we regard $ST(\mathcal{T})$ as a set of paths. The k-th principal component derived from Eq. (2) is

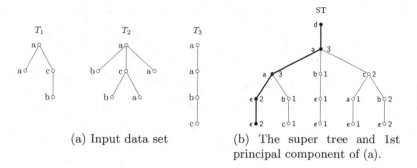

(a) Input data set

(b) The super tree and 1st principal component of (a).

Fig. 1. (a) An example of input data set $\mathcal{T}$. The characters, a,b and c, are labels of vertices. (b) The super tree of the input data (Fig. 1a). A label ϵ represents the terminal and d represents the dummy vertex. An integer which describes the right-hand side of each vertex is the weight of Eq. (4). The heavy line is the first principal component.

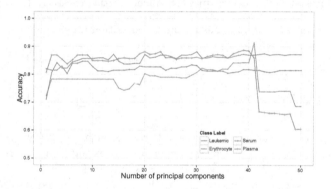

Fig. 2. The accuracy of classification based on a top-down indel distance between cumulative principal components and the given data tree.

$$PC_k = \arg \max_{P \in ST(\mathcal{T})} \sum_{v \in P} \sum_{T \in \mathcal{T}} w_k(v, T, P), \qquad (3)$$

$$\text{where } w_k(v, T, P) = \begin{cases} 1 & \text{if } v \in M_{\text{TD}}^{\{P,T\}}|^P \text{and } v \notin P_1 \cup \cdots \cup P_{k-1}, \\ 0 & \text{otherwise.} \end{cases} \qquad (4)$$

Therefore, the k-th principal component is the path whose sum of weights is the maximum. Then, we can extract k principal components, converting w_k on principal components to 0. The time complexity of extracting k principal components is $O((\text{Leaf}(ST(\mathcal{T}))\text{dep}(ST(\mathcal{T})) + k|V|)|\mathcal{T}|)$ where $|V| = \max\{|V_1|, |V_2|\}$.

4   Experiment

In our experiment, we use glycan structure data from the KEGG/GLYCAN database [5] and their annotations are from the CarbBank/CCSD database [4].

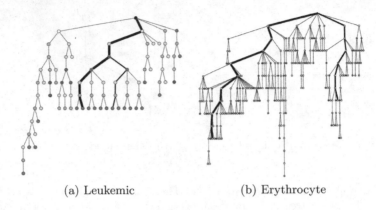

(a) Leukemic   (b) Erythrocyte

Fig. 3. The super trees of Leukemic and Erythrocyte, respectively, where the each root vertex is a dummy vertex and not including ϵ vertices. The heaviest and second heaviest lines are 1st and 2nd principal component paths, respectively.

Table 2. Comparing the classification accuracy.

Comparing	Leukemic	Erythrocyte	Serum	Plasma
20 principal components	0.802	0.871	0.827	0.881
Other global structure	0.914	0.841	0.843	0.787

We regard a glycan structure as a labeled tree as with [6]. A glycan structure is often regarded as an ordered tree, however, in our experiment, we regard it as a labeled unordered tree because we focus on only paths.

For a glycan structure, many studies have been proposed e.g. [6,13]. Each glycan structure is assigned to a blood component class among Leukemic, Erythrocyte, Serum, and Plasma, and the number of each class data used in our experiment is 140, 127, 78 and 60, respectively.

First, we visualize super trees and the first and second principal components of the Leukemic and Erythrocyte in Fig. 3. The heaviest and the second heaviest lines in Fig. 3 are the first and second principal component paths, respectively. Next, we try to classify the input data set into the 4 classes by using principal components. First, we classify a given data set with a top-down indel distance, that is the number of vertices not including the largest common prefix structure. The results are shown in Fig. 2. The accuracy of all class labels is mostly over 0.8. We compare the accuracy of classification by using principal components from 1st to 20th with the one of measuring a global edit distance. The result is shown in Table. 2. The classification of the Erythrocyte and Plasma using principal components is higher while the one of the other classes using the global edit distance is higher. According to the results, we could conclude Leukemic has specific global structures while Erythrocyte and Plasma have specific local structures because principal components are local dominant structures of the given data. Moreover, the classification using principal components runs faster because the time complexity of computing the global edit distance between two trees is $O(|V|^3)$.

5   Concluding Remarks

We introduced a general schema for defining PCA for labeled rooted trees in Sect. 3. Because of lack of space, we gave only one instance of the schema. Another instance can be given with bottom-up mappings [10]. We should select a mapping depending on the given data set.

In [5,6], glycan data are classified by using kernels, but in this method, we cannot know the similar structures explicitly. By extracting principal components, we can observe the dominant structures and the similar structures and classify the given data by using the principal components. Moreover, the time complexity of classification by using principal components is lower than the one by measuring a global edit distance between two unordered trees, NP-hard problem.

Acknowledgments. The authors would like to thank both the anonymous reviewers and Kouichi Hirata, Kyushu Institute of Technology, Japan for their valuable comments. This work was partially supported by the Grant-in-Aid for Scientific Research (KAKENHI Grant Numbers 26280085, 26280090, and 24300060) from the Japan Society for the Promotion of Science.

References

1. Alfaro, C.A., Aydin, B., Valencia, C.E., Bullitt, E., Ladha, A.: Dimension reduction in principal component analysis for trees. CSDA **74**, 157–179 (2014)
2. Aydin, B., Pataki, C., Wang, H., Bullitt, E., Marron, J.S.: A principal component analysis for trees. Ann. Appl. Stat. **3**(4), 1597–1615 (2009)
3. Chartrand, G., Lesniak, L.: Graphs and Digraphs, 3rd edn. Chapman and Hall/CRC, London (2000)
4. Doubet, S., Albersheim, P.: CarbBank. Glycobiology **2**(6), 505–507 (1992)
5. Hashimoto, K., Goto, S., Kawano, S., Aoki-Kinoshita, K.F., Ueda, N.: KEGG as a glycan informatics resource. Glycobiology **16**, 63–70 (2006)
6. Kuboyama, T., Hirata, K., Aoki-Kinoshita, K.F., Kashima, H., Yasuda, H.: A gram distribution kernel applied to glycan classification and motif extraction. Genome Inform. **17**(2), 25–34 (2006)
7. Kuboyama, T.: Matching and learning in trees, Ph.D. thesis, Univ. Tokyo (2007)
8. Pearson, K.: On lines and planes of closest fit to systems of points in space. Philos. Mag. **2**(6), 559–572 (1901)
9. Tai, K.C.: The tree-to-tree correction problem. J. Addociation Comput. Mach. **26**(3), 422–433 (1979)
10. Valiente, G.: An efficient bottom-up distance between trees. In: Proceedings of the 8th SPIRE, pp. 212–219. IEEE Comp. Science Press (2001)
11. Wang, H., Marron, J.S.: Object oriented data analysis: set of trees. Ann. Stat. **35**(5), 1849–1873 (2007)
12. Wang, J.T.-L., Zhang, K.: Finding similar consensus between trees : an algorithm and a distance hierarchy. Pattern Recogn. **34**, 127–137 (2001)
13. Yamanishi, Y., Bach, F., Vert, J.P.: Glycan classification with tree kernels. Bioinformatics **23**(10), 1211–1216 (2007)

Enumerating Maximal Clique Sets
with Pseudo-Clique Constraint

Hongjie Zhai[1]([✉]), Makoto Haraguchi[1], Yoshiaki Okubo[1], and Etsuji Tomita[2]

[1] Graduate School of Information Science and Technology, Hokkaido University,
N-14 W-9, Sapporo 060-0814, Japan
{zhaihj,makoto}@kb.ist.hokudai.ac.jp
[2] The Advanced Algorithms Research Laboratory,
The University of Electro-Communications,
Chofugoka 1-5-1, Chofu, Tokyo 182-8585, Japan

Abstract. It is an important task in Data Mining and Social Network Analysis to detect dense subgraphs, namely pseudo-cliques in networks. Given a positive integer k designating an upper bound of the number of disconnections, some algorithms to enumerate k-plexes as pseudo-cliques have been proposed based on the anti-monotonicity property similar to the case of cliques. Those algorithms are however effective only for small k, since every vertex set with its size less than $k + 1$ is trivially a k-plex. Moreover, there still exist non-dense k-plexes with their sizes exceeding k. For these reasons, it has been a hard task to design an efficient k-plex enumerator for non-small k. This paper aims at developing a fast enumerator for finding densely connected k-plexes for non-small k, avoiding both of the small k-plexes and non-dense medium k plexes. For this purpose, we construct a clique-graph from the original input graph and consider meta-cliques of overlapping cliques satisfying several constraints about k-plexness and overlappingness using bond measure for set-theoretic correlation. We also show its usefulness by exhaustive experiments about the number of solution k-plexes, computational costs and even the quality of output k-plexes.

Keywords: k-plex · Clique graph · Meta-clique · Bond measure

1 Introduction

A social network is often represented by a graph of vertices representing actors in a social network. The edges represent ties between actors [6]. The actors are typically people, and the ties are their relationships as friendship, the collegiality, the acquaintance, or other types of associations. In recent years, with the increasing availability of network data, the structures of networks have drawn great interest. A key feature of networks is that vertices are often organized into *communities* [9]. As a formal model of communities, a clique notion has been used from the beginning of study [5]. A clique satisfies three important properties of communities, familiarity, reachability, and robustness. These properties

© Springer International Publishing Switzerland 2015
N. Japkowicz and S. Matwin (Eds.): DS 2015, LNAI 9356, pp. 324–339, 2015.
DOI: 10.1007/978-3-319-24282-8_28

are represented by vertex degree, path length, and connectivity, respectively in graph theoretic terms [17].

However, the clique approach has been criticized for its restrictive nature for analysis of real data [22]. Most communities are in fact not cliques, allowing exceptional disconnection between actors. This motivates us to the study of clique relaxation models [20]. Luce [16] proposed a distance-based model called k-clique and Alba [2] introduced a diameter-based model called k-club. These models were also studied along with a variant called k-clan which was introduced by [18]. In a word, the major parameter k of these models controls admissible distances among vertices. As is well known by the study of Small World Network [26], when we allow a longer distance parameter, large dense subgraphs appear which are almost cliques even when the subgraphs w.r.t. the original edge connection are not dense. On the other hand, k-plex model [22,27] discusses the density w.r.t. the original graph. A vertex set X is called a k-plex if, for any vertex v in X, the number of vertices not connected to v is at most k including v itself. We also concern in this paper the density of vertex sets in the original graph. However, the definition of k-plex is clearly weak in detecting dense vertex sets whose sizes are not small. As we target larger dense vertex sets, more number of missing edges as exceptions must be allowed in the set. This means that we have to set larger k accordingly. Since every vertex set with its size less than $k + 1$ is trivially a k-plex, there exist many trivial k-plexes for such a k. Moreover, k-plexes with their sizes exceeding k are not necessarily densely connected. These simple facts prevent us from designing efficient k-plex enumerator for densely connected vertex sets.

A hint to overcome this difficulty would be found in the study of communities and networks. Various indices for evaluating clusters of vertices have been proposed to show how the vertices in a community cluster together. A clustering coefficient [26] is often used as such an index. As more number of small cliques, namely triangles, in a cluster, it has a higher index value. Some of those small cliques are directly connected via overlapped common vertices. In this paper, we focus on finding clusters consisting of cliques overlapping each other to keep the density as a whole. We call such a cluster a meta-clique in a graph of cliques in the original graph. As a vertex set, the cluster is simply a set union of member cliques. The cluster thus defined may not be a k-plex when the member cliques less overlap. We require that the clusters must be k-plexes, and call them k-clique sets. We also use a set-theoretic correlation measure, similar to the bond used for transaction databases [19], to guarantee the degree of connection among member cliques. Typical meta-cliques targeted in this paper are illustrated in Fig. 1.

A notion of overlapping cliques has been already studied in [12], where its search process is much complicated. The procedure firstly tries to enumerate the intersections called cores, and then try to find out member cliques by adding additional vertices to the cores. Thus, the process invokes clique enumerator each time candidate core is found by a clique enumerator. On the other hand, the procedure *MetaClique* developed in this paper firstly enumerates possible cliques that can be member cliques and their connections constrained by the

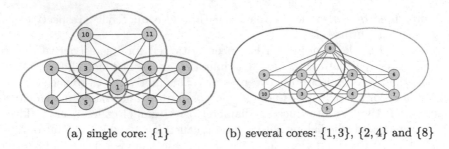

(a) single core: {1}   (b) several cores: {1,3}, {2,4} and {8}

Fig. 1. Target meta-clique

bond measure beforehand. Then the remaining task is simply computing cliques in the clique graph under k-plex constraint. This reduces the overall computational cost drastically. Since the clique enumeration is fast enough for sparse graph as has been demonstrated in [8], *MetaClique* for finding overlapping cliques is working quite well even for medium k, as we demonstrate by several experiments in Sect. 5.

In our experimentation for several synthetic and real world networks, we compare with an extended maximal k-plex enumerator and verify the effectiveness of our approach. For a protein-protein interaction network, we also evaluate the quality of extracted communities by $F1$-score defined with recall and precision for a given answer set. Particularly, we observe that *MetaClique* gives the best score for a parameter setting which it is hard for GCE [14], known as the most effective system, to detect solutions with high quality. This implies that our *MetaClique* has an ability to detect many meaningful communities missed by GCE.

The remainder of this paper is organized as follows. In the next section, we introduce some terminologies used throughout this paper. In Sect. 3, we define our problem of enumerating overlapping cliques as meta-cliques. Our algorithm *MetaClique* for the problem is presented with a pseudo-code in Sect. 4. Section 5 presents our experimental results for several networks. In Sect. 6, we conclude this paper with a summary and future work.

2 Preliminaries

A *simple* graph is denoted by $G = (V, E)$, where V is a set of vertices and $E \subseteq V \times V$ a set of undirected edges. For a pair of vertices, $v_i, v_j \in V$, v_i is said to be adjacent to v_j if $(v_i, v_j) \in E$. For a vertex v, the set of vertices adjacent to v is denoted by $N_G(v)$, where $|N_G(v)|$ is called the *degree* of v in G and is referred to as $deg_G(v)$. If any pair of vertices $v, v' \in V (v \neq v')$ are adjacent each other, then G is said to be *complete*.

For a graph $G = (V, E)$, a *subgraph* of G induced by $V' \subseteq V$, donoted by $G[V']$, is defined as $G[V'] = (V', E \cap (V' \times V'))$. A complete subgraph is called a *clique* in G. We often refer to a clique $G[V']$ as simply V'. A clique is said to be *maximal* if it is not a proper subset of any other cliques. The problem of enumerating all the maximal cliques is called *"Maximal Clique Problem"*.

As previously discussed, although a clique is a perfect structure for community, it is difficult to find them in real world. Therefore, a notion of k-plex has been introduced as one of the relaxation model [22].

Definition 1 *(k-plex).* *For a set of vertices $S \subseteq V$, the subgraph $G[S]$ is called a k-plex if $deg_{G[S]}(v) \geq |S| - k$ for every vertex v in S.*

A k-plex $G[S]$ is also referred to as simply S.

Similar to the case of cliques, a k-plex is said to be *maximal* if it is not a proper subgraph of any other k-plexes. A k-plex is regarded as a pseudo-clique in the sense that it can be obtained by deleting *less than k-edges* from each vertex in a clique. Thus, a clique is a special case of k-plex with $k = 1$. We should point out that Definition 1 does not emphasize the requirement of connectivity. From the definition, it is clear to see that k-plex is anti-monotonic, that is, any subset of a k-plex is also a k-plex.

3 k-Maximal Clique Set

As has been mentioned, existing algorithms for enumerating maximal k-plexes will detect a large amount of sparse patterns and work well with only relatively small k values. In this section, we present a new subclass of k-plexes to address this problem.

3.1 k-Maximal Clique Set: k-MCS

The new class of pseudo-clique we define is called a k-*Maximal Clique Set* (k-MCS). It is formally defined as follows.

Definition 2 *(k-Maximal Clique Set).* *Given a graph G, let Comp be a set of maximal cliques in G, where each $c \in$ Comp is considered as a primitive component. Then, a set of components $T \subseteq$ Comp is called a k-MCS (Maximal Clique Set) iff $\bigcup_{c_i \in T} c_i$ is a k-plex in G.*

A k-MCS T is defined as a set of component maximal cliques whose union form a k-plex in G. In what follows, the k-plex defined as the union of components in T is referred to as $Ext(T)$, that is, $Ext(T) = \bigcup_{c_i \in T} c_i$.

Roughly speaking, by combining maximal cliques, we can expect to efficiently obtain a densely connected community with reasonable cost, because several practical algorithms for enumerating maximal cliques are recently available [24].

From the anti-monotonicity of k-plex, we are particularly concerned with a *maximal k-MCS*, that is, a *maximal set of components* whose union can give a k-plex. It should be noted here that in some cases, we could have different maximal k-MCSs T and T' such that $Ext(T) = Ext(T')$.

Figure 2 shows a simple example of k-MCS. In the figure, c represents the term "clique". $\{c_1, c_2, c_3, c_4, c_5\}$ is a maximal k-MCS consisting of five components. It is clear that maximal k-MCSs are not always maximal k-plexes. In the example, a maximal k-plex must contain some of the vertices in c_6. However, the maximal k-MCS cannot contain those vertices since we consider only maximal cliques as primitive components.

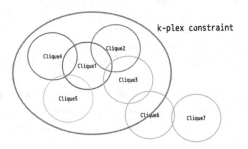

Fig. 2. Image of maximal k-MCS

From now on, we use the term "k-MCS" as the meaning of "maximal k-MCS" in a given graph G unless otherwise stated.

3.2 Restricting k-MCS with Size and Overlappingness of Components

In Definition 2, each k-MCS is required to form a k-plex. Although the constraint can exclude many useless combinations of components which result in obviously sparse communities, we often have undesirable k-MCSs when we assume a relatively larger k value. For example, we could obtain a k-MCS whose components have little shared vertices. Needless to say, such a community would not be so meaningful because it is not reasonable to combine components with low commonality into one group. Moreover, in many real world networks, there exist a large amount of small maximal cliques consisting of just a few vertices. However, those small cliques (components) exponentially increase the number of useless combinations, making computational efficiency worse. Therefore, we impose the following additional constraints on size of component maximal cliques in order to exclusively find densely connected pseudo-cliques with certain amount of size.

More precisely speaking, given a graph G and a positive integer $minsize$, we consider

$$C_{minsize}(G) = \{c \mid c \text{ is a maximal clique in } G \text{ such that } |c| \geq minsize\}$$

as our set of components and try to find maximal k-MCSs consisting of components in $C_{minsize}(G)$. The constraint on component size can effectively exclude sparse communities mainly consisting of component cliques with small size.

In addition to the constraint, we take overlappingness of components in a maximal k-MCS into account. That is, in order to obtain densely connected communities, we require our maximal k-MCS to consist of components overlapping each other. Formally speaking, we evaluate our degree of overlappingness by *bond measure* [19] which is regarded as an extension of *Jaccard Coefficient*.

Given a pair of sets A and B, the bond value for A and B, $Bond(A, B)$, is simply defined as $Bond(A, B) = \frac{|A \cap B|}{|A \cup B|}$. From the definition, it is easy to see that if A and B are highly overlapping, then the bond value $Bond(A, B)$ is close to 1.0.

3.3   Problem of Finding Maximal k-MCS

From the above discussion, we can now define our problem of enumerating maximal k-MCSs in a given graph.

Definition 3 *(Maximal k-MCS Problem). Let G be a graph, $minsize$ a minimum size of components and $minbond$ a threshold for minimum bond value. Then, a Maximal k-MCS Problem is to find every maximal set of components $T \subseteq C_{minsize}(G)$ satisfying the following conditions, where $C_{minsize}(G) = \{c | c$ is a maximal clique in G such that $|c| \geq minsize\}$:*

- $Ext(T) = \bigcup_{c_i \in T} c_i$ *is a k-plex in G.*
- *For any pair of c_i and c_j in T, $Bond(c_i, c_j) \geq minbond$.*

In the next section, we present our algorithm for this problem.

4   Enumerating Maximal k-Maximal Clique Set

In order to enumerate all maximal k-MCSs for given $minsize$ and $minbond$, we have to exhaustively examine possible combinations of components in $C_{minsize}(G)$. For efficient computation, we first introduce a notion of k-*Clique Graph* by which we can exclude quite useless combinations of components in our search process.

4.1 k-Clique Graph

Our maximal k-MCS T to be detected must consist of pairwisely overlapping components whose union forms a k-plex. This implies that any pair of components in T, c_i and c_j, gives a sufficient bond value and the union $c_i \cup c_j$ forms a k-plex from the monotonicity of k-plex. Let us here consider an auxiliary graph $\widetilde{G}$ in which each vertex is a component in $c \in C_{minsize}(G)$ and an edge means the connected vertices (that is, components) give a sufficient bond value and form a k-plex. Then, any maximal k-MCS can be extracted as a clique in $\widetilde{G}$. That is, in order to enumerate our maximal k-MCSs, it is sufficient to examine cliques in $\widetilde{G}$. Thus, the auxiliary graph $\widetilde{G}$ can provide useful information about candidate combinations of components we have to check. Since each vertex in

$\widetilde{G}$ is a component maximal clique for our problem, we call $\widetilde{G}$ a *k-clique graph* which is regarded as an extension of clique graph [11]. It is formally defined as follows.

Definition 4 (*k-Clique Graph*). *For a given graph G, k, minsize and minbond, a graph $\widetilde{G}^k_{minsize,minbond} = (\widetilde{V}, \widetilde{E})$ is called the k-clique graph for the maximal k-MCS problem, where*

$$\widetilde{V} = C_{minsize}(G) = \{c \mid c \text{ is a maximal } k - plex \text{ in } G \text{ such that } |c| \geq minsize\}$$

$$\widetilde{E} = \{(c_i, c_j) \mid c_i, c_j \in \widetilde{V}, Bond(c_i, c_j) \geq minbond \text{ and } (c_i \cup c_j) \text{ is a } k - plex\}$$

Figure 3(a) and (b) present a simple instance of k-clique graph with $k = 4$. The k-clique graph can often simplify the input graph.

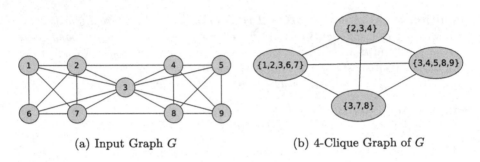

(a) Input Graph G (b) 4-Clique Graph of G

Fig. 3. k-Clique graph

As is just mentioned above, since a maximal k-MCS can be detected as a clique in $\widetilde{G}^k_{minsize,minbond}$, that is, a clique of component cliques, we often call it a *Meta-Clique*.

With the help of the k-clique graph $\widetilde{G}^k_{minsize,minbond}$, we can efficiently extract all maximal k-MCSs. We discuss below our algorithm in details.

4.2   Basic Search Strategy

Given a graph G, k, *minsize* and *minbond*, we first identify the set of components $C_{minsize}(G)$ which consists of all maximal cliques in G with size no less than *minsize*. We then construct the k-clique graph, $\widetilde{G}^k_{minsize,minbond}$, by examining whether each pair of components in $C_{minsize}(G)$ have a sufficient degree of overlap and forms a k-plex. Finally, we try to enumerate (meta-)cliques in $\widetilde{G}^k_{minsize,minbond}$ and check their maximality and k-plexness. Our process of enumerating meta-cliques is based on CLIQUES [24], an excellent algorithm for listing all maximal cliques. As a basic procedure, in our algorithm, a meta-clique X is tried to expand by adding a component $c \in Cand(X)$ to obtain a larger meta-clique $X \cup \{c\}$, where $Cand(X)$, called the candidate set for X, is the set of components each of which is adjacent to all components of X in $\widetilde{G}^k_{minsize,minbond}$. Starting with $X = \emptyset$ and $cand(X) = C_{minsize}(G)$, such an expansion process is iterated in depth-first manner until no meta-clique can be expanded. At each expansion step, the meta-clique X is examined whether the union of all components in X (that is, $Ext(X)$) forms a k-plex or not. If it is true, X is tried to further expand. On the other hand, if it is false, we can safely stop expanding X and then backtrack to the next alternative because any expansion of X can never be our solution from the monotonicity of k-plex.

4.3   Pruning Useless Search Branches

As has been mentioned, for a pair of component sets X and X', we often observe $Ext(X) = Ext(X')$. In our search process, for a meta-clique X, if the candidiate set $Cand(X)$ includes a component c such that $c \subset Ext(X)$, expanding X with c is quite useless because we must have $Ext(X) = Ext(X \cup \{c\})$. In other words, we do not need to *actually* examine the search branch corresponding to such an expansion. Just adding c to X is sufficient in our search process.

Based on this idea, at each expansion step for X, we first detect a meta-clique $M \subseteq Cand(X)$ defined as $M = \{c \mid c \in Cand(X) \text{ and } c \subseteq Ext(X)\}$ and then try to expand $(X \cup M)$ with a component in $Cand(X \cup M)$. As the result, we can prune $|M|$ search branches from the search node of X without loss of completeness.

4.4   Algorithms

Summarizing the above discussion, we present an algorithm for enumerating all maximal k-MCSs based on k-clique graph as well as several pruning strategies. A pseudo-code is presented in Algorithm 1.

Procedure ENUMERATECLIQUE identifies the set of primitive components (maximal cliques in G) with size no less than $minsize$. Procedure BUILDKCLIQUE-GRAPH then builds a k-clique graph. Finally, Procedure ENUERATEKMCS tries to detect all maximal k-MCSs. All pruning mechanisms presented above are incorporated into the algorithms. Due to the space limitation, the pseudo-code for the component identification is omitted.

5   Experimental Results

We present our experimental results in this section. Because our method, referred to as *MetaClique*, points to find a subclass of k-plex, we here compare our method against a maximal k-plex enumerator *MaxKPlex*.

MaxKPlex tries to enumerate maximal k-plexes except obviously undesirable and trivial ones. More concretely speaking, the task of *MaxKPlex* is to extract all *non-pass connected k-plexes each of which contains more than k vertices*.

In order to compare *MetaClique* and *MaxKPlex* under similar conditions, *MaxKPlex* is imposed an additional constraint on size of solution maximal

Algorithm 1. Algorithms for finding maximal k-MCSes

```
1: procedure MAIN(G, k, j, b)                                    ▷ G = G(V, E) is the input graph
2:     C ← ENUMERATECLIQUE(G, j)                                ▷ Enumerate maximal cliques larger than j
3:     F ← BUILDKCLIQUEGRAPH(C, G, k, b)
4:     return ENUMERATEKMCS(F, G, k)
5: end procedure

1: procedure BUILDKCLIQUEGRAPH(C, G, k, b)
2:     V_F = C
3:     E_F = ∅
4:     N_F = {n|n = |c|, c ∈ C}
5:     M_F = ∅
6:     F = F(V_F, E_F, N_F, M_F)                                 ▷ Initialize k-clique graph
7:     for each c ∈ C do
8:         for each c' ∈ C \ {c} do
9:             if bond(c, c') > b and c ∪ c' is k-plex in G then
10:                E_F ← E_F ∪ {(c, c')}                          ▷ Add edge
11:                M_F ← M_F ∪ {|c ∩ c'|}                         ▷ Calculate overlap of cliques
12:            end if
13:        end for
14:    end for
15:    return F
16: end procedure

1: procedure ENUMERATEKMCS(F, G, k)
2:     C ← V(F)                                                  ▷ Initial candidate set takes all vertices of F
3:     comp ← ∅
4:     not ← ∅
5:     DOENUMERATEKMCS(comp, C, not)
6: end procedure

1: procedure DOENUMERATEKMCS(comp, C, not)
2:     if C = ∅ and not = ∅ then
3:         Print comp                                            ▷ Maximal k-MCS detected
4:     end if
5:     while C ≠ ∅ do
6:         u ← one of the largest cliques in C
7:         C ← C \ {u}
8:         not ← not ∪ {u}
9:         C_connected ← N(u) ∩ C
10:        not_connected ← N(u) ∩ not
11:        C_kmcs ← ∅
12:        not_kmcs ← ∅
13:        for each d ∈ C_connected do
14:            if d ∪ comp is a k − plex then
15:                C_kmcs ← C_kmcs ∪ {d}
16:            end if
17:        end for
18:        for each d ∈ not_connected do
19:            if d ∪ comp is a k − plex then
20:                not_kmcs ← not_kmcs ∪ {d}
21:            end if
22:        end for
23:        D_bridge ← the set of all bridges in C_kmcs
24:        for each r in ENUMERATECLIQUE(G[D_bridge], 1) do
25:            DOENUMERATEKMCS(comp ∪ {u} ∪ r, C_kmcs ∩ N(r), not ∪ N(r))
26:        end for
27:    end while
28: end procedure
```

k-plexes to be extracted, because size of k-plexes found by *MetaClique* is at least *minsize*, the minimum size of components. It should be noted here that the additional constraint on size is beneficial for computational efficiency of *MaxKPlex* because its branch-and-bound pruning can work more powerfully.

Datasets: We observe computational performance of *MetaClique* compared with *MaxKPlex* for various benchmark graphs, including synthetic random graphs and real world networks. Because of the limit of space, we here present just a small part of our experiments.

Our random graphs have been generated by the method described in [25]. We have selected some degree distributions under power law as well as average degrees to generate different types of graphs.

CA-GRQC [15] is a collaboration network constructed from the e-print arXiv and covers scientific collaborations. among authors whose papers submitted to General Relativity and Quantum Cosmology category.

GEOM-0 [3] is an authors' collaboration network produced from Computational Geometry Database *geombib*.

COM-AMAZON [15] is based on the feature *customers who bought this item also bought* provided by Amazon website. It is one of the large scale datasets to test the computational performance of our algorithm.

Moreover, in order to verify ability of *MetaClique* to identify meaningful communities, we have made an experiment on a protein-pretoin interaction network referred to as PPI [7] and compared *MetaClique* with existing community detectors for PPI networks.

Detailed information of those networks are summarized in Table 1.

Table 1. Information of graphs

Name	# of Vert	# of Edges	Density	Avg. Clique Size	Max. Clique Size
RANDOM-10000-1	10000	40497	0.00081	2.3	4
RANDOM-10000-2	10000	95474	0.00191	3.3	6
GEOM-0	7343	11898	0.00044	3.1	22
CA-GRQC	5242	14484	0.00105	3.1	44
COM-AMAZON	334863	925872	0.00002	2.9	7
PPI	1622	9070	0.00690	17.9	33

Computational Performance: We observe computation times and numbers of solutions detected by *MetaClique* and *MaxKPlex*. For each graph, we have executed both systems with *minsize* of the range from 3 to 7 and various k-values.

Figure 4 shows computation times for each graph. In these figures, missing data points mean that the algorithms have failed to return within 3 h. Computation times by *MetaClique* include those for enumerating component maximal cliques and constructing k-clique graphs. For each graph, however, it takes less than 1 % of the total.

For $k = 2, 3$, both *MetaClique* and *MaxKPlex* achieve good efficiency. However, when k becomes larger than 3, our *MetaClique* greatly outperforms *MaxKPlex*. Particularly, for CA-GRQC, *MetaClique* can detect solutions $100,000$ times faster than *MaxKPlex*.

In Fig. 5, we also present numbers of solutions. From the figure, *MetaClique* extracts a much less number of solutions than *MaxKPlex* for each graph. As a remarkable point the authors would like to emphasize, the figures show that numbers of solutions by *MetaClique* are not strongly affected by the parameter k. For each j as *minsize*, we have almost the same numbers of solutions even for different k values. This means that the number of solutions by *MetaClique* could be a useful index for characterizing networks.

Figure 6(a) and (b) show the behavior of *MetaClique* and *MaxKPlex* according to the parameter k. We can see that for small k values, *MaxKPlex* gets better performance than *MetaClique*. As k becomes larger, however, our *MetaClique* outperforms *MaxKPlex*. The reason of this behavior is that when k is smaller, enumeration cost for maximal k-plexes is expected to be not high. It would often be almost the same as that for enumerating maximal cliques. Thus, *MaxKPlex* efficiently works for small k-values. On the other hand, even for small k-values, *MetaClique* must first enumerate maximal cliques to obtain primitive components and then construct a k-clique graph. We have to accept the construction cost as an overhead in our framework. After the construction of k-clique graph, however, *MetaClique* can efficiently extract solutions. Particularly, for a larger value of k, since k-clique graphs usually become sparse, our solutions can be obtained much more efficiently and the overhead for graph construction can be disregarded.

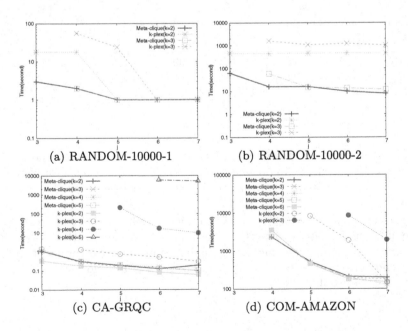

(a) RANDOM-10000-1   (b) RANDOM-10000-2

(c) CA-GRQC   (d) COM-AMAZON

Fig. 4. Computation times ($bond = 0$)

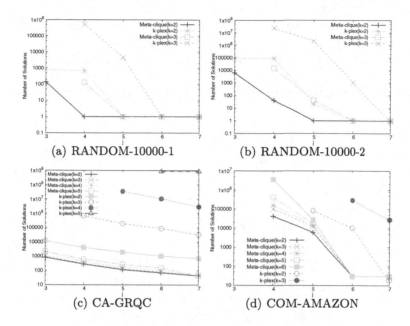

(a) RANDOM-10000-1   (b) RANDOM-10000-2

(c) CA-GRQC   (d) COM-AMAZON

Fig. 5. Number of solutions ($bond = 0$)

We have found that bond values do not affect both runtime and number of solutions greatly when $bond < 0.75$. When its value becomes larger than 0.75, times for building k-clique graphs decrease greatly. In such a case, since our k-clique graph tends to be sparse, we can obtain solutions very efficiently.

Quality: In order to verify quality of communities extracted by *MetaClique*, we compare our system with existing systems for community detection in PPI networks [10,14].

A PPI network [7] treats proteins as vertices and protein-protein interactions as edges. For a given PPI network, we define pre-discovered multiprotein complexes as the "ground-truth" communities, because proteins in multiprotein complexes are usually linked by protein-protein interactions. These "ground-truth" communities are considered as an answer set to be detected. Then we apply different community detection algorithms on the PPI network and compare their outputs with "ground-truth" communities.

In general, communities detected by those algorithms usually do not perfectly match the "ground-truth" communities. Moreover, we may detected more than one clusters which match one "ground-truth" community. The general definition of precision and recall can not be used directly here. Furthermore, when we consider the size constraint j, any ground-truth community with less than j vertices is out of our targets.

Given a threshold w, for a community c detected by an algorithm, if there exists a "ground-truth" community $\tilde{c}$ such that $\frac{|c \cap \tilde{c}|}{min(|c|,|\tilde{c}|)} > w$, we consider that c matches $\tilde{c}$. We also say that $\tilde{c}$ is detected by the algorithm and c is a correct community. Based on this judgement, we define precision $p = \frac{\text{\# of correct communities}}{\text{\# of detected communities}}$ and recall $r = \frac{\text{\# of detected ground-truth communities}}{\text{\# of ground-truth communities}}$. To measure quality of communities detected by each algorithm, we here use $F1-score$, defined as $F1-score = 2 \times \frac{p \times r}{p+r}$. In our preliminary experiment, we have found that $w = 0.8$ is an adequate value to measure the correctness of detected communities allowing some divergence.

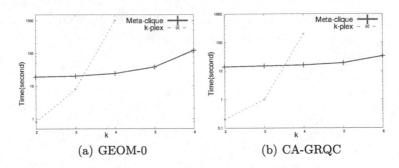

(a) GEOM-0   (b) CA-GRQC

Fig. 6. Computation times $(j = 1, bond = 0)$

In our experimentation, we have used the Combined-AP/MS data [7] from Yeast Interactome Project as interaction network as a graph dataset. For ground-truth communities, we use CYC2008 [21] which contains the complexes that have actually been discovered. We have applied different algorithms on the PPI networks then measured quality of detected communities.

Figure 7 shows $F1$-scores of *MetaClique*, Greedy Clique Expansion (GCE)[1] [14], LINK[2] [1], Order Statistics Local Optimization Method (OSLOM2)[3] [13] and Community Overlap PRopagation Algorithm (COPRA)[4] [10] on the PPI network, where GCE works best for PPI network as far as we know. Each vertical line shows $F1$-score and horizontal line $minsize$-value for our minimum component size. Because COPRA, LINK and OSLOM are not clique-based systems, we can not apply constaints on cliques. Thus, they give horizontal lines through j. Even though Meta-Clique generally detects much more clusters than other algorithms, our algorithm still achieves the highest score in all the listed methods.

When j increases, the quality of solutions by *MetaClique* worsens because only a small number of (larger) components are available. Even in such a case, however, *MetaClique* shows the highest score among all systems. Particularly, although GCE shows the best score at $j = 6$, our *MetaClique* can have a score clearly higher than that. It is emphasized here that *MetaClique* gives the best

---

[1] https://sites.google.com/site/greedycliqueexpansion/.

[2] https://github.com/bagrow/linkcomm.

[3] http://www.oslom.org/software.htm.

[4] http://www.cs.bris.ac.uk/~steve/networks/software/copra.html.

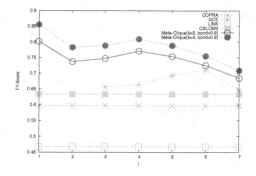

Fig. 7. Comparison of F1-score on protein-protein interaction network

score at $j = 4$ at which it is difficult for GCE to detect solutions with high quality. This means that our *MetaClique* has an ability to detect many meaningfull communities which are unfortunately missed by GCE.

We find that *MetaClique* outputs many highly-overlapped clusters which usually only have one vertex different from each other. In our evaluation, these highly-overlapped clusters are all marked as correct. However, in the task of community detection, highly-overlapped clusters may be unwelcomed. Thus, we need to reduce the overlapness of Meta-Clique's output in future works.

6   Conclusions and Future Works

In this paper, we discussed Maximal k-MCS Problem for finding densely connected communities in a given graph. In order to restrict the solutions to what we want, we introduced constraints on component size and overlappingness.

Our problem of enumerating k-MCSs can be transformed into a problem of finding meta-cliques in a k-clique graph. Therefore, our algorithm has been designed based on CLIQUES [24], an efficient maximal clique enumerator. We empirically verified effectiveness of our method for several synthetic and real world networks from the viewpoints of computational performance and quality of solutions. In order to make our method more appealing, we need to have a theoretical analysis of our algorithm as an important future work. Since the literature [24] has already investigated the worst-case time complexity for CLIQUES, it would provide many helpful and constructive suggestions in our theoretical analysis.

To get denser communities, we cut off small cliques. However, by removing them, we lose the ability of clustering the vertices contained only by these small cliques. Even though the number of small cliques is huge, most of them are highly overlapped. From this, we can surmise that the cover set of these vertices can be small and able to be considered without slowing down the algorithm. As a solution, a clique coverage heuristic process [14] can be introduced.

Very recently, it has been argued that in many networks, a lower bound for the typical community size is given as 5 to 7 [23]. Therefore, it is natural to assume a minimum size of components in order to detect meaningfull communities. In this sense, our approach in this paper would be quite reasonable. However, one might claim that our target k-plex as a combination of component cliques seems too restrictive. In order to make the restriction weaken, we can consider our components to be pseudo-clique so that we can obtain various dense communities more flexibly. Then it would be worth designing an efficient algorithm for such a relaxed maximal k-MCS problem.

As another approach in this direction, it would be promissing to impose a constraint on connectivity of vertices in communities. More concretely speaking, for a community, we may require each vertex to be adjacent to a certain number of vertices in the community. This kind of community can be formalized with the notion of j-core [4] and is currently under investigation as j-cored k-plexes.

References

1. Ahn, Y.Y., Bagrow, J.P., Lehmann, S.: Link communities reveal multiscale complexity in networks. Nature **466**(7307), 761–764 (2010)
2. Alba, R.D.: A graph-theoretic definition of a sociometric clique. J. Math. Soci. **3**, 3–113 (1973)
3. Batagelj, V., Mrvar, A.: Pajek datasets (2006). http://vlado.fmf.uni-lj.si/pub/networks/data/
4. Batagelj, V., Zaversnik, M.: An O(m) algorithm for cores decomposition of networks. Adv. Data Anal. Classif. **5**, 129–145 (2003)
5. Bron, C., Kerbosch, J.: Algorithm 457: finding all cliques of an undirected graph. Commun. ACM **16**(9), 575–577 (1973)
6. Carrington, P.J., Scott, J., Wasserman, S.: Models and Methods in Social Network Analysis, vol. 28. Cambridge University Press, New York (2005)
7. Collins, S.R., Kemmeren, P., Zhao, X.C., Greenblatt, J.F., Spencer, F., Holstege, F.C., Weissman, J.S., Krogan, N.J.: Toward a comprehensive atlas of the physical interactome of saccharomyces cerevisiae. Mol. Cell. Proteomics **6**(3), 439–450 (2007)
8. Eppstein, D., Strash, D.: Listing all maximal cliques in large sparse real-world graphs. In: Pardalos, P.M., Rebennack, S. (eds.) SEA 2011. LNCS, vol. 6630, pp. 364–375. Springer, Heidelberg (2011)
9. Girvan, M., Newman, M.E.: Community structure in social and biological networks. Proc. Nat. Acad. Sci. **99**(12), 7821–7826 (2002)
10. Gregory, S.: Finding overlapping communities in networks by label propagation. New J. Phys. **12**(10), 103018 (2010)
11. Hamelink, R.C.: A partial characterization of clique graphs. J. Comb. Theor. **5**(2), 192–197 (1968)
12. Haraguchi, M., Okubo, Y.: A method for pinpoint clustering of web pages with pseudo-clique search. In: Jantke, K.P., Lunzer, A., Spyratos, N., Tanaka, Y. (eds.) Federation over the Web. LNCS (LNAI), vol. 3847, pp. 59–78. Springer, Heidelberg (2006)
13. Lancichinetti, A., Radicchi, F., Ramasco, J.J., Fortunato, S.: Finding statistically significant communities in networks. PloS one **6**(4), e18961 (2011)

14. Lee, C., Reid, F., McDaid, A., Hurley, N.: Detecting highly overlapping community structure by greedy clique expansion. In: Proceedings of the 4th Workshop on Social Network Mining and Analysis held in Conjunction with the International Conference on Knowledge Discovery and Data Mining (SNA/KDD 2010), pp. 33–42 (2010)

15. Leskovec, J., Krevl, A.: SNAP Datasets: Stanford large network dataset collection, June 2014. http://snap.stanford.edu/data

16. Luce, R.D.: Connectivity and generalized cliques in sociometric group structure. Psychometrika **15**(2), 169–190 (1950)

17. McClosky, B., Hicks, I.V.: Combinatorial algorithms for the maximum k-plex problem. J. Comb. Optim. **23**(1), 29–49 (2012)

18. Mokken, R.: Cliques, clubs and clans. Qual. Quant. Int. J. Methodol. **13**(2), 161–173 (1979)

19. Omiecinski, E.R.: Alternative interest measures for mining associations in databases. IEEE Trans. Knowl. Data Eng. **15**(1), 57–69 (2003)

20. Pattillo, J., Youssef, N., Butenko, S.: Clique relaxation models in social network analysis. In: Thai, M.T., Pardalos, P.M. (eds.) Handbook of Optimization in Complex Networks, pp. 143–162. Springer, New York (2012)

21. Pu, S., Wong, J., Turner, B., Cho, E., Wodak, S.J.: Up-to-date catalogues of yeast protein complexes. Nucleic Acids Res. **37**(3), 825–831 (2009)

22. Seidman, S.B., Foster, B.L.: A graph-theoretic generalization of the clique concept*. J. Math. Sociol. **6**(1), 139–154 (1978)

23. Slater, N., Itzchack, R., Louzoun, Y.: Mid size cliques are more common in real world networks than triangles. Netw. Sci. **2**(03), 387–402 (2014)

24. Tomita, E., Tanaka, A., Takahashi, H.: The worst-case time complexity for generating all maximal cliques and computational experiments. Theoret. Comput. Sci. **363**(1), 28–42 (2006)

25. Viger, F., Latapy, M.: Efficient and simple generation of random simple connected graphs with prescribed degree sequence. In: Wang, L. (ed.) COCOON 2005. LNCS, vol. 3595, pp. 440–449. Springer, Heidelberg (2005)

26. Watts, D.J., Strogatz, S.H.: Collective dynamics of small-world networks. Nature **393**(6684), 440–442 (1998)

27. Wu, B., Pei, X.: A parallel algorithm for enumerating all the maximal k-plexes. In: Washio, T., Zhou, Z.-H., Huang, J.Z., Hu, X., Li, J., Xie, C., He, J., Zou, D., Li, K.-C., Freire, M.M. (eds.) PAKDD 2007. LNCS (LNAI), vol. 4819, pp. 476–483. Springer, Heidelberg (2007)

Author Index

Printed in the United States
By Bookmasters